D1300854

Ecotoxicology

Ecotoxicology

The Study of Pollutants in Ecosystems

3rd edition

F. MORIARTY

ACADEMIC PRESS

San Diego   London   Boston
New York   Sydney   Tokyo   Toronto

Academic Press
24–28 Oval Road, London NW1 7DX, UK
http://www.hbuk.co.uk/ap/

Academic Press
a division of Harcourt Brace & Company
525 B Street, Suite 1900, San Diego, California 92101-4495, USA
http://www.apnet.com

ISBN 0-12-506763-1

A catalogue record for this book is available from the British Library

Typeset by Saxon Graphics Ltd, Derby
Printed in Great Britain by MPG, Bodmin, Cornwall

99 00 01 02 03 04 MP 9 8 7 6 5 4 3 2 1

Contents

Preface

I have retained the structure of the previous editions, with two main sections. After the introduction, the next three chapters discuss the relevant aspects of populations, communities and genetics, and indicate the implications for ecotoxicology. The succeeding chapters then discuss topics of more immediate relevance. The first two describe effects on habitats (with an extensive account of global warming) and on individual organisms, and are followed by critical appraisals of methods used to predict and to monitor effects, four case-studies and a final chapter with a few general comments. I have not tried to write a textbook, nor have I attempted a comprehensive review of the literature: I doubt whether that would be feasible or worth while, for the literature is both enormous and, in large part, trivial. This book is best regarded as a series of closely linked essays, in which I have selected examples from the literature to illustrate ideas that may have some more general relevance.

There is a bias towards an historical account, which may help counter the tendency to overlook older work. Such ignorance leads to redundancy at best and to confusion at worst. Such knowledge helps set newer work in perspective, as with the Gaia theory, and may help avoid confusion, as with interactions. I have tried to use the original definitions of new terms, although, as I have sometimes indicated, usage in the literature can be inconsistent. Even the term ecotoxicology has now been defined as part of environmental toxicology (Eaton and Klaassen, 1996), which is perhaps an example of the part being greater than the whole.

I do not expect all of the arguments advanced in this book to be accepted by all who chance to read them, but I do hope that those who may disagree with any of my suggestions will be sufficiently stimulated to produce the contrary evidence or scientific argument.

F. Moriarty
Cambridge, UK

Acknowledgements

I am indebted to the following holders of copyright for permission to reproduce some of the figures and tables in this book: Academic Press Inc. for Figs 2.13, 2.14 and 7.7; Academic Press Inc. (London) for Figs 6.6, 6.17 and 6.21; the American Chemical Society for Fig. 7.2; the editors of the *Annals of Applied Biology* for Fig. 2.10 and Table 6.3; ASTM, 1916 Race Street, Philadelphia, Pennsylvania 19103 for Fig. 8.11; the editor of *Bird Study* for Fig. 2.11; Blackwell Scientific Publications for Fig. 3.3; Dr. J.D. Burton and The Royal Society of London for Figs 1.1 and 1.2; Cambridge University Press for Figs 2.1, 6.5, 6.9, 8.2 and 8.3, and Table 3.3; the Council of the Linnaean Society of London for Table 7.1; Dr. J. Cullen and Springer-Verlag for Fig. 7.6; Dr. J.M. Davies and The Royal Society of London for Fig. 7.9; Dr. J.P. Dempster for Fig. 2.6 and Table 2.2; Dr. W. Dickson for Fig. 5.13; Dr. J.M. Elliott and Oxford University Press for Fig. 2.7a; Elsevier Science for Figs 6.10 and 8.9; the editor of *Environmental Pollution* for Figs 8.6 and 9.3, and Tables 6.1, 6.6, 6.7 and 7.4; the editor of *Food and Cosmetics Toxicology* for Fig. 6.16; Dr. D.L. Hawksworth for Fig. 8.13 and Table 8.3; Heldref Publications for Fig. 6.22; the editor of *Heredity* for Figs 4.7 and 4.8; the editor of the *Journal of Animal Ecology* for Fig. 3.5; the editor of the *Journal of Applied Ecology* for Figs 4.6 and 6.7; the editor of the *Journal of Ecology* for Fig. 3.12; the editor of the *Journal of the Fisheries Research Board of Canada* for Fig. 7.3; Dr. C.D. Keeling and MacMillan Magazines Ltd for Fig. 5.5; MacMillan Journals Ltd for Fig. 8.4, and Tables 4.6 and 6.2; C.V. Mosby Co. for Fig. 6.20; the National Research Council of Canada for Fig. 8.15; Mr. I.A. Nicholson for Fig. 5.12; North-Holland Publishing Co. for Fig. 3.8; Oxford University Press for Figs 4.5 and 6.1; Pergamon Press Ltd for Fig. 8.12; Plenum Publishing Corp. for Fig. 7.8; Springer-Verlag for Fig. 2.9 and Tables 6.9, 7.6, 7.7 and 7.8; the University of Chicago for Fig. 2.8; Prof. R.P. Wayne and Oxford University Press for Fig. 3.13; The Wildlife Society for Table 7.2; J. Wiley & Sons Ltd for Fig. 6.15; and The Zoological Society of London for Figs 7.5 and 9.2. Fig. 5.8 and Table 5.2 are by courtesy of the Hadley Centre for Climate Prediction and Research.

I particularly thank Dr. D.J. Jefferies for the loan of a print for Fig. 6.6, and Dr. D.A. Ratcliffe and the publishers T. & A.D. Poyser for copyright permission and the loan of the original drawing for Fig. 9.1. I am also grateful to Dr. H. Blanck for letting me see a paper that was in press, before publication. The copyright of Fig. 3.9 and Table 3.4 was held by the late Professor R.H. Whittaker, whose untimely death was a great loss to the scientific community.

I have, of course, a general debt to the scientific community, for work past and present, much of which is not specifically mentioned in this book but which has influenced my thinking in various ways. I would also like to thank some individuals who have had a more immediate and direct impact: I am grateful to Professor A.D. Bradshaw, FRS, Dr. J.P. Dempster, Dr. S. Dobson, Mr. A.V. Holden, Dr. D.J. Jefferies, Dr. E.E. Kenaga, Professor K. Mellanby, Professor N.W. Moore and Dr. M.B. Usher for information, opinion and advice during the preparation of the first edition. Similarly I would like to thank Dr. I. Denholm, Dr. A.L. Devonshire, Dr. H.F. Evans, Dr. M.O. Hill, Dr. A.J. Southward, Dr. E.C. Southward, Mr. J.L. Vosser and Dr. D. Wainhouse for their help with the second edition. Numerous individuals have helped with details of this third edition, and I am particularly grateful to Professor P. Brimblecombe and Professor F.I. Woodward.

I also owe much to my wife, for help and encouragement at all stages of what has been, on occasion, an onerous task. To combine marriage and writing without a forbearing wife would be a hazardous enterprise.

"One curious aspect of Earth history is the continuity of life during the past 3.8 b.y." "It remains to be seen whether this long record will be sustained in the presence of modern man."

The Chemical Evolution of the Atmosphere
and Oceans
H.D. Holland, 1984

1 | Introduction

The term ecotoxicology was coined by Truhaut in 1969 (see Truhaut, 1977) as a natural extension from toxicology, the science of the effects of poisons on individual organisms, to the ecological effects of pollutants. However, the transition from the study of single organisms to that of ecosystems brings additional complexities. The immediate effects of pollutants arise either from direct toxicity on individual organisms or from altering the environment, which again affects the individual organism although there may be no direct toxicity. Either way the ecological significance, or lack of it, resides in the indirect impact on the populations of species and not in the effects on individual organisms. Toxicology is concerned with effects on single organisms, but ecotoxicology is concerned with effects on ecosystems. The fact that a pollutant kills, say, half of the individuals in a species population may be of little or no ecological significance. whereas a pollutant that kills no organisms but retards development may have a considerable ecological impact. Pollutants that alter the environment can also have considerable ecological consequence. We will see that this indirect link between cause—impacts on individual organisms—and effect—impacts on populations and communities—underlies many of the problems in ecotoxicology.

The widespread concern about possible ecological effects of chemicals developed during the 1950s and 1960s. when some agricultural pesticides were found to affect wildlife (Carson, 1962; Rudd, 1964; Moore, 1966a; Sheail, 1985). In retrospect, given that pesticides are non-specific poisons and are released into the environment deliberately, these effects were perhaps not too surprising, but both the form that some biological effects took, and the subsequently discovered effects of other pollutants on wildlife, were completely unexpected by most people. One striking example, not strictly ecological, must suffice.

Tomato crops grown in glasshouses in Essex showed symptoms of damage from herbicide, for no apparent reason, in 1973. Eventually, it was established that the damage was caused by the herbicide 2,3,6-TBA, present in both mains water and river water used for watering the plants (Williams *et al.*, 1977). The source was a factory near Cambridge, 130–170 km by river and artificial channel from the glasshouses. Effluent from the factory had contained small amounts of TBA for the previous 15 years, with no previously reported damage to crops, but this source of water had only just come into use for the glasshouses, and tomatoes are particularly sensitive to this herbicide.

This example illustrates a pathway, persistence and sensitivity that were not anticipated. Many such unforeseen incidents have occurred with wildlife, but, for reasons developed in this book, the determination of cause and effect is usually more complicated and difficult than in this horticultural example.

The focus of interest is not always on the possible effects of pollutants in the environment on all species, but on the implications for our own species. Effects on other species are then appraised principally for the possible impacts on ourselves. Contamination of food species is one obvious example. Studies of this type may be seen more accurately as part of environmental toxicology: the effects of chemicals in the environment on human beings (see Guthrie and Perry, 1980; Klaassen, 1996).

It is difficult to state precisely how many compounds have been synthesized. Some occur once in one laboratory in small quantities, and over 9 million are listed in the Chemical Abstract Service's registry of chemicals (Cairns and Mount, 1990). A provisional list of substances in commercial use within the European Community during the decade 1971–1981 contained about 100 000 names (EINECS, 1987), of which about 10 000 are manufactured in quantities greater than 10 tonnes/year (Haigh, 1995). About 200 new chemicals came into use within the Community during the next seven years (OECD, 1989a). World-wide, 3000 compounds account for almost 90% of the total weight of chemicals produced by industry (IRPTC, 1983).

The need to predict the ecological effects of chemicals is now accepted as so important that many countries have introduced legislation. First was the Toxic Substances Control Act (ToSCA), enacted in the United States in 1976 (Harwell, 1989). The emphasis in this and other similar legislation was initially on the potential effects of new substances, but pre-existing chemicals in commercial use are now also being assessed.

This was possibly the first time in history that legislation posed the problems for scientific research. I say problems for research because one thing is abundantly clear: we do not know how to predict the effects of chemicals on ecosystems, nor how best to monitor for these effects. Indeed, it is arguable that we never will be able to predict with absolute certainty, but there is scope for improvement on our present performance. Certainly, we know enough to tackle some relevant questions: How useful are acute toxicity tests? What are the relevant measurements for a species or a community in its usual environment? Do model ecosystems tell us anything that we cannot discover more easily in other ways? How should we monitor for the effects of pollutants?

The rest of this chapter considers two topics, the nature of pollutants and of ecosystems, which will provide the context of the ensuing chapters.

*

A pollutant is defined, in this book at least, as a substance that occurs in the environment at least in part as a result of man's activities, and which has a deleterious effect on living organisms. Less restricted definitions are sometimes used, which then embrace disparate phenomena.

It is sometimes useful to distinguish between a contaminant and a pollutant: a substance released by man's activities is a contaminant, unless there is reason to suppose that it is having biological effects, although the term pollutant is frequently used loosely to cover both situations. Often, of course, it is an open question whether or not a contaminant is having any biological effects.

There are many ways of classifying pollutants: by chemical or physical nature, properties, source, use (e.g. industrial solvents), place of occurrence in the environment. The appropriate choice depends on the purpose (Holdgate, 1979). Useful though these classifications can be, a complete understanding requires that the properties of each compound or element be considered separately. This can be a difficult task, for instance with sewage. It is not the aim of this book to characterize every pollutant—Walker et al. (1996) introduce that topic, developed further by Connell and Miller (1984), and Manahan (1993, 1994) may also be helpful—but several points need to be emphasized.

Some pollutants, in the amounts produced, do not have any apparent direct effects on living organisms, but do so alter the physical and chemical environment as to affect the ability of organisms to survive. For example the enrichment of inland waters and coastal seas with inorganic nutrients derived from sewage and agricultural fertilizers—eutrophication—alters the distribution of species by altering the chemical characteristics of the environment in which they live.

For pollutants that are toxic (sometimes called toxicants to distinguish them from those that alter the environment) a major theme of this book is that effects depend on exposure and dose. The environment can as a first approximation be divided into three sectors—air, soil and water—and the sector has some influence on exposures, because both transport and diffusion of pollutants are most rapid in gases, less so in liquids and least in soils. It is important to know the rate at which a pollutant is released and the pathways it takes, but exposure also depends on many other variables, both physical and chemical, and some of these will depend on the fine details of an organism's microhabitat. In practice though, it can be difficult to measure the variation in amount of pollutant with both time and part of the environment (see Harrison and Perry, 1986, for an account of the difficulties with air pollutants).

The speciation of pollutants, that is the occurrence of well-defined chemical entities, poses considerable problems for the analyst (Krull, 1991; Ure and

Davidson, 1995; Caroli, 1996). Difficulties can include lack of adequate sensitivity and selectivity in the analytical techniques, the transitory state of some chemical species and the risk of altering the system during sampling, pretreatment and measurement. These problems can limit the relevance of extrapolations from experimental studies on the biological effects of pollutants to field situations (e.g. p. 188). It is useful to distinguish between chemical species (distinct chemical entities, be they ions, molecules or complexes), forms (all species of a similar type, such as all ionic species) and fractions (those species that are detected by a particular analytical technique, such as the dissolved fraction). In other words, chemical fractions are operationally defined, indicate the behaviour rather than the identity of the chemical species involved, and are therefore not mutually exclusive.

Heavy metals in water exemplify this particularly well: they are commonly measured as concentrations of total metals, but occur in a variety of forms (Fig. 1.1) (see also Förstner and Wittmann, 1981; Stumm and Morgan, 1995). The chemical form and species of a metal can affect greatly all aspects of the metal's behaviour and biological effects (Florence, 1982), and different natural water systems contain different proportions of the various species (Burton, 1979; Whitfield and Turner, 1986).

For instance, humic compounds in river water may form complexes with almost all of the dissolved copper, but when salinity increases as the river water passes through an estuary to the sea, calcium and magnesium ions

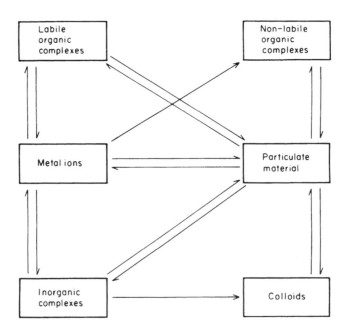

Fig. 1.1 The principal forms and conversion pathways of trace metals in natural waters. (From Burton, 1979; adapted from Whitfield, 1975.)

displace the copper from the humic complex, so that an increasing proportion of the copper occurs as a range of inorganic species (Fig. 1.2). It is noteworthy that the data in Fig. 1.2 are to some extent theoretical, and assume that chemical equilibrium exists, although in practice some equilibria are attained very slowly. Sometimes the kinetics of a reaction are more important than the chemical equilibrium.

The polychlorinated biphenyls (PCBs) illustrate a related problem. These are distinct chemical compounds, whose composition differs only in the number and position of chlorine atoms substituted onto the molecule of biphenyl. Individual PCBs differ in their effects on, and persistence in, animals, but until recently they were rarely considered separately. Commercial PCBs are mixtures of many individual compounds, and analysis of environmental samples by gas–liquid chromatography invariably detects many of them. In the past, the usual procedure for quantifying the results from a sample was to compare the total area of peaks recorded on the chromatogram with that for the peaks from a known concentration of a commercial mixture whose composition is similar. A more sophisticated approach was to convert all components of the PCB mixture to a single derivative before analysis (Armour, 1973). Either way, any differences between individual PCBs in their toxicity or rate of intake or loss by organisms were ignored: the same implicit assumptions that are made when metals are expressed as total metal content. Some recent work does take account of the individual compounds, with analyses using capillary columns

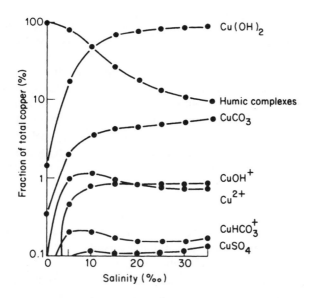

Fig. 1.2 The proportions of the different species of dissolved copper, and the significance of humic acids, in the range of salinities to be found in an estuary. The values are for chemical equilibrium, and are derived from a mathematical model of an estuary. (From Burton, 1979; original data from Mantoura et al., 1978.)

and mass spectrometers. However, interpretation of the biological significance of analyses for field specimens is that much more difficult when the effects of more than one compound have to be considered.

One difference between PCBs and heavy metals is important for questions of environmental contamination. The toxity of a PCB depends on the structure of the whole molecule. A metal, such as copper, is toxic in its own right, although the toxicity can vary greatly with the molecular form. Once released into the environment, the potential for pollution by copper does not disappear with a change of chemical form. However, the toxicity of a PCB depends on the structure of the whole molecule, and once the molecule has been disrupted into its constituent elements, there is no longer any possibility of pollution from the biphenyl.

It has been assumed so far that the precise cause of toxicity, the biologically active molecule or species, is known. The active form is not always readily apparent. Many organophosphorus insecticides become effective only after they have been metabolized, or activated, to a related form (O'Brien, 1967; Thompson, 1992). The organochlorine insecticide p,p'-DDT is toxic in its own right, but one of its biological effects, on eggshells in birds, is due to one of its common metabolites, p,p'-DDE, and DDT itself may not affect eggshells at all. Sometimes the toxicity of a compound comes from a contaminant: the teratogenic effect of 2,4,5-T is caused by dioxin (2,4,7,8-tetrachlorodibenzo-p-dioxin, or TCDD), an impurity which can be formed during manufacture of 2,4,5-T.

The American Chemical Society (Anon., 1978) has reviewed in considerable detail the chemical aspects of pollution, but a few points are particularly relevant for assessment of biological effects. Two simple statistical comments should be made first. Both the precision, the degree of agreement between replicates, and the accuracy, the degree of bias, need to be estimated whenever possible. In most circumstances the biological significance of any analytical result is difficult to determine anyway, and unless one has some indication of both precision and accuracy it can become almost impossible. One also needs to know how selective the analytical method is. For example, concentrations of sulphur dioxide in air are sometimes measured as the amount of acidity in a known volume of air. Other acids, and alkalis, could bias results, particularly in rural areas where concentrations of sulphur dioxide are low (Derouane et al., 1982).

Analysts are well aware of these difficulties. Ideally, there is a standard method, one which has been tried and tested over a considerable period of time and found to give consistent reliable results (see Fifield and Haines, 1995). However, there is a great demand, for many pollutants, for rapid development of methods, when collaborative tests, to compare results from different laboratories on the same material, become very important if not essential.

All measurements contain some random error, so analytical results close to the limit of detection are not easy to interpret. One may conclude falsely that no contaminant is present, or falsely that the sample does contain a low level of contaminant. Repeated analyses of the same sample enable one to determine a detection limit that gives an acceptably low level of probability (e.g. 5%) for both types of error (Black, 1991).

The period of time to which measurements relate also deserves consideration. Results of analyses of plants and animals usually depend on the past history of exposure to a pollutant (pp. 157–163). Samples of air or water may be used for more or less instantaneous determinations (e.g. Fig. 5.12), or for integrated measurements of the total amount of pollutant that has been present during some time, as with a standard method of sulphur dioxide measurement in which air is drawn through a sampler at a known rate. This variability with time needs to be distinguished from questions of precision and bias, and its magnitude assessed.

One final point. Analytical results are usually expressed as concentrations, not in units of mass. The implications for analyses of plants and animals are discussed later (pp. 162–163), but for studies on the sources and extent of environmental contamination the important measurement is of mass of pollutant (Warner, 1979).

Pollutants are but one small part of the total impact that man has on wildlife. Agriculture, forestry, recreation, the control and supply of water, and similar activities all have significant, and often major, effects. One can distinguish between three types of vegetation: natural vegetation is that which occurs as it does without having been affected by man's activities, semi-natural vegetation is derived from natural vegetation but has been influenced by man's activities, and some vegetation is planted deliberately by man. It is possibly true to say, for somewhere like Great Britain, that there is no truly natural vegetation. So changes caused by pollutants occur against a background of changes caused by other human activities. Deliberately or unconsciously, we determine the nature of the fauna and flora around us.

It is time to introduce some of the terms that help ecologists to analyse and understand the living world. Loosely, ecology can be described as scientific natural history (Elton, 1927), or more precisely as the scientific study of the interactions that determine the distribution and abundance of organisms (Krebs, 1994). Pollutants are then seen as one small part of the total, and ecotoxicology as a subdiscipline of ecology.

The individual organism is the basic unit of study. Individual plants and animals are, usually, physically distinct from each other, and, with practice, it is usually possible to allocate like individuals of plant or animal to individual species. Linnaeus laid the foundations for the current system of classification into species, in which each organism is given a generic and specific name. Man is thus placed in the genus *Homo*, and classified as *Homo sapiens*.

Linnaeus supposed that each species is a distinct type, which the individual specimens reflect, in much the same way that individual crystals of a mineral may not all have precisely the same form, but all do reflect the distinctive crystalline form that is characteristic of that mineral. The concept of a species has developed considerably since the time of Linnaeus (1707–1778). The current understanding of what constitutes a species is an entirely novel concept: "Species are groups of interbreeding natural populations that are reproductively isolated from other such groups" (Mayr, 1963, 1970).

This definition results from the evolutionary view of life, and reconciles the historical conflict between the view of species as fixed unchangeable entities and the evolutionary view that species do change with time. Two species are distinguished essentially not by any difference in form or structure, but by not interbreeding: species are distinct rather than different. Thus species, the entities to which individual organisms are allocated, are not constant in their properties, unlike the chemist's elements or the physicist's abstractions from natural phenomena.

Ecological studies involve several levels of organization, for which special terms are commonly used, although a cautionary note is appropriate. Many of the terms used by ecologists have had, and still have, a range of meanings, and there is still disagreement about the meanings to be attached to some of them.

Within one species, those individuals that occur within a defined area can be called a population. Alternatively, a population may be defined as those individuals of a species "so situated that any two of them have equal probability of mating with each other and producing offspring", provided of course that the needs are met for opposite sexes, maturity and equivalence for sexual selection (Mayr, 1963). For observation and experiment, a population should ideally consist of a group of individuals that is clearly isolated, on a remote island, a mountain summit or in a pond. In practice, the boundaries of populations can be difficult to define, are likely to change with time, and are commonly defined somewhat arbitrarily, albeit with an attempt to delimit a distinct interbreeding group of individuals. Some populations, sometimes called meta-populations, consist of numerous separate relatively ephemeral groups (see Shorrocks and Swingland, 1990).

No one population exists alone, and the populations of different species that exist in the same area form a community. Whittaker (1975) defined a natural community—a community, such as redwood forest, that has not been significantly affected by man's activities—as an "assemblage of plants, animals, bacteria and fungi that live in an environment and interact with one another, forming a distinctive living system with its own composition, structure, environmental relations, development and function". In practice, only a limited range of species is usually studied, and rarely if ever are all of the species within a community studied. Like populations, communities are not always easy to delimit, and MacFadyen (1963) emphasized the empirical aspect when he described the community as a working hypothesis, a group of organisms that presumably interact with each other and are separable by survey from other groups.

This series, of organism, population and community, takes no account of the organism's surroundings, for which three terms in particular are useful. The term environment denotes all of an individual organism's surroundings, both the inanimate components such as the air, soil and water, and other plants and animals, including other members of its own species. The inanimate, or abiotic, components alone, for individual, population and community, constitute the habitat (Whittaker *et al.*, 1973), although the term biotope is sometimes used to distinguish the environment in which a community exists (Udvardy, 1959). Finally, a community with its habitat constitutes an ecosystem (Tansley, 1935), and it is the effects of pollutants on ecosystems that form the subject of ecotoxicology.

The nature of ecological relationships influences the impacts that pollutants have on wildlife, and the next three chapters consider some of the relevant aspects of ecology before turning to the impact of pollutants on habitats and directly on organisms, and of how to predict and monitor the biological effects of pollutants.

2 | Population Dynamics

No population remains constant in number for ever. This should be a self-evident statement, but it tends sometimes to be neglected when the ecological impact of pollutants is considered. This chapter is about one consequence of this statement: the effect that pollutants exert on population size is influenced greatly by many other factors. The introduction, or increase in amount, of a pollutant may coincide with the decrease in population size of one or more species, but this does not by itself indicate that the pollutant has necessarily affected that species. Nor does the certain proof that a pollutant has killed some individuals necessarily indicate that the population has been affected. Conversely, a pollutant that kills no individuals but has sublethal effects on a significant proportion of individuals could have a severe impact on the population.

Although many factors can affect population size, their combined total impact can be determined by four measures: birth rates, death rates, gains from immigration and losses from emigration. The rates of these four processes must determine the subsequent size of any population. Pollutants are simply one factor amongst many that can influence one or more of these measures, and before we can understand the ecological effects of pollutants, we need to consider the nature of the fluctuations in population sizes—population dynamics.

It will help if first we discuss briefly the "carrying capacity" of any particular habitat for any of the populations that occur within it. Every organism needs certain things from its environment: food, shelter, light, a restricted range of salinities for aquatic organisms—a very long list of possible needs can be made, but the essential point is that there is a limit to the amounts of these resources that any habitat can provide. The carrying capacity can then be defined as the maximum number of a species that a habitat can support throughout the complete life cycle.

To take one simple example, it appears that the size of the breeding population of peregrine falcons (*Falco peregrinus*) in Great Britain had, until 1940, remained relatively constant for possibly some centuries (Ratcliffe, 1963). In the 1930s there were probably about 820 breeding pairs (Ratcliffe, 1993). The limiting factor in their environment was suitable sites for their nests. Each summer there would be an average of about 2.5 fledged young reared by each breeding pair, but by the next breeding season there would again be about 820 breeding pairs of falcons. This limiting resource, of suitable nesting sites, would clearly remain fairly constant from year to year, and is therefore in this instance a sufficient explanation for a stable population size.

Populations of other species are often limited by more variable resources, such as food. But in all cases the environment's carrying capacity limits the size to which a population can possibly increase. If, as frequently happens, the number of adults plus offspring exceeds this capacity, then there will be competition between individuals for this resource, be it nesting sites, food or whatever.

The degree of competition for a resource that any individual will have to contend with increases as the population increases. Thus if one pair of peregrine falcons acquire a breeding site, this reduces the chances for other pairs to find breeding sites. One can then readily extend this idea of competition into a semi-quantitative form: the degree of competition that exists for a limited resource will be density-dependent. The larger the population the more intense will be the competition and the lower the proportion of individuals that will obtain an adequate amount of the limited resource. It is important to note that as the population increases a density-dependent factor increases not only the absolute number of individuals that disappear, but also the proportion or percentage that disappears. There are three possibilities for the surplus individuals: some may die, some may emigrate, or reproductive success may decrease.

In general, it is probably true to say that competition only exists above a certain density of individuals, and that below this critical or threshold level competition for a resource has no effect on the population density. For example, mortality in larvae of the almond moth (*Ephestia cautella*) increases suddenly above a critical density, probably because of food shortage (Fig. 2.1).

Many other factors besides that of competition can affect a population, and not all factors that do affect a population act in a density-dependent manner. There are also situations in which the density-dependence is inverse, that is, a higher percentage is lost at low densities. Bess (1961) studied populations of the gypsy moth (*Porthetria dispar*) in the USA for 8 years. This species has one generation a year and survives the winter as eggs. There are two main species that parasitize the eggs (*Anastatus disparis* and *Oenocyrtus kuwanae*), and both species are believed to be almost specific parasites on eggs of the gypsy moth. The proportion of parasitized eggs in experimental plots was found to be inversely density-dependent (Fig. 2.2). Some situations combine both types of interaction. Way (1968) found that the bean aphid (*Aphis fabae*) has an optimum density of adults feeding on bean plants for a maximum multiplication rate (Fig. 2.3). Up to this optimum density an increase in

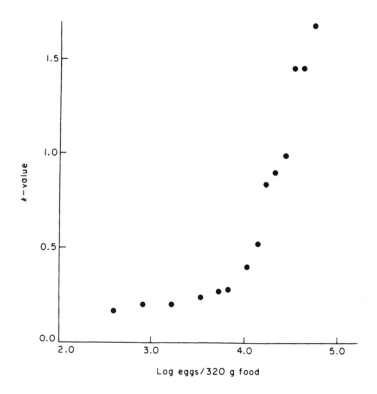

Fig. 2.1   Mortality in eggs and larvae of the almond moth *Ephestia cautella* reared at different initial egg densities on a fixed amount of food. Mortality is expressed as a *k*-value, where *k* = ln (initial number of eggs/number of larval survivors). (From Benson, 1973; original data by Takahashi, 1956.)

numbers appears to stimulate reproduction, and the flow of nutrients within the plant towards the aphids' feeding site may become more vigorous. At higher densities adverse factors override these beneficial effects, and the aphids may be said to become overcrowded.

There may also be situations where factors that affect a population are quite independent of population size, but it would be difficult ever to prove complete independence. This led Andrewartha and Birch (1954) to argue that density-independent factors do not exist, and the question is still debated (see Tamarin, 1978; Sinclair, 1989). It is probably a somewhat academic point, and for practical purposes we may suppose that some factors that affect numbers act independently of population size. Inclement weather is a good example, which may as a first approximation be expected to act independently of density, although it can be argued that even here effects are density-dependent. Solomon (1969) gives a lucid discussion of this topic.

The carrying capacity sets an upper limit to population size, but is not sufficient by itself to explain why populations change in numbers in the way

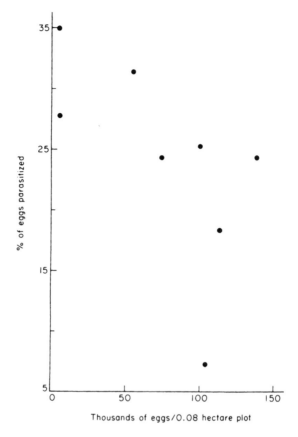

Fig. 2.2 The percentage of gypsy moth (*Porthetria dispar*) eggs parasitized in eight consecutive generations near Freetown, Massachusetts, 1937–1944. (Data from Bess, 1961.)

they do. Change in number must be the sum of gains from reproduction and immigration and of losses from death and emigration. Let us concentrate for a while on two of these terms: the balance between the rates of reproduction and of death. It is a relatively simple matter to keep experimental populations of small organisms such as bacteria, protozoa or insects in containers so that there is no immigration or emigration. It is invariably found with such populations, if there is no limiting factor such as lack of food or space, that initially the population increases in geometric (exponential) manner. During this phase one could meaningfully estimate the time needed for the population to double in size, regardless of the initial number of individuals. At some stage though, as numbers continue to increase, one or more factors start to reduce the rate of increase, until the population reaches and maintains a relatively constant number (Fig. 2.4). Data of this sort have stimulated the development of many mathematical models (May, 1974). Some would go so

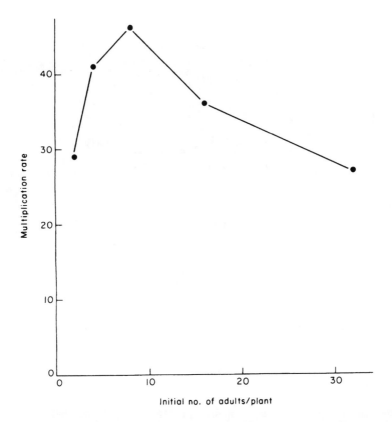

Fig. 2.3 The multiplication rate after 8 days for different numbers of bean aphids (*Aphis fabae*) kept on single plants of the field bean, *Vicia faba*. Each multiplication rate was significantly different ($P<0.05$) from the adjacent ones. (Data from Way, 1968.)

far as to say that there is a surfeit of such models and that the prime need is for more detailed studies of populations in their normal habitats (Dempster, 1975a), but we can usefully abstract one or two helpful ideas from a somewhat superficial appraisal of some of the simpler models.

The classic logistic equation, first proposed by Verhulst in 1838, states in its differential form that

$$\frac{\mathrm{d}N}{\mathrm{d}t} = rN\left(1 - \frac{N}{K}\right) \tag{2.1}$$

where N is the population size, t is time, r is the rate of population growth when there are no environmental constraints such as shortage of food or space, and K is the carrying capacity of the environment; $\mathrm{d}N/\mathrm{d}t$ is, of course, the population's growth rate at time t. For those who are unfamiliar with differential equations, this statement can be expressed verbally as

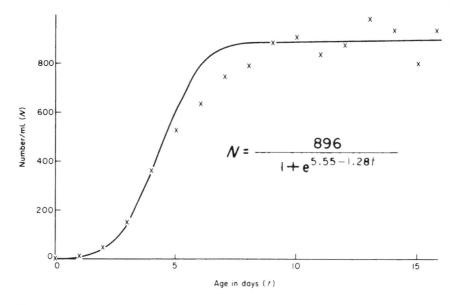

Fig. 2.4 The increase in population density of a culture of the protozoan *Paramecium aurelia* kept in 5 ml of buffered salt solution at 26°C. The salt solution was renewed every 2 days, and aliquots from a pure bacterial culture were added daily for food. The curve is calculated from the logistic equation. (Data from Gause, 1934.)

The population's actual growth rate	=	the population's potential growth rate	×	the proportion by which the population is below the environment's carrying capacity

It can also be restated in its integrated form as

$$N = \frac{K}{1 + e^{(a - rt)}} \tag{2.2}$$

where e and a are constants.

As with all mathematical models, we need first to know what assumptions have been made: this model assumes that the population has a stable age structure with complete overlap between generations. This qualification is needed because the probability of both reproduction and death varies with an individual's age. For our purposes we can neglect those situations where generations do not overlap—the equations are different in form, but they still derive from and illustrate the same general ideas (May, 1981a). It should also be noted that the equation predicts that the population stabilizes at a fixed level, K, but K is often taken to denote, confusingly, not the limit or ceiling value that N can take, but an equilibrium level around which N fluctuates (Pollard, 1981; Dempster, 1983).

These equations and Fig. 2.4 illustrate three essential features:

(1) Population growth is not unlimited, and will reach a stable maximum. This maximum value is measured by one parameter, K, which may be equated with the habitat's carrying capacity. For the simple example illustrated in Fig. 2.4, the carrying capacity is determined by the food supply, which was kept at a near-constant level by frequent renewal.

(2) The speed with which a population approaches its maximum value is determined solely by its rate of increase, r (increase in number/unit time/unit of population), which, for a population with a stable age structure, is the birth rate minus the death rate, when there are no environmental constraints.

(3) The effective rate of population growth, $r(1 - N/K)$, is density-dependent: the larger the population, the slower it increases in size. This equation also supposes the effect of population size on the recruitment of additional individuals to be instantaneous.

I doubt whether anyone would argue that this simple model gives a complete and totally satisfactory picture of the population changes in even such simple systems as can be kept within small containers in the laboratory. However, leaving these considerations aside, one much more fundamental point needs to be raised: the logistic equation describes populations that grow in a sigmoid or S-shaped fashion (e.g. Fig. 2.4), and there are quite a few equations that will fit such data (e.g. Medawar (1945) discusses a range of equations that describe the analogous process of growth in individual organisms). An adequate mathematical model must, of course, give a reasonable fit to existing data. Sometimes it may also focus attention on the important biological features of the system, and enable predictions to be made. Perhaps we can best see how useful the logistic equation is in these respects by considering in some detail an intensive study on a "natural" population. This will also introduce us to some more ideas that may help us to understand the dynamics of populations.

The cinnabar moth (*Tyria jacobaeae*) occurs naturally throughout Europe and adjacent parts of Asia. It is in many ways a good animal to study. The larval (caterpillar) stages have only one important host plant, the ragwort (*Senecio jacobaea*), the larvae are relatively easy to find, and in Britain there is only one generation per year. The life cycle starts with the egg, which hatches to produce a larva that moults four times, thus giving five larval instars. The life cycle is completed by the pupal (chrysalid) and adult stages. It is the pupae that survive the winter, in the surface layers of the soil, and the adult moths emerge the next summer in May, when they lay their eggs on the underside of leaves of their host plant. Caterpillars hatch from the eggs 2–3 weeks after laying, and the caterpillars feed and grow during the next 2–3 months before they are ready to pupate and so survive the next winter.

Dempster (1975a) studied a population at Weeting Heath in Norfolk for 8 years. The study area was 19 ha of sandy grass-heath, dominated by lichens, with clumps of a few grass species, and a few herb species, of which ragwort was an abundant component. The numbers of each stage in the life cycle were estimated each year from 1966 to 1973, and the causes of death at each stage were also estimated. The crude data for survivors to each stage for each year (Fig. 2.5) show several interesting features:

(1) There is not much sign of a stable maximum size of population: the number of eggs, for example, varied from 62 in 1969 to 21 699 in 1971, a difference of 350-fold.

(2) In a general sort of way the numbers of each stage in the life cycle did depend on the number of individuals in the immediately preceding stage, but even for these simple graphs there are some striking discordancies. For example, there were far fewer pupae in 1968 than one might have anticipated, and likewise the number of fifth instar larvae in 1971 appears to be rather low.

(3) The percentage mortality varied markedly with both the stage of the life cycle and the year (Table 2.1).

Clearly, there were many interacting variables that combined to determine the number of adult moths that emerged each year. A somewhat complicated analysis of the data, whose validity depends to a considerable extent on detailed biological knowledge of the population, is needed before one can decide which factors were most significant in determining population trends. To put it another way, one wants to determine the key factors—those factors which best predict the numbers in future generations.

Key factors can be detected by measuring the mortality (k) at specific stages of the life cycle for several generations (Varley and Gradwell, 1960). k-Values (ln initial number/final number of individuals in each stage) are calculated for each successive stage in the life cycle and their sum (K) measures the total mortality for each generation. One can then see graphically, or statistically, which factors contribute significantly to the variation in total mortality (e.g. Fig. 2.6).

Some care is needed when interpreting key factors. One is not strictly establishing a cause and effect relationship between the key factor(s) and changes in population. In strict logical terms it is a correlation, which may be very effective for predicting future population size, but does not necessarily mean that it is the factor(s) that control(s) population size (see Eberhardt, 1970). A brief digression may help make this clear.

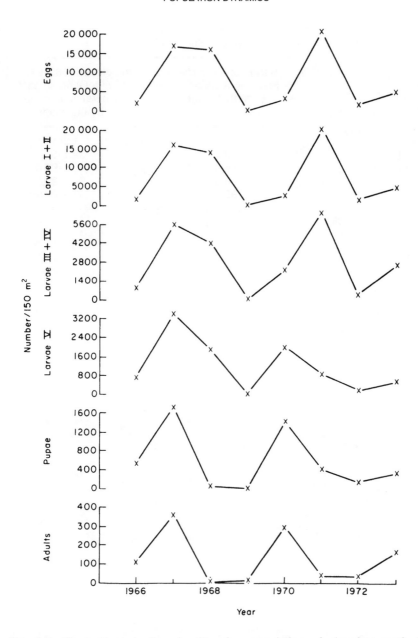

Fig. 2.5 Fluctuations in the density of a population of cinnabar moth (*Tyria jacobaeae*) from 1966 to 1973 at Weeting Heath, Norfolk. (Data from Dempster, 1975a.)

Table 2.1   The percentage mortality of cinnabar moth (*Tyria jacobaeae*) at different stages of the life cycle during the years 1966–1973

	Range of mortalities during the years 1966–1973	
Stage of life cycle	% reaching a stage that died in that stage	% of total no. in each generation that died by that stage
Eggs	1.2–13.1	1.2–13.1
Larvae I + II	23.2–83.8	32.1–85.3
Larvae III + IV	0.0–86.6	32.3–96.0
Larvae V	9.6–96.0	55.1–99.6
Pupae	51.4–98.0	80.6–99.99

From Dempster (1975a).

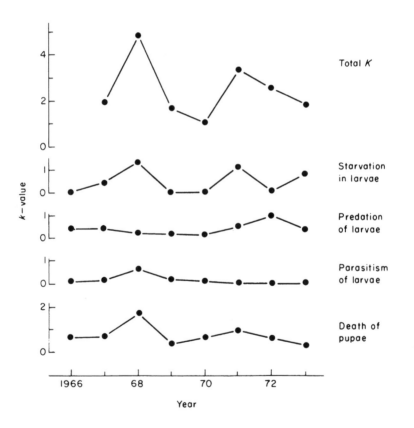

Fig. 2.6   The annual fluctuations in the degree of mortality from various causes, and their relationship with total mortality, for a population of cinnabar moth, *Tyria jacobaeae*, at Weeting Heath, Norfolk. Mortality is measured by *k*-values (defined in the legend to Fig. 2.1). (From Dempster, 1975a.)

Hypericum perforatum (common St John's wort, or Klamath weed) is a common but agriculturally unimportant plant in Europe. It has become widely distributed in other temperate parts of the world, where it has sometimes become a serious weed. It became such a nuisance on the west coast of the USA that attempts were made at biological control. One of the candidate species for control was a beetle, *Chrysolina quadrigemina*, imported from Australia. Like the larvae of cinnabar moth on ragwort plants, the larvae of this beetle feed on the leaves of Klamath weed, and after 3 years the plant's food reserves become exhausted and the plant dies. Control with this beetle was very successful, but control was most complete in sunny areas—the adult female tends to avoid shade when laying her eggs. Huffaker and Messenger (1964) point out that we might well have concluded from the present situation, without the benefit of hindsight, that this beetle has little influence on the population density of Klamath weed. Harper (1977) developed this thought further by suggesting that an ecologist would probably attempt to understand this plant's "shade-loving" character by growing it in a range of controlled light intensities until he discovered some process that the plant managed best at low light intensities. This would then become the explanation for the plant's observed distribution!

To return to the cinnabar moth, Dempster (1975a) concluded that in this particular population starvation amongst the larvae, and predation of the pupae, were the key factors that determined the trends from one year to the next. In 4 of the 8 years of study (1967, 1968, 1971 and 1973) there were sufficient caterpillars present to completely defoliate their food plants, so that many died from lack of food. As a rough guide, about 1 g of ragwort plant per larva at the time of hatch is needed to avoid starvation. The amount actually present ranged from 0.04 to 162.07 g/larva (Table 2.2). Thus there is a cycle of abundance for both the moth and its food plant: the number of moths crashes when the ragwort plants are eaten out, and the rate of recovery then depends on how quickly the ragwort plants recover.

The ragwort is normally a biennial: it develops into a rosette plant during the first summer, flowers in the second season, and then dies. If prevented from flowering, it will, like many other biennials, become a perennial. Defoliation causes regeneration from rootbuds, especially in young rosette plants. These facts of the plant's physiology, plus variations in rainfall, suffice to explain, in general terms at least, the observed fluctuations in density of ragwort plants at Weeting Heath (Table 2.2).

Thus the summer of 1968 was very wet, and so the severe defoliation stimulated a considerable amount of regeneration. The other years of complete defoliation were much drier, and there were correspondingly fewer plants in the following year.

One might argue from Fig. 2.5 that starvation did not cause a crash in numbers in 1967, which in fact had more adults than did any of the other 7 years. This contradiction is more apparent than real. Total mortality from egg to adult was high in 1967, with only 2.1% of the eggs producing adults, but

Table 2.2 The number of ragwort plants per m² and the weight of ragwort per larva at the time of hatching

	1966	1967	1968	1969	1970	1971	1972	1973
No. of rosette plants	5.3	4.0	5.5	59.7	33.9	18.1	17.8	19.2
No. of flowering plants	4.4	4.1	0.1	8.4	18.4	7.7	0.3	3.1
Total no. of plants	9.7	8.1	5.6	68.1	52.3	25.8	18.1	22.3
Weight (g) per larva	1.36	0.14	0.04	162.07	4.79	0.25	0.91	0.78
Complete defoliation	↑	↑				↑		↑
Rainfall (mm) from July to September	164	127	296	129	113	153	182	173

From Dempster (1975a).

in 1968 and 1971—the other 2 years of crash that started from high numbers of eggs—total mortality was even more intense, with only 0.01 and 0.02% of the eggs producing adults. It is important to appreciate that population trends are not necessarily determined by the major cause or causes of death. Figure 2.6 shows clearly that the fluctuations in total mortality correlate closely with death from larval starvation and death of pupae. Parasitism and predation of the larvae sometimes cause more deaths, but their changes from year to year do not correlate with, and therefore cannot explain, the trends in population size.

This account of the factors controlling a population of cinnabar moth is quite satisfying, as far as it goes. It does explain the causes and extent of the fluctuations in numbers from year to year. The population of ragwort plants is a major component in the environment's carrying capacity: starvation in cinnabar larvae clearly relates to the number of eggs laid and to the physiology and population dynamics of the ragwort plant. Pupal mortality is thought, at Weeting, to be due principally to being eaten by moles (*Talpa europaea*), which depends presumably on the behaviour and population dynamics of that species.

However, this, like most scientific studies, raises perhaps as many questions as it answers. What would happen if the degree of predation on larvae (Fig. 2.6) were different: either more or less intense? Would larval starvation still be a key factor? And why does the amount of predation on larvae take the pattern it does from year to year? All of these specific questions can be put into the general form of: how do the interactions between populations of different species affect those populations? It is difficult if not impossible, when dealing with a single population, to discuss data in terms of simple cause and effect. Not only does the environment affect the population, but the population also affects its environment: not only does the ragwort population affect the cinnabar population, but the population of the cinnabar also affects that of the ragwort. This is an example of feedback, or what Hutchinson (1948) called circular causal systems. It seems quite certain that, at Weeting, lack of sufficient ragwort plants causes population crashes of cinnabar moth. Other populations of this species are not necessarily controlled by the same factors. A population of cinnabar moth at Monks Wood, in Cambridgeshire, never ate out its food supply during several years' observation. Predation of larvae was more intense, and pupal mortality was very high, probably because the soil became waterlogged during the winter. Flooding in the autumn of 1968 exterminated the population.

To generalize, the key factor or factors that predict population trends may act at any stage in the life cycle, and these factors may be competition between or within species for resources (e.g. food, nesting sites), interactions between species (e.g. parasitism, predation), or effects of the abiotic environment, such as drought, flood, heat or cold. These factors may also differ from one population to another of the same species.

It should perhaps be stressed that in practice it can be very difficult to determine the precise nature of the interactions that do occur between a population and its environment. Evans (1976) illustrates this with an account, based on work by Ashton, of *Eucalyptus regnans*, a tree native to south-east Australia and Tasmania. The mature trees are very large, and often occur in groves of even age with no younger trees present. The trees flower and fruit regularly and abundantly. It is estimated that in a typical area there should be about 400 000 seedlings per hectare each year. In fact, most of the seeds are carried by ants too deep into the soil for the seedlings to be able to surface. Those that do germinate on the surface succumb to a range of factors, and if they do grow as far as the four-leaf stage, they are grazed by wallabies. Thus none of a year's seed survives, and the even age of the groves shows that this situation has existed for a long time. One might deduce that the present ecosystem prevents this species from reproducing itself, and that in due course it would die out and be replaced by other tree species. However, one factor is missing from this account—fire. Occasionally, in dry summers fires break out naturally. After a fire in 1939 seedlings appeared in a study area at a density of 2.5 million per hectare. The fire had greatly reduced the impact of other species on the seedlings, and enough survived to form a new grove.

This example also illustrates another theme. We have implicitly assumed so far that a population's dynamics are deterministic, that given enough knowledge of a population and its environment at one time we could predict its future history. In fact, ecologists are coming to realize that sometimes, especially in small populations, what happens to a population depends on chance events. These are what mathematicians call stochastic situations, in which events are not defined precisely by earlier events, but events have a degree of probability attached to them (Renshaw, 1991). Thus one could estimate the annual probability of fire occurring in an area and so permitting new groves of *Eucalyptus regnans* to establish themselves. More generally, because many of the factors that control population size have an element of chance, it follows that the smaller the population the greater the probability of extinction by chance (Haldane, 1956b; Shaffer, 1987).

Because of all these complexities there is no general agreement about the control of population size. The "balance of nature"—the idea that populations are in stable equilibrium and constant in size—was already implicit in the writings of Plato and Herodotus (Egerton, 1973) and has influenced ideas on population ever since. It is a fact that many populations survive for a long time, which led Nicholson (1933; see also Nicholson and Bailey, 1935) to suggest that population size is regulated by density-dependent factors. Although any factor that alters the rate of population gain or loss, for example the effect of temperature on reproductive rate, helps *limit* population size, only density-dependent factors can help *regulate* a population by returning numbers towards a stable equilibrium. This does not necessarily imply that there is a stable equilibrium density of population. There may well be oscillations tuned to fluctuations in environmental factors, but in the long run these oscillations will be about a mean. This proposal has aroused much furious discussion, with many alternative theories being proposed (Sinclair, 1989). The weakest point in Nicholson's argument is perhaps that, so far as our present knowledge goes, density-dependent factors are only density-dependent in the intensity of their effect over a limited range of densities. Of critical importance, there is a threshold value below which the effect of a factor is thought to be independent of density (Fig. 2.1). Moreover, one cannot assume automatically that, because a factor acts in a density-dependent manner up to a certain density, the factor will continue to act in such a way at higher densities.

Haldane (1956b) suggested that in favourable habitats with high population densities numbers are controlled largely by density-dependent factors but that density-independent factors become more important in less favourable habitats with lower densities. Studies on two populations of brown trout (*Salmo trutta*), 3 km apart in the English Lake District, support this suggestion (Elliott, 1994). During the years 1966–1990 the k-values showed that for the denser population, in Black Brows Beck, only mortality during the first summer after hatch contributed significantly to the variation between years in total mortality. Such predictive mortality factors are not always density-dependent, but plots of k-values against the initial population density on which they acted will distinguish them: a significant regression indicates dependence provided that some statistical requirements are met (Varley and Gradwell, 1968; Southwood et al., 1989). First-year mortality was density-dependent (Fig. 2.7a). By contrast, data from 1966 to 1972 for a sparser population in Wilfin Beck with a trout density about 10% of that in Black Brows Beck found no evidence of a significant regression nor therefore of density-dependence (Fig. 2.7b).

Apart from density, the regularity with which individuals occur within the habitat may also be important. "Meta-populations", populations that are subdivided to some extent, may not have the same population dynamics as populations with a relatively uniform distribution (Kareiva, 1990).

This debate about regulation, or limitation by a ceiling, of population size is not yet concluded. Thus, key-factor analysis may not reveal existing density-dependence if the study population is heterogeneous (Hassell *et al.*, 1987), and

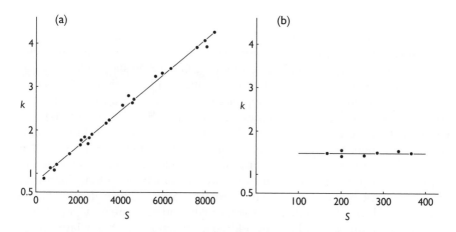

Fig. 2.7 Plots for two populations of brown trout (*Salmo trutta*) of *k*-values, for spring mortality in the first year of life, on initial egg density (*S* is the number of eggs laid per 60 m² of river bed). (a) Linear regression for trout in Black Brows Beck, 1966–1990. (From Elliott, 1994.) (b) Mean value for trout in Wilfin Beck, 1966–1972. (Data from Elliott, 1994.)

conversely detection of density-dependence does not of itself prove that regulation occurs (Dempster, 1983). We lack sufficient knowledge at present to discriminate between these and other theories.

This chapter has been dominated so far by examples from animal populations. This is a fair reflection of the way the study of populations has developed, although plants are in some ways more amenable to study—they do not move, and so are much easier to observe in the field. There are two main reasons for the paucity of studies on plant populations. First, individual plants of the same age can vary enormously in size, which can mean, for example, that more resources yield fewer individual plants. Conditions conducive to growth may produce bigger plants with fewer individuals per unit area. Secondly, the genetic and the physiological unit may not be identical. Plants are very prone to what is normally called vegetative reproduction— one genotype of the grass *Holcus mollis* appeared to extend in this way for over a kilometre in grassland vegetation (Harberd, 1967). There are thus two types of individuals, called genets (genetic units) and ramets (physiological units) (Kays and Harper, 1974). Both of these reasons can make it difficult to decide what to count in a plant population. However, despite the great differences of detail that exist between studies of plant and of animal populations, the same essential principles do appear to govern both (Harper, 1977).

✳

We started this consideration of changes in population size to see how relevant is the logistic equation. Not all species behave like the cinnabar moth, but although the logistic equation does contain the two parameters r and K, both of which are meaningful biologically, it is clearly a rather unsatisfactory model for populations that fluctuate appreciably in size. It is at this stage that what has come to be called "theoretical ecology" (May, 1981b) starts to extend ecological theory beyond that based on observation and experiment. It is a matter of observation that often a population's response to an environmental change is not instantaneous: equation (2.1) assumes that the density-dependent regulatory mechanism (simulated by the factor $1 - N/K$) acts immediately. In practice, there is often a delay before a density-dependent mechanism takes effect. For example a predator population may respond to an increased population of prey by better survival and reproduction, so that there is an increased number of predators in the next generation. The result may then be an increased level of predation on the next generation of the prey population. Such features can be simulated by adding a constant T for the delay before a density-dependent regulatory effect operates (Hutchinson, 1948):

$$\frac{dN_t}{dt} = rN_t\left(1 - \frac{N_{(t-T)}}{K}\right) \tag{2.3}$$

The differential equation now states that the rate of increase for a population at time t depends on the size of the population not at the time t, but at the earlier time $t-T$. The consequences of such a seemingly small alteration are considerable. The important feature is the relationship between T and r. As the ratio r/T increases, so one has populations that change from a stable equilibrium, via populations that oscillate in a consistent pattern, to ones that fluctuate in a seemingly random, or chaotic, manner (May, 1981a). It is at about this stage that the theory outstrips observational data (e.g. Hassell and Anderson, 1989). It is perfectly true to say that populations have been observed to follow all of these patterns, but detailed studies are needed to determine the critical factors that predict the size of real populations. Indeed, random events may mean predictions cannot be deterministic, but are probabilistic, and may obliterate many of the predicted patterns (Renshaw, 1991).

Before we can usefully extend our ideas on population dynamics any further, we must digress to a philosophical point. One tends almost inevitably on occasion to explain biological facts in terms of purpose. "Giraffes have long necks so that they can reach up to the leaves on the trees" might serve as an example. This is what is known technically as a teleological statement—the fact that giraffes have long necks is explained as the result of a purpose or design, that giraffes should be able to eat leaves from trees. It is indeed true that giraffes eat leaves from trees, but the usually accepted ideas of evolution do not see such facts as part of a conscious design. Perhaps there is a conscious design, but even if there is, there are, for scientific purposes, more effective ways of approaching such questions. Southwood (1976) expressed it very well "... the organism's habitat may be viewed as a templet against which evolutionary forces fashion its bionomic or ecological strategy". The key word here is "strategy". It does not imply any sense of conscious purpose or intent. Rather, it implies that an organism has evolved one way rather than another of coping with the vagaries of its environment: different competing species have different strategies.

MacArthur and Wilson (1967) promoted the distinction between two types of strategy for the survival of populations when they were considering why islands contain the numbers and types of species that they do. In more general terms, what makes a species a successful colonizer of a new habitat, be it a small isolated island in the middle of the Pacific or a tidal pool on a rocky shore? This can be regarded as the extreme form of an important question for ecotoxicology: how does a population react when a large proportion has been killed by a pollutant? If the pollutant eliminated the population, the question would become: how does a species recolonize an area? MacArthur and Wilson (1967) constructed sophisticated mathematical models that were based on a hypothetical species birth rate (λ), death rate (μ) and equilibrium population size (K). $\lambda - \mu$ is, of course, equal by definition to r. These models could not be tested except in a few restricted instances, but MacArthur and Wilson deduced that the probability of colonizing success is about r/λ, and that the greater the value of K the longer the population is likely to persist. Currently, there is much enthusiasm for this distinction, and certainly many biological facts can plausibly be ascribed in terms of an r- or K-strategy for survival. It has also been used successfully to test hypotheses about the differences between populations and species. Southwood (1976) expanded these ideas and suggested that all species can be placed on an r–K continuum of strategies. r-Strategists are the successful colonizers, and continually invade new temporary habitats. They have high rates of increase, high migrating ability, and poor competitive ability. Every gardener will have encountered some of these species—they are the weeds that usually arrive first on a bare piece of soil. Insect pests too tend to this end of the spectrum. The aphid *Aphis fabae*, for example, has the potential to increase its population 1000-fold in 6 weeks in a crop of field beans (Way and Banks, 1967). In practice, the extent to which this potential for increase is realized would depend on several factors. The most important for the aphid are probably the condition of the food plant, drought, heavy rainfall and predators. In contrast, K-strategists have stable habitats, in which they maximize their use of the available resources, thus maximizing the size population that the environment can carry. Their populations tend to remain at this equilibrium level, with long-lived individuals and low reproductive rates. They are likely to have high interspecific competitive ability, with an appreciable investment in defence mechanisms.

Plants too can be placed on an r–K continuum of strategies. This was illustrated in a study on six populations of four species of *Solidago* (golden rod), which were studied in three sites (Abrahamson and Gadgil, 1973). The sites were at three different stages of development, from an open dry disturbed site to a wet meadow site with dense growth, and in contrast a woodland site with little disturbance where a K-strategy would be more appropriate than in either of the other two sites. *Solidago speciosa* and *S. rugosa* both occurred in the woodland site, and in one of the other two earlier successional sites. The proportion of plant biomass allocated to seed production was less in the woodland site than in the other two sites (Fig. 2.8). Two other species, *S.*

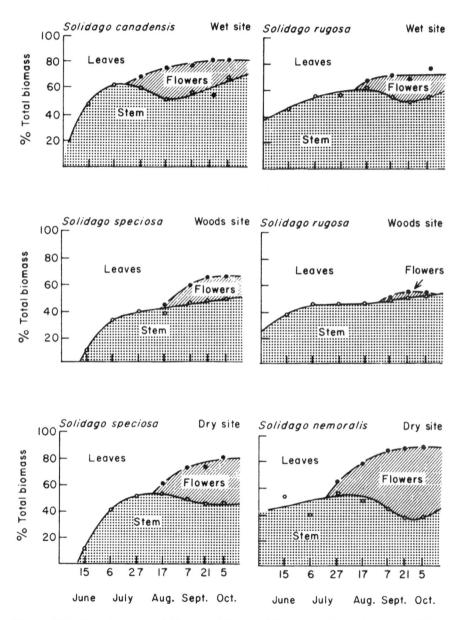

Figure 2.8 The allocation of biomass (dry weight) among stems, leaves and flowers in six populations of four species of *Solidago* (golden rod) at three sites at different stages of community development. (From Abrahamson and Gadgil, 1973.)

canadensis and *S. nemoralis*, only occurred in one of the earlier successional sites. Both of these species had a larger proportion of biomass for seed production than the other species in those sites that also occurred in the woodland site. This example raises an interesting and important question to which we will return in Chapter 4. We do not know whether the differences between populations within the two species of *Solidago* that occur in two types of site are determined by their genetic constitution or by the environment in which they are living.

In general, *r*-species produce small seeds, whereas *K*-species will tend to have larger seeds and devote a larger proportion of their resources to survival. The most developed form of *K*-strategy for plants is perhaps to become a large, long-lived tree. A tree's woody tissues use a major part of the plant's resources. This gives the plant a long-lived advantage in height, so that leaves may shade those of neighbouring plants, flowers may be more obviously displayed for pollination by wind or insect, and seeds may be dispersed greater distances.

Differences between *r*- and *K*-strategists may be quite subtle. Cates and Orians (1975) made an interesting demonstration of one aspect. They argued that plants that first colonize bare ground will have short life-spans with relatively low risks of being found by herbivores, whereas those species that tend to arrive later will be slower growing and much more liable to attack by herbivores. They therefore suggested that the early colonists would commit fewer resources to defence against herbivores, and would therefore be more acceptable as food. They tested 80 species of plants for palatability to two species of slug (Table 2.3). There was, indeed, a trend towards lower palatability, although the standard errors for the later colonists show that not all of these species were less palatable.

One particular aspect of *r*- and *K*-strategies should be noted. Fenchel (1974) showed for 42 species ranging in size from the virus T-phage to the cow that *r*, the potential rate of increase, is related to body weight (*W*) by the

Table 2.3 The palatability of plants to two species of slug

	Number of species tested	Palatability index[a] (±SE)	
Type of plant		*Arion ater*	*Ariolimax columbianus*
Early successional annuals and biennials	18	0.99±0.14	0.96±0.09
Early successional perennials	45	0.69±0.36	0.77±0.30
Later successional and climax plants	17	0.40±0.39	0.46±0.45

Data from Cates and Orians (1975).

[a]Palatability index = $\dfrac{\text{log. of amount of test material eaten}}{\text{log. of amount of control material eaten}}$

equation $r=aW^n$, where a and n are constants. There is also an additional component that differentiates among unicellular organisms, cold-blooded animals, and mammals (Fig. 2.9). Some caution needs to be applied to this relationship—in particular, many of the species were standard laboratory species, many of these are pest species, and on both these counts these species may tend to have higher values for r than other species of similar weight. However, there is undoubtedly a general trend for larger organisms to have lower values of r.

There are conflicting views on the relevance of r- and K-strategies (e.g. Stearns, 1977; Boyce, 1984). The idea of different strategies can be quite productive of testable hypotheses (e.g. Grime and Hunt, 1975), but it cannot be accepted as a fundamental law—a spectrum in which all organisms have their due place. Environmental conditions differ from place to place and from time to time. Moreover, individuals may have different strategies at different stages of their life cycle, although this criticism can be met by restricting the distinction between the two types of strategy to two criteria, by which r-strategists are defined as species with a short expectation of life and that devote a large proportion of their resources to reproduction, and K-strategists are the converse. Again though, as Fig. 2.8 shows for resource allocation, the way organisms develop is determined to some extent by interaction with their environment. More fundamentally, it is probably unrealistic to suppose that species fall somewhere along the continuum between

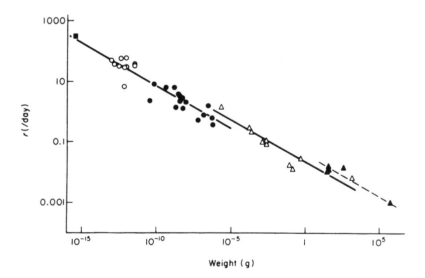

Fig. 2.9 The relationship for 42 species between the potential rate of increase (r) and body weight (W). Both axes are on logarithmic scales. (■) T-phage; (○) bacteria; (●) protozoa; (△) cold-blooded metazoa; (▲) mammals. (From Fenchel, 1974.) Blueweiss *et al.* (1978) argue that there is no significant difference between the three regression lines.

only two extreme conditions. Grime (1979, 1988) has recently argued that there are three strategies for plants, in which plants have either high competitive ability (C-selection), high endurance of stress (S-selection) or have high colonizing ability of disturbed sites such as marsh and arable land (R-selection, or ruderal species). Grime then points out that, of course, these three strategies and the gradations between them do not form a neat linear sequence, but that plants may be adapted to various degrees to all three of these factors. Whittaker (1975) refers to r, K and L selection, or species that are adapted to temporary, stable or adverse environments.

For ecotoxicology, the important, obvious and simple deduction is that different species are adapted to survive in different ways: they have different needs for resources, can have very different types of life cycle and react to the challenges of life in different ways. One may then suggest that the net effect of a pollutant on populations of two species will be quite different even though the individuals of the two populations are affected to the same extent.

*

It should be obvious that population dynamics is a rapidly developing subject, with much unknown, and considerable discussion as to the correct conclusions to be drawn from what is known. Nevertheless, there are important implications for ecotoxicology. The LD_{50} (that dose which is lethal to 50% of a test population) and similar measures conventionally form a major part of the assessment of ecological risks from potential pollutants. Indeed, there is a school of thought that considers the acute lethality test to be the most useful of all available tests (Cairns et al., 1978). We will accept for the moment the assumption that death is the effect of paramount importance. It is necessary then to realize that, although probably no potential pollutant is a completely specific poison that affects only one species, differences between species in their susceptibility can be important (Blanck et al., 1984; Slooff et al., 1986).

A striking illustration of unexpected differential toxicity in the field occurred after dieldrin had been banned from use in Great Britain as a seed dressing to protect winter wheat from wheat bulb fly, because of the risk to seed-eating birds. The organophosphorus insecticide carbophenothion was one of the substitutes that replaced dieldrin, but several incidents were reported from 1971 to 1975 of deaths of greylag geese (*Anser anser*) and of pink-footed geese (*Anser fabalis brachyrhynchus*) that had been feeding in fields sown with dressed grain. One of the notable features of these incidents was that only geese of the genus *Anser* were found dead, although other species of grain-eating birds had also been present on these fields. In laboratory tests an oral dose of 25 mg carbophenothion/kg body weight was sufficient to kill these two species of geese, whereas Canada geese (*Branta canadensis*) with an estimated LD_{50} of 29–35 mg/kg were unafffected, as were pigeons (*Columba livia*) and chickens (*Gallus domesticus*) (Westlake et al., 1978). This example illustrates the point that we cannot predict with complete confidence from toxicity data for one species to the toxicity of the same chemical for another species, even if closely related. It also needs to be emphasized that the environment can have a considerable effect on the intake of a pollutant (e.g. Graney *et al.*, 1984) and its activity (e.g. McClusky *et al.*, 1986, and Fig. 2.10), and we will return to this point later (Chapter 6).

If we can suppose that we are aware of a pollutant's relative toxicity to different species, we come to the question: if a pollutant kills an appreciable number of individuals, what is the effect on the population? Some bird-kills from insecticides have been very dramatic, with many corpses visible, but it is not easy to decide what effect such incidents may have on the population. Many field observations contradict the simplistic view that a large number of sudden deaths will permanently damage a population. In this context the cause of death is immaterial, and a brief episode of deaths from pollution might simply mean that the population would suffer fewer deaths from other causes. Murton *et al.* (1974) showed, in an analogous situation, that the shooting of wood pigeons (*Columba palumbus*) during the winter in agricultural habitats had little effect on population size. Rather, it reduced the number that would have died from starvation. Moreover, populations do sometimes suffer natural disasters such as hard winters, when large numbers die, and from which they usually recover (Fig. 2.11). Conversely, populations are sometimes culled if they are thought to be getting too large, and this has to be repeated from time to time if the population is to be restrained at the lower level, unless of course the environment is modified in such a way that its carrying capacity is reduced.

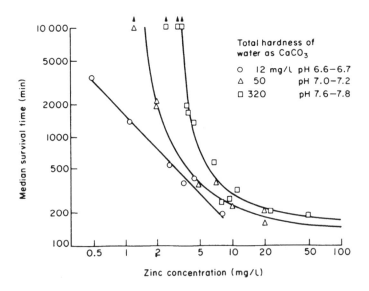

Fig. 2.10 The effect of water quality on the survival of rainbow trout (*Salmo gairdneri*) exposed to zinc sulphate. (From Lloyd, 1960.)

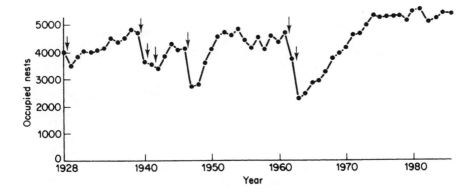

Fig. 2.11   The approximate number of nests occupied by herons (*Ardea cinerea*) in England and Wales each year from 1928 to 1985. The arrows indicate severe winters. (From Reynolds, 1979, and Reynolds, in prep.)

A more sophisticated variant of this approach is to estimate the maximum sustainable yield, that is to say the maximum yield that can be taken or harvested regularly from a population without jeopardizing future yields (Clark, 1990). This question usually arises for species that are used for food or for sport—fish and game-birds provide obvious examples—and a simplistic model (Fig. 2.12) may serve to illustrate the basic idea. If the carrying capacity remains constant then the population size will fluctuate about a mean unless a harvest is taken. Regular harvesting will permit the survivors to increase in number. If the number harvested each year equals the potential increase in numbers, then the population will remain constant and a sustained yield is possible.

In practice sustainable yields can be difficult to determine (Getz and Haight, 1989), but two points from our simple model are relevant:

(1) There are two population sizes that will produce any size of sustained yield, except for the population size that gives the maximum sustained yield.
(2) If a constant number is harvested each year from a population that was initially at its stable size, then the population will decline in size until it reaches the upper size for which that constant harvest is the sustained yield. If this harvest exceeds the maximum sustained yield, then the population will decline and become extinct.

We may conclude that deaths from a pollutant (or any other cause) do not necessarily reduce a population.

A less simple variation of the assumption that a population is damaged by a large number of sudden deaths is to ask whether, in the most general terms, populations of different species react in similar ways to a sudden drop in numbers. The simple answer is, no.

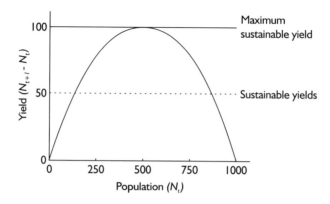

Fig. 2.12   A simple model to illustrate the sustainable annual yield from an exploited population with one generation per year, where $r = 0.2$, $K = 1000$ (see equation 2.1). N_t is the number of adults in one year after all density-dependent effects on birth, death and migration have occurred, and N_{t+1} is the number of adults the next year. (Adapted from Caughley and Sinclair, 1994.)

We can illustrate several possible types of response from the effects of insecticides. It is usually considered impracticable to eliminate an insect pest with insecticides. All one can reasonably aim for is a spray regime that keeps the pest population below the level at which its effects are unacceptable. One reason for this will be that even a 100% kill in the sprayed area will not eliminate the pest, because it will simply recolonize the area from adjacent regions. This, however, is not the whole explanation, and the mosquito vectors of malaria provide a good example. Considerable efforts have been made to eliminate these species of mosquito from many parts of the world, with but very limited success. Not only is it almost impossible to achieve a 100% kill, but the consequence of a very high percentage kill is that one rapidly selects out those genotypes (genetic constitutions, see p. 68) that are resistant to the insecticide. Mosquitoes can, like many insect pests, multiply rapidly. This then enables them, in a regime of frequent exposures to insecticide, to rapidly develop populations that are resistant to insecticides.

Sometimes an insecticide also reduces populations of a pest's predators or competitors, when the pest population may recover to reach higher levels than before (resurgence). New pest species may also appear: species whose predators and competitors had previously kept them at sufficiently low density to avoid notice (Dempster, 1975b).

To return to the peregrine falcon (*Falco peregrinus*), the British population illustrates a very different situation. The peregrine falcon had very poor reproductive success from the late 1940s onwards, most probably as a result of organochlorine insecticides, but the population remained fairly stable until the mid-1950s, when this bird virtually disappeared from most parts of Britain, probably from acute poisoning from dieldrin ingested with prey

(Ratcliffe, 1970). If there had been a deliberate policy to exterminate this species completely from Britain, it would have been a relatively simple matter, and then restoration of a normal-sized population by recolonization and reproduction would have been a slow process measured in years.

Compared with the peregrine falcon, mosquitoes show at least some of the characteristics of an r-strategy, with high reproductive rates and great opportunism about breeding sites—temporary pools of water are readily utilized for their eggs and larvae. Peregrine falcons tend to a K-strategy, with much lower reproductive rates, longer-lived individuals and more stable populations.

The golden eagle (*Aquila chrysaëtos*) illustrates how long-lived K-strategists could be affected by pollutants in a way that could not occur with short-lived r-strategists. Individual eagles commonly live for as long as 30 years. Their breeding success in Scotland during the years 1963–1965 was greatly reduced: in only 31% of the nests that contained eggs did the adults successfully rear any young (Lockie et al., 1969). This was attributed to dieldrin residues obtained from sheep carrion. Despite this reduced breeding success the population did not decline, and one explanation could be that the population simply got older (Dempster, 1975b). In due course an ageing population of this sort would decrease in size, but a population of a typical r-strategist whose reproduction was inhibited would decrease much more rapidly because of the short life-span. Any survivors of a brief exposure, or emigrants to another habitat, would rapidly build up in numbers again, whereas the typical K-strategist whose population had been reduced to low numbers would take much longer to recover, and there would be a greater risk of extinction if the population was very small.

We have been considering situations where it is known, or assumed, that a pollutant is killing part of a population. It should be obvious by now that rarely will simple estimates of changes in population size with time be adequate to detect effects of pollutants unless the effect is immediate and obvious, such as oil pollution on a beach. To explain why a population is the size it is and fluctuates in the way it does is difficult for many reasons: not all of the relevant factors may be known; species interact with each other; it may take literally years to establish the key factors; sometimes chance events can be significant; and different populations of the same species may not be controlled in the same way.

It is unwise, moreover, to focus all of our attention on the lethal effect of pollutants. Even for human toxicology, where interest is centred on the individual rather than the whole population, the LD_{50} is of doubtful value (Frazer and Sharratt, 1969). The LD_{50} test was designed for the biological assay of drugs that could not be determined chemically. It is ideal for that purpose, giving reproducible results under standardized conditions, but the reverse process of using an LD_{50} as the measure of a chemical's toxicity is a much more dubious affair. The LD_{50} varies with species, strain, sex, age and environmental conditions. Moreover, it takes no account of sublethal effects.

Clearly, an increased death rate will reduce the value of r, the potential rate of increase. However, another equally effective way of reducing the rate of increase is to reduce the birth rate. Birth rate will be reduced by anything that reduces the individual's ability to survive and so reduces longevity. Other sublethal effects such as longer developmental period, shorter reproductive period, reduced fecundity or fertility will all have the same effect. Seemingly small changes may be quite significant.

Lewontin (1965) considered this topic in some detail. He argued that one could reasonably use a triangle to represent the way in which the reproductive rate of a cohort of similarly aged organisms will change with increasing age (Fig. 2.13). The total number of offspring (S) produced by the cohort is given by the area of the triangle. Lewontin assumed that population growth conformed to the logistic equation (equations 2.1 and 2.2) and that the population was at an early stage, so that growth was exponential. He then computed the effects of changes in A, T, W and S on the value of r, the rate of population increase (increase in numbers/individual/unit time). There are many possible variations, and Fig. 2.14 shows one of the simplest cases, where the intervals of time between A, T and W remain constant, although the values of all three can change by equal amounts. In addition, the maximum value of $V(x)$, at time T, can vary, thus varying the value of S. The lines on the graph of Fig. 2.14 indicate changes in the time between birth and reproductive age (A) that compensate exactly for the effect of changes in the total number of offspring (S) on the rate at which the

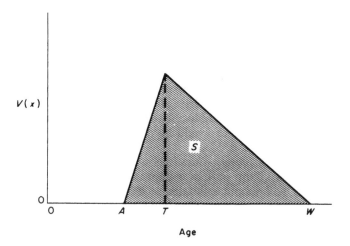

Fig. 2.13 A generalized representation of reproduction by a cohort of similarly aged individuals. A represents the age at which reproduction first occurs, T the age at which reproduction is maximal, and W the age at which reproduction ceases. The vertical axis $V(x)$ indicates how reproductive rate changes with age in the cohort. It is calculated for any age by multiplying the number that have survived to that age by the mean reproductive rate/survivor at that age. S is the total area of the triangle, and represents the total number of offspring. (From Lewontin, 1965.)

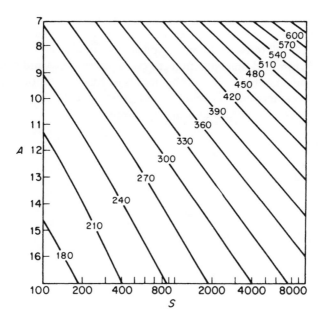

Fig. 2.14 Graph to show how the potential reproductive rate (r) changes with total number of offspring (S) and the time between birth and the start of reproduction (A). It is assumed that the intervals of time between A, T and W (defined in Fig. 2.13) are constant for all values of A (W–A was set at 43 units of time, T–A at 11 units of time). Each line on the graph represents a constant value of r, and the values of r are given × 10^3, e.g. 510 corresponds to $r = 0.510$. (From Lewontin, 1965.)

population increases. For example, at the upper end of the range of values for r, halving the total number of offspring from 10 000 to 5000 can decrease the value of r from 0.565 to 0.510. An identical decrease could be obtained by increasing the value of A from 7.5 to 8.6 units of time—increasing the time to start of reproduction by one-seventh is equivalent to halving the total number of offspring. For lower reproductive rates, halving the number of offspring from 300 to 150 would decrease r from 0.205 to 0.180, equivalent to an increase in A of nearly one-quarter from 13.2 to 16.2 units of time. Clearly, these ideas are most relevant to r-strategists, but larger organisms with smaller numbers of offspring are also likely to be sensitive to relatively minor changes in their breeding performance. The timing of breeding in many vertebrates, for example, is geared very closely to environmental variables, and a small shift could have a significant effect on reproductive success.

Of course, the same qualification applies to the effect of these sublethal changes on population size as for lethal effects. There is no simple relationship between a change in r and the effect if any on population size.

We have seen in this chapter that population dynamics are complicated, that populations of different species have different strategies for survival, that different populations of the same species are not necessarily controlled by the same factors and that sublethal effects of pollutants could be as important as lethal effects. It should also be apparent that no population lives in isolation, but interacts with populations of other species. We will now consider some aspects of communities of populations.

3 | Communities

We discussed in the previous chapter why populations are the size they are and why they fluctuate in size. We can ask similar questions about communities: why do some species but not others occur within a community, and what aspects of the community structure influence the sizes of those populations that do occur?

A community contains populations of different species that interact with each other, and these interactions among species are a major determinant of community composition. If we consider just two species within a community, the effect of each on the other may be beneficial, negligible or adverse, although it can be difficult to determine the net effect of the population of one species on the population of another species. For example, a predator or parasite may reduce the population of a prey or host species, but this may benefit the prey or host population by reducing its total demand for food and so increasing the net survival. To return to the example of the cinnabar moth (pp. 17–23), a parasite that reduced the number of moth larvae could conceivably, in some years, increase the number of subsequent adults.

We have seen that intraspecific competition—competition between members of one species—can be an important factor in the control of population size. Likewise, interspecific competition—competition between species for a limited resource—in which both species populations are smaller than they would otherwise be, appears to be an important factor in the structure of communities (Keddy, 1989). We will consider one example of interspecific competition to introduce some general ideas about community structure and function.

Park (1954) described a classically simple but revealing experiment on competition between two species of flour beetle, *Tribolium confusum* and *T. castaneum*, pests of stored flour that can easily be maintained in simple, standardized conditions in the laboratory. These two species are rather similar in size and in their birth and death rates under optimal conditions, with about one generation per month. Park set up a whole series of cultures. Each consisted of a vial with 8 g of food (95% flour, 5% brewers' yeast) to which were added four adult male and four adult female beetles. The number of beetles in each vial was then counted every 30 days, after which the beetles were returned to a fresh batch of medium. This process was repeated for 750 days, about 25 generations in all, which was taken to be sufficient time to reveal all likely trends. Milled cereals form the usual habitat of these beetles, so that despite the extreme simplicity of the experimental units, they bore some resemblance to the insects' normal environment. Cultures were started with either four males and four females of one species, or two males and two females of both species, and each culture was allocated to one of six treatments formed by the combination of three temperatures (24, 29 and 34°C) and two relative humidities (30% and 70% relative humidity).

The cultures of single species persisted in all treatments but one: all 15 replicates of *T. castaneum* in the cold, dry treatment (24°C, 30% relative humidity) became extinct. Where the two species were cultured together, one or other species always became extinct, although the outcome was not the same for every single replicate (Table 3.1). These results illustrate two important points. First, there was for some treatments an element of chance as to which species would be exterminated—we have already mentioned the stochastic element of population dynamics in the previous chapter (p. 23). Secondly, no treatment permitted the two species to survive together. This inability to coexist was first generalized as the "principle" of competitive exclusion by Gause in 1934, although the idea was recognized much earlier. Many now take it as axiomatic that two species cannot coexist in the same habitat if they need identical resources for survival (Pianka, 1981). An alternative and in some ways more fruitful statement is that there must be some ecological difference between coexisting species in a community. This ensures that species within a community exploit different components of the available resources: to give one simple example, different species of predator may hunt at different times—owls at night, hawks by day.

We have seen that it is a moot point how much competition actually occurs within populations. There can be little doubt that the possibility of

Table 3.1 Survival in two species of the flour beetle *Tribolium* in mixed cultures exposed to six different treatments

Relative humidity (%)	Temperature (°C)	Favoured species	% of cultures in which favoured species survived	% of cultures in which unfavoured species survived	No. of cultures
30	24	*T. confusum*	100	0	20
	29	*T. confusum*	87	13	30
	34	*T. confusum*	90	10	29
70	24	*T. confusum*	71	29	28
	29	*T. castaneum*	86	14	28
	34	*T. castaneum*	100	0	29

Data from Park (1954).

competition between populations of different species that have somewhat similar requirements has influenced the demands that populations of such species actually make on their environment, but again it is an open question how much of ecological differences is due to competition, and how much to other factors, particularly predation (Schoener, 1974).

The logistic equation can be adapted to illustrate the competitive situation, albeit with due regard for the simplifying assumptions that this equation embodies (Gause and Witt, 1935). If N_1 and N_2 are the numbers of the competing species 1 and 2, then we can write

$$\frac{dN_1}{dt} = r_1 N_1 \left(1 - \frac{N_1 + \alpha_1 N_2}{K_1}\right) \tag{3.1}$$

and

$$\frac{dN_2}{dt} = r_2 N_2 \left(1 - \frac{N_2 + \alpha_2 N_1}{K_2}\right) \tag{3.2}$$

where K_1, K_2 indicate the carrying capacity of the environment for species 1 and 2 respectively, in the absence of the other species. α is the coefficient of competition, assumed in this model to be constant, and measures the extent to which numbers of one species are depressed by members of the other species. Thus the term $\alpha_1 N_2$ expresses the retarding effect that competition from species 2 has on the population growth of species 1. In other words, N_2 individuals reduce population growth of species 1 to the same extent as would $\alpha_1 N_2$ additional individuals of species 1.

Thus, equation (3.1) can be expressed in words as

The actual growth rate of the population of species 1	=	the population's potential growth rate	×	the proportion by which the population, plus the decrement due to competition from species 2, is below the environment's carrying capacity

and equation (3.2) conveys a similar meaning.

If there is no competition, $\alpha = 0$ and the equations reduce to the original logistic equation. If there is competition between the two species, then the consequences can be depicted graphically by plotting values of N_2 against the corresponding values of N_1 (Fig. 3.1). There are four graphs (a–d), to illustrate all four of the possible consequences of competition between two species. Each graph contains two continuous lines. Those lines that start at the value K_1 on the axis N_1 indicate, for different values of N_2, the carrying capacity of the environment for species 1. The intercept of this line on the axis N_2 is, by definition from the model, K_1/α_1, and the equation for this line is $K_1 - N_1 - \alpha_1 N_2 = 0$. Similarly, the other line indicates, for different values of N_1, the carrying capacity of the environment for species 2. It follows that, if the line that indicates the carrying capacity of the environment for species 2 indicates higher values for N_2 at all values of N_1, then species 1 will become extinct (Fig. 3.1a): species 2 becomes extinct in the reverse situation (Fig. 3.1b). If the values of K and α are such that the two lines intersect, the two

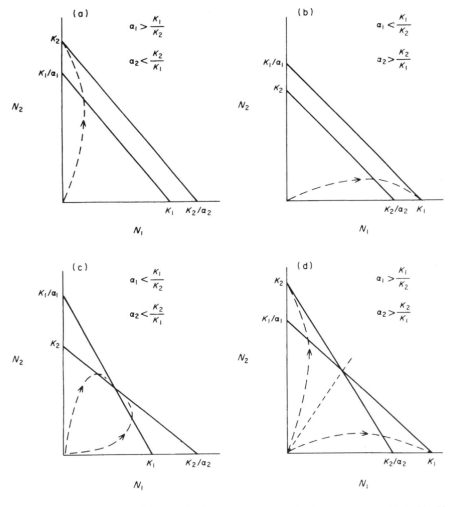

Fig. 3.1 The four possible results from competition between two species. N_1, N_2 indicate the numbers of species 1 and 2, respectively. The relative magnitudes of α and K are indicated for each type of result, and the broken line indicates possible changes with time in the numbers of the two species. Further details are given in the text.

species can only coexist indefinitely if $\alpha_1 < K_1/K_2$ and $\alpha_2 < K_2/K_1$ or, in non-mathematical terms, density-dependent, intraspecific competition halts growth in both species before the other species is eliminated (Fig. 3.1c). Finally, if $\alpha_1 > K_1/K_2$ and $\alpha_2 > K_2/K_1$, then one or other of the species will disappear. The outcome depends on the proportions of the two species present initially, and the critical ratio is indicated by the dotted line (Fig. 3.1d).

Both species can survive only when both species limit their own population growth more than they limit that of the other species (Fig. 3.1c). How then can two competing species coexist within a community? The balance of advantage between the two species may change from time to time. Species can coexist in a uniform environment if the rate at which one species excludes another is of the same order as the rate at which the environment alters from, for example, seasonal change, when the competitive advantage will also change (Hutchinson, 1961). Alternatively the balance of advantage may change from place to place, and diversity of habitats is the very stuff of which communities are made. Figure 2.4 illustrates data for a protozoan cultured in 5 ml of a pure bacterial culture. Gause found that a significant diversity of habitats could occur even in such a simple system: particles of some types of nutritive medium settled to the bottom of the flasks if unshaken. He then showed that, in cultures with two species of protozoa, this sediment provided a refuge in which one species could escape from a second competing species.

The idea that the degree of competition between two species can be represented by simple coefficients is probably illusory. The population consists of individuals, which differ, and so they are likely to exert different degrees of effect on the other species. The use of coefficients also assumes that interactions between a pair of species are independent of what is happening to all the other species in the community. Neill (1974) showed this assumption to be untrue in laboratory experiments with a small number of species. However, these equations are still of considerable heuristic value (Pianka, 1981), although more complex models have been developed (Renshaw, 1991). Whatever the detailed mechanisms may be for the nature of the interactions between a pair of species, the relevant points for us are that populations of two species cannot coexist indefinitely in exactly the same habitat unless their demands for resources differ at least to some extent, and that if there is any degree of competition between them, this will tend to slow down the rates at which both populations can increase in numbers.

It has proved difficult to actually demonstrate the effects of competition in the field: many experimental studies are severely flawed in their design, implementation and interpretation (Underwood, 1986). There have however been some successful field experiments (Connell, 1975). For example, two species of barnacle occur on rocks in the intertidal zone of the Isle of Cumbrae in Scotland. Adult *Chthamalus stellatus* are normally confined to the upper part of this zone, whereas adult *Balanus balanoides* occur in the lower zone. *Chthamalus* do settle successfully in the lower as well as the upper zone if *Balanus* are removed and kept out of an area, but *Balanus* grows quicker so that when not excluded it smothers, crushes or lifts up *Chthamalus*. *Chthamalus* survives in the upper drier zone because it withstands desiccation better (Connell, 1961). We do not really know how frequently interspecific competition occurs. Probably the usual situation is that after a long period of adaptation communities of species have become so arranged that the populations avoid overt competition (Law and Watkinson, 1989).

Fine subdivisions of habitat are commonly found when looked for in field situations. Up to five species of warbler occur together during the breeding season in spruce (*Picea* spp.) and balsam fir (*Abies balsamea*) trees in Maine and Vermont. These birds are closely related, all being of the genus *Dendroica*; they are of about the same size and shape, and all feed mainly on insects. Ecologists had concluded that any differences in the species' requirements were at best obscure, and that perhaps they were an exception to the idea that species must either have different habitats or ranges, or be limited by different factors.

MacArthur (1958) studied the feeding habits of coexisting populations of these five species to discover what prevents all but one of the species being exterminated by competition. He observed the feeding behaviour within individual trees, which were divided into 16 zones (Fig. 3.2). The trees were 50–60 ft tall, and were divided into 10-ft vertical zones measured from the top of the tree. Each branch was divided into three zones: a bare or lichen-covered base (B), a middle zone of old needles (M) and a terminal zone of young needles or buds (T). There were clear differences in feeding zones, although there was some overlap. There were also differences in detailed feeding behaviour. For example, *D. tigrina* and *D. fusca* both occur on the

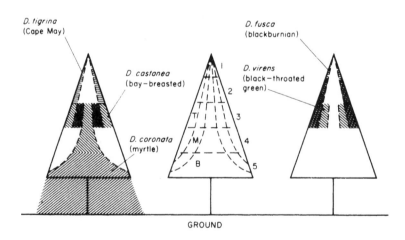

Fig. 3.2 Division of coniferous trees into 16 zones by height (divisions 1–5) and distance from trunk (basal (B), middle (M) and terminal (T)). The two lateral representations of trees indicate those zones within which at least 50% of the feeding activity occurred for each of five species of warbler (*Dendroica* spp.). (Data from MacArthur, 1958.)

upper terminal zones, but *D. tigrina* catches more insects on the wing and tends to move vertically rather than radially in the tree, so that its feeding zone is more restricted to the outer region of the tree. Clearly, there is, for these insectivorous species, a great and significant degree of variation within the habitat of a single tree canopy.

The key idea in this approach to community structure is that of a species niche. Species can coexist if they occupy different niches within the habitat. Several shades of meaning can be attached to this term niche. The word was first used by Grinnell (1917), when he noted that the structure and physiology of the Californian thrasher (*Toxostoma redivivum*) implies definite requirements from the environment, that this bird can only occur where these requirements are available, and that these requirements delimit the species niche. Grinnell's analysis denied much significance to interactions with other species, and defines the fundamental niche of a species as the limits set by the habitat on a population's abundance and distribution. Elton (1927) then used the term to emphasize an organism's interaction with other species—food and enemies—the realized niche. Since then the concept has developed considerably (Vandermeer, 1972; Giller, 1984; Schoener, 1989) and the emphasis has shifted from the organism to the environment, which is then assessed quantitatively. It is commonly used as a measure of the amount of a set of resources that is used by a population, and Fig. 3.3 illustrates the use of one such resource, food, by a group of potentially competing species

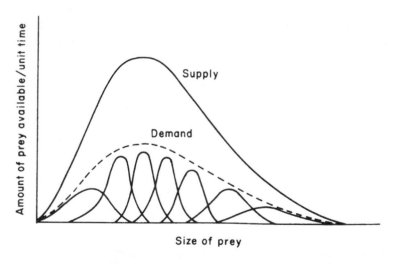

Fig. 3.3 Model to illustrate the differences between seven hypothetical predatory species with overlapping food requirements. Each species is distinguished by its particular selection of size of prey from within the relatively wide range of size of prey that is available. The curve for demand is the sum of the curves for the seven individual species, and should therefore coincide with the individual curves at both ends of the size range: it is depicted separately to distinguish it from the individual curves. (From Pianka, 1981.)

(Pianka, 1981). The relative positions of the supply and demand curves indicate the severity of competition. The use of other resources can be illustrated in the same way, and between them these graphs of resource utilization delimit a species multidimensional niche, or hypervolume, an idea first developed by Kostitzin in 1935 (see Hutchinson, 1978). This approach can become mathematically quite sophisticated (Hutchinson, 1957), and can distinguish between the fundamental niche, just described, and the realized niche, that lesser part which is still occupied after interactions with other species. The essential point remains that there is only one species per niche.

The idea of a niche implies that species do not occur at random within a community, but that there is a definite pattern, imposed by the interactions that a population has both with the physical and chemical conditions of its habitat and by its interactions with populations of other species. The search for food is one of the most obvious forms that these interactions can take.

A plant may be eaten by one animal, which is in turn eaten by another animal, and so on: Elton (1927) coined the term food chain to describe such a sequence of events, and Table 3.2 illustrates one of his examples, with five trophic levels. In all food chains the first level contains primary producers, or autotrophs, which unlike all subsequent trophic levels do not obtain their energy from organic food. All green plants are autotrophs: they are green because of their chlorophyll pigments, which convert the sun's radiant energy into the chemical energy of organic compounds by photosynthesis. Carbon dioxide from the ambient air (or ambient water for submerged aquatic plants) is converted first into simple sugars and then into the whole range of organic compounds:

$$n\text{CO}_2 + n\text{H}_2\text{O} \xrightarrow{\text{radiant energy}} \underset{\text{carbohydrate}}{(\text{CH}_2\text{O})_n} + n\text{O}_2$$

Table 3.2 Example of a food chain

	Trophic level		
Primary producer *(autotroph)*			
Pine trees	1st	producers	primary producer
Secondary producers *(heterotrophs)*			
Aphids	2nd	herbivores	primary consumer
Spiders	3rd	carnivores	secondary consumer
Tits and Warblers	4th	carnivores	tertiary consumer
Hawks	5th	carnivores	quaternary consumer

From Elton (1927).

Energy for all trophic levels comes from the converse process, respiration:

$$(CH_2O)_n + nO_2 \longrightarrow nCO_2 + nH_2O + energy$$

Some bacteria are also autotrophs, but obtain their energy from inorganic chemical compounds in their environment, not from the sun. They are normally considered to be a relatively insignificant source of food in food chains except in habitats on the ocean floor where hot springs (deep-sea hydrothermal vents) from underlying volcanic activity release high concentrations of hydrogen sulphide. Here abundant bacteria live by oxidizing the hydrogen sulphide and form the first trophic level of thriving food chains (Schmidt-Nielsen, 1990). Subsequent trophic levels of all food chains contain only heterotrophs, which cannot synthesize complex compounds from simple molecules and so obtain all of their energy from organic food. All animals, fungi and a few other plants, and most bacteria are heterotrophs. Their food comes directly or indirectly from the photosynthetic activities of green plants and, to a minor extent, the chemosynthetic activities of some bacteria.

Elton (1927) noted that the herbivores at the base of a food chain are relatively numerous, whereas carnivores in the top trophic level are relatively few, thus forming a pyramid of numbers as one passes up the trophic levels. More specifically, as one progresses from herbivores to top carnivores one tends to find fewer species, smaller populations, lower reproductive rates, increased body size and an increased feeding area within which food is obtained from a more diverse range of habitats. These changes parallel to some extent those characteristic of the shift from r- to K-strategies.

This pyramid of numbers does not always occur: one host can carry many parasites, and one large tree can support many herbivores. A pyramid is still appropriate for these examples if we use the biomass (weight of living organisms) of each trophic level instead of numbers of individuals. It is important though to distinguish between the standing crop of a trophic level—the biomass of that level at a given time—and the rate at which it is produced. It is conceivable that one would not necessarily always obtain a pyramid for the amounts of biomass at successive trophic levels. The standard example is from the English Channel off Plymouth, where Harvey (1950) found that a column of water of 1 m^2 cross-section contained 4 g dry weight of phytoplankton, whereas the consumer levels had a combined dry weight of 21 g/m^2. Provided that there is no significant input of energy from elsewhere, this must mean that the phytoplankton had a relatively high production rate ($g/m^2/day$), with most of the production rapidly being eaten first by the zooplankton or after death by animals on the sea bed. Indeed, Harvey found that the conventional pyramid reappears when the estimates of biomass are converted to estimates of production rates (Table 3.3). One qualification is needed, perhaps. These data are mean annual values, and it may well be that in the spring, when the biomass of phytoplankton is highest, the conventional pyramid of biomass does occur. Figure 3.4 illustrates this seasonal change for Lago Maggiore, northern Italy.

Table 3.3 Biomass and production rates in the English Channel off Plymouth[a]

	Biomass (dry weight (g) of organic matter/m²)	Production (g/m²/day)
Phytoplankton	≈4	0.4–0.5
Zooplankton	1.5	0.15
Pelagic fish	1.8	0.0016
Bacteria in water column	0.04	–
Demersal fish	1–1.25	0.001
Benthic fauna	17	0.03
Bacteria in sediments	0.1	–

From Harvey (1950).
[a] Bacterial production rates were not determined.

There was initially perhaps an undue emphasis on live food, or the grazing food chain (Odum, 1962). Organisms that die before they are eaten are still food: the jackal may symbolize scavengers, but most organisms that feed on dead organic matter—the detritivores or decomposers—range in size from moderate, such as earthworms, to the very small, such as most fungi, bacteria and protozoa (see Anderson and MacFadyen, 1976). They occur mainly in the sediments of aquatic habitats and in the litter and soil of terrestrial systems. These decomposer food chains are self-evidently important but have been relatively little studied (see Usher et al., 1979), in part perhaps because of the difficulties of studying small organisms in habitats such as soil and sediments. There are almost insuperable problems in determining even population size for most species, and for practical purposes one usually has therefore to describe decomposition processes in chemical rather than biological terms (Swift et al., 1979; Ritz et al., 1994). Of particular relevance to our concerns, the decomposition of organic matter in the soil releases both car-

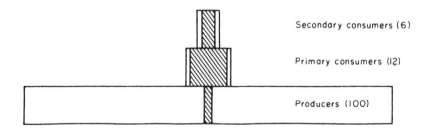

Fig. 3.4 Seasonal changes in the biomass of three trophic levels in the water of Lago Maggiore, northern Italy, for organisms captured in a plankton net. The numbers in brackets give the dry weight (mg) per cubic metre of water in the spring, and the cross-hatched areas indicate the values for winter. (Adapted from Odum, 1971; original data from Ravera, 1969.)

bon, as carbon dioxide, and nitrogen, an important plant nutrient, which may then be taken up by plants, remain in the soil or be lost from the ecosystem by leaching or to the atmosphere.

Feeding relationships are found to be more complicated than the simple food chain suggests. Some species are omnivores and feed on species in at least two trophic levels. Some animals change their food requirements during their life history. Predators tend to feed upon a range of prey species on the same trophic level, and many herbivores too will feed on a range of plant species. In other words, trophic levels describe functions, not species. Food web then becomes a better description than food chain (Fig. 3.5).

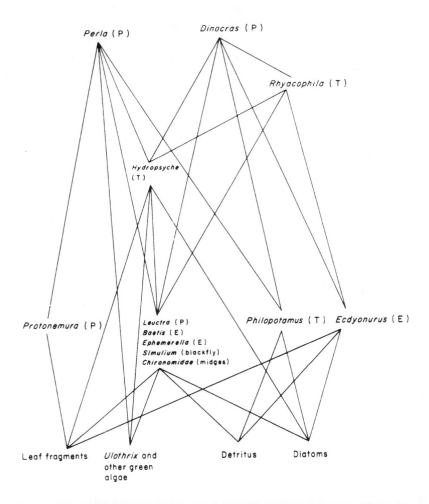

Fig. 3.5 The main links in the food web for the dominant insects in a Welsh stream system. P, a member of the Plecoptera (stone flies); T, a member of the Trichoptera (caddis flies); E, a member of the Ephemeroptera (mayflies). (From Jones, 1949.)

Two of the most obvious functions of food are to supply energy and nutrients, and although both aspects have been used for the analysis of food
webs, food as energy has received the more attention. As Lindeman (1942)
put it: "The basic process in trophic dynamics is the transfer of energy from
one part of the ecosystem to another". In practice, the more that is known
about a food web the more complicated it becomes, and it is useful sometimes to ignore the species composition of a food web and to consider it
simply as a series of trophic (or feeding) levels, in which the individuals in
one particular level feed on the organisms in the level below. One is then able
to utilize more general ways of looking at and comparing food webs.

MacArthur (1955) predicted that the stability of communities—defined as
the degree to which a change in numbers of one population has little effect
on the size of other species populations—increases with both number of
species and the number of species that they eat. Maximum stability occurs
when there is only one species per trophic level and the species at any level
eats the species at all lower trophic levels. Conversely, minimum stability
occurs when one species feeds on all other species, with only two trophic levels (Fig. 3.6). Studies of energy flow should then yield predictions on a community's stability and persistence, and facilitate comparisons with the
structure and function of other communities but first we must consider some
of the details of energy within ecosystems.

✳

The most generally useful unit of energy in ecological energetics is the joule (a
unit of work equivalent to the amount of heat needed to raise the temperature

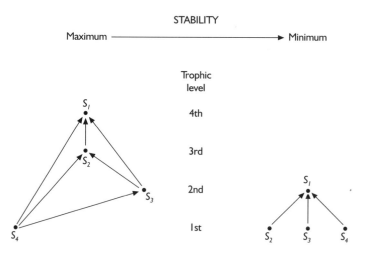

Fig. 3.6 The effect of feeding relationships on community stability. Two communities are illustrated, both of four species (S_{1-4}) with different feeding relationships.
Arrows indicate the direction of energy flow: species S_1 feeds on all three other
species. (Based on MacArthur, 1955.)

of approximately 0.24 g of water from 14.5 to 15.5°C). Biomass can then be converted into units of energy by measuring the heat produced when the organisms are completely oxidized (Phillipson, 1966). We have already seen that plants store chemical energy. The total amount of radiant energy that is retained per unit area of ground per unit time is called the gross primary production: short-term maxima for crop plants grown in ideal conditions range from 3 to 10% of the incoming light energy (summarized by Begon *et al.*, 1996, from Cooper, 1975). Not all of it is potential food for herbivores: some of this chemical energy is used by the plant itself. Energy is released by respiration, the reverse process to photosynthesis. The amount of food available for subsequent trophic levels is therefore the resultant of gross primary production minus respiration, called net primary production, which was estimated to range from near 3% for forests to very low values for desert in the USA (Webb *et al.*, 1983). Similarly, for consumers, some of the energy ingested in the food will be used for respiration. Much of the food's energy will not even be assimilated, but will pass out with the faeces. It is therefore impossible for there to be more energy (expressed as joules produced/unit area/unit time) in one trophic level than there is in the level below it. The energy ingested by predators of one trophic level is commonly thought to be about 10–20% of the energy that has been taken in by their prey (Odum, 1971), although this generalization should not be taken too precisely (Slobodkin, 1972; May, 1979). With efficiencies of this sort it is not surprising that food webs rarely have more than five trophic levels.

This poses the question, what happens to the other 80–90% of the energy taken into a trophic level? To answer this adequately one needs to make a budget for the flow of energy through an ecosystem. Estimates are needed for the gross and net production of autotrophs, for input of energy from outside, and for consumption of autotrophs by primary consumers. Estimates are also needed, for each consumer level, of changes in standing crop or biomass, and for losses to the next trophic level and by respiration. A significant amount of energy may also be exported from the ecosystem. Even for the simplest ecosystem such a budget takes a great deal of work. Because of this not many budgets have been made, but it does seem likely that in many ecosystems more energy flows through the detritivores than through the grazers. Odum (1957) found an example of this in his study of a freshwater spring, Silver Springs, in Florida (Fig. 3.7), where 21 185 kJ/m^2/year flows through the detritivores compared with 14 101 kJ/m^2/year through the grazing food chain. The detritivores in most freshwater habitats will probably have an even greater proportion of the total energy flow, because a major part of the chemical energy (from organic matter) usually comes from outside the system (Hynes, 1972). There are, however, few good measurements for the productivity of decomposers: even their biomass is difficult to estimate, and their biomass is small relative to their productivity (Swift *et al.*, 1979).

A very great deal of research has been devoted to production studies, stimulated in particular by the International Biological Programme (see Worthington,

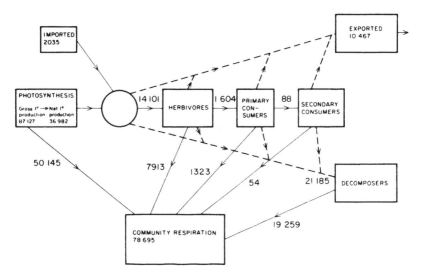

Fig. 3.7   The flow of energy in the headwaters of Silver Springs, Florida, which extend for over 1 km. (➤-) Routes for which estimates of energy flow were made; (->-) routes for which rates of energy flow were not made. All numbers are for kJ/m²/year. (Data from Odum, 1957.)

1975), but although it is of interest and value as a description of how ecosystems function, it appears so far to have had little predictive value. It is obvious that pollutants that affect decomposers could have appreciable effects on an ecosystem, but it is difficult to know how these effects would manifest themselves. The effect of a reduction or elimination of some decomposer species populations on other decomposer species is not usually at all certain.

There is the additional complexity that populations of some species may increase the availability of energy for other species. Edwards and Heath (1963) placed leaves of oak and beech in nylon mesh bags before burying them in soil, thus excluding different groups of organisms according to the size mesh that was used. It was found that the exclusion of earthworms reduced several-fold the rate at which leaves decomposed (Fig. 3.8). This is because micro-organisms attack litter from its surfaces, and although earthworms do not appear to digest much from leaves they do break them down into small pieces. So although little of the energy contained within fallen leaves passes through the earthworms, they facilitate energy flow through micro-organisms. Conversely, considerable energy flow between two species may have no relevance to community structure (Paine, 1980).

Nor does the account of energy flow describe all of the important functional aspects of an ecosystem. Many of the chemical elements are essential to life, and they tend to circulate in characteristic cycles between the biota and the non-living environment. Some organisms are particularly important in these cycles: the nitrogen-fixing bacteria are a classic example. These

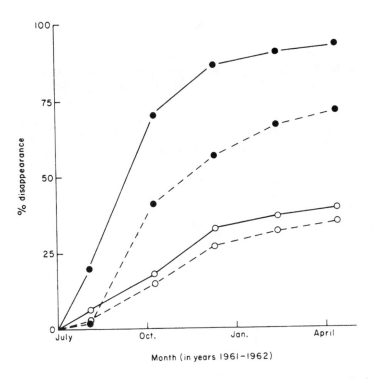

Fig. 3.8 Decomposition of leaves in nylon mesh bags buried in soil. (—) Oak leaves; (– –) beech leaves; (○) earthworms excluded; (●) earthworms not excluded. (From Edwards and Heath, 1963.)

bacteria convert nitrogen from the air into nitrate, which is then available to plants, and some will then be used for protein synthesis.

Phosphorus is another essential element, vital for example in both respiration and photosynthesis. Kuenzler (1961) studied the effect of a population of mussels, *Modiolus demissus*, on phosphorus cycling in a Georgia salt marsh. Every 2–6 days these mussels had removed from the water by their filter-feeding particulate phosphorus equivalent to the average amount present at any one time in suspension in the water. This phosphorus was then deposited onto the bottom sediments, where it was available for other organisms, such as snails, worms and bacteria. Clearly, this must be a major pathway for phosphorus in this ecosystem, but the amount of energy consumed by mussels was one to two orders of magnitude less than the energy used by bacteria, quite apart from energy used by other organisms. This illustrates the conclusion that the amount of energy that flows through one component of an ecosystem does not necessarily indicate the ecological significance of that component.

There have been attempts to explain the structure of ecosystems in terms of energy flow and nutrient cycles (also called biogeochemical cycles), but

Rigler (1975) has suggested that these attempts are misguided. It is technically difficult to estimate accurately the flow of energy or nutrient from one trophic level to another. Moreover, production fluctuates at all trophic levels, which can be important ecologically but difficult to quantify and interpret (Kokkinn and Davis, 1986). More importantly, it is not practicable, except for primary producers, to assign sufficient individual organisms to specific trophic levels—too many fall into a spectrum between two or more levels. It then becomes impossible to test any statement about energy or nutrient flow between trophic levels, because it is impossible to identify the members of any trophic level, apart from primary producers, with adequate precision.

For studies on sediments, where interest focuses largely on the flow of energy and nutrients, ecosystems are best divided into three subsystems (Fig. 3.9), with the activity of each subsystem then measured by its inputs and outputs of nutrients and energy, without too much consideration of the processes within each subsystem.

*

So far we have ignored the practical difficulties that may arise in delimiting a community, nor have we considered the possibility that community structure may change with time. Conflicting views have been expressed on both topics, and the literature is confusing. Many terms have been introduced, which then acquire different shades of meaning with different authors, and often the precise meanings are unclear. There is, however, one fundamental dichotomy of views: that a community is either some kind of supra-organism, or that it is an aggregate of interacting populations.

In much of what follows we will restrict our attention principally to plants. This is historically where most emphasis has occurred in studies of community structure and function, and there are several good reasons for this. Plants are the prime source of energy in most ecosystems. They also determine the physical structure in and around which the animals live: the simplest classifications, such as desert, grassland, scrub and forest, exemplify

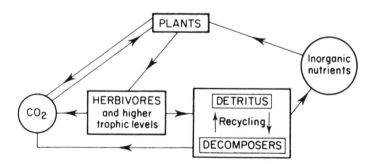

Fig. 3.9 The pathways of organic and inorganic matter within an ecosystem. (Adapted from Swift *et al.*, 1979.)

this. Moreover, many species of animal are very mobile, so that although it is meaningful to consider the interactions between plant and animal species within a plant community, the limits within which the populations of some of the animal species occur may extend far beyond the boundaries of the plant community. Nevertheless, ideas had developed sufficiently by 1877 for Möbius to coin the term biocoenosis to include both plants and animals in a single complex of organisms, although Tansley (1935) could still emphasize that we cannot usefully lump plants and animals together as members of one community, because they are so very different: "Animals and plants are not common members of anything except the organic world", and one may more usefully refer to different animal communities that live in or on a plant community. There is a long history of ideas on the nature of animal communities (Whittaker, 1962) and there is still no consensus on how to define and delimit them (McIntosh, 1995). In practice, the traditional academic disciplines still tend to maintain a certain distinctness between studies in plant and animal ecology (e.g. MacFadyen, 1963), although plants and animals can interact in many ways, and ideas about plant communities are frequently extended to animals as well (e.g. Mills, 1969).

Clements (1916) took it as axiomatic that plant formations (communities with a definite physiognomy, e.g. oak woodland) are organic entities comparable to individual organisms. Both formations and individuals have distinct boundaries, arise, grow, mature, die, and can reproduce themselves. This became the dominant view until quite recently and we will see how it has influenced ideas on classification and succession.

Community classification

In the same way that population studies have been predominantly zoological, attempts to classify communities have been based principally on the plant component. The traditional view of communities argued from the fact that they are living systems to the presumption that communities possess similar attributes to those of individual organisms, with the implication that individual species were necessary parts of this structure (see Whittaker, 1962, 1975). In other words, species form distinct groups, these groups are characteristic of distinct communities, and communities are separated from each other by clear boundaries. As Braun-Blanquet (1932) put it, "The community has an existence altogether independent of the individual".

This traditional view did not pass entirely unchallenged. Ramensky in Russia first published a clear opposition in 1924, and in the USA Gleason developed similar views independently in 1926 (see Whittaker, 1962, 1975). Ramensky argued that the presence and abundance of each species depends on its interactions with its environment, and that these will be different for each species. Therefore, different species will have different distributions. It then follows that communities that occur along continuous environmental gradients will merge continuously one into the next. This does, in fact, seem to be the more realistic approach: when subjective bias is avoided there is little evidence for distinct groups (Whittaker, 1962), and Fig. 3.10 shows an example for changes in two forests along moisture gradients. Obviously, this view still allows for the possibility of separate communities with distinct boundaries when there are sharp transitions in the abiotic environment.

Even when one accepts that individual species merge into each other along gradients one does need to distinguish types of community as an aid to comparison and discussion. This problem is not, of course, unique to ecology. There is, for example, a continuous range of wavelengths in the visible spectrum, arbitrarily but usefully classified into seven colours.

Several systems have been developed for classifying communities (Whittaker, 1975). Some may have only limited use, but no one system is most suitable for all applications. All systems are abstractions, and which is the most appropriate depends on the purpose of the classification. One of the simplest is to name a community by the one or two dominant plant species (those which are the most abundant and/or cover the largest area of ground) combined with their physiognomy or growth form, e.g. pine forest, hawthorn scrub. This has the limitations that sometimes, as in tropical rain forest, there are so many species that it is difficult to define a few dominant species, or alternatively the dominant species may occur in several different kinds of community.

At the other extreme, Braun-Blanquet (1932) classified communities, or associations, by considering all of their plant species—the floristic approach. This is more complicated but one might expect the full species composition of a community to give a better index of the species relationships to each other and to the environment than dominance or any other less complete

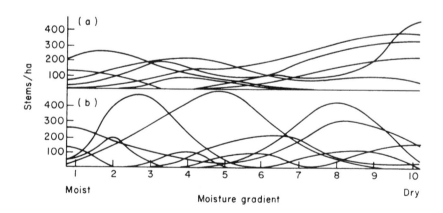

Fig. 3.10 The densities of different tree species plotted against position along a moisture gradient. Densities are calculated for the number of tree stems at least 2 cm in diameter per hectare. Both graphs are for slopes that face south-west; (a) from the Siskiyou Mountains, Oregon; (b) from the Santa Catalina Mountains, Arizona. (From Whittaker, 1975.)

description. We will illustrate this approach by studying an example from Whittaker (1975), based on data for Polish forests obtained by Frydman. Twelve sample areas were examined, and all species present were listed and scored for abundance. The data were then tabulated to bring together those species with similar distributions among the sample areas, and to bring together those sample areas with similar species composition (Table 3.4). Some sets of species were characteristic of some sample areas. These diagnostic species are enclosed by boxes in Table 3.4, and are used to distinguish between three types of community, on the assumption that these diagnostic species indicate relationships more clearly than do other species, and that these diagnostic species can be used to construct a hierarchical classification of communities. Mathematical analyses for data of this type can become quite sophisticated (Pielou, 1984; Digby and Kempton, 1987).

Table 3.4 The relative abundance of plant species in 12 samples in Polish forests[a]

| | | Fir forests | | | | Pine-bilberry forests | | | | | | | |
| | | | | | | Moist | | | | Dry | | | |
Group	Species Sample no....	1	2	3	4	5	6	7	8	9	10	11	12
A	Abies alba	4	2	2	2	+	+	+	+	+	+		+
	Pinus sylvestris	+	+	+	+	4	3	2	4	4	1	2	3
	Picea excelsa	+	+	2	+		2	+	+	+			+
	Vaccinium myrtillus	+	2	+	+	5	4	2	+	1	+	+	2
	Vaccinium vitis-idaea	+		+	+	+	+		+	1	3	3	2
B	Lycopodium selago	+		+	+								
	Circaea alpina	+	+		+								
	Pyrola secunda		1	+	+								
	Pyrola minor		+	+	+	+							
C	Lycopodium annotinum	+		+	+	+		+	+	+			
	Ptilium crista-castrensis	2		4	+	2	+	3	3				
	Dicranum undulatum	4	+	2	2	+	+	+	+				+
	Entodon schreberi		+			5	1	5	2		+		
D	Pyrola chlorantha					+			+	+	+		
	Melampyrum vulgatum					1	+	1	2		+		
	Calluna vulgaris						+	+		2	+	+	
	Cladonia sylvatica					2	+			3	+	3	+
	Cladonia rangiferina					1				1	2	+	4	+
E	Quercus sessilis				+				+	+	+	+	+
	Betula verrucosa									+	+	+	+
	Thymus ovatus							+		+	2	+	+
	Lycopodium clavatum									+	+	+	1
	Total number of species	35	37	38	37	20	17	24	25	39	41	32	34

From Whittaker (1975): original data from Frydman (in Frydman and Whittaker, 1968.)
[a] +, Rare; 1–5, increasing degrees of abundance. The species of groups B–E are diagnostic.

When more community types are compared, these diagnostic species can be seen to fall into two groups: those which are more or less peculiar to one type of community, the character-species, and those which occur in some but not all types of community, the differential-species. In other words, character-species can be used to distinguish between communities by their abundance, which is greatest in the communities they characterize, whereas differential-species distinguish between communities by the limits of their distribution. Table 3.4 does not tell us whether the diagnostic species are character- or differential-species—we need to know their distribution in a wider range of communities.

There is another aspect to this way of classifying plant communities. Various mathematical techniques can be used to quantify the degree of similarity between samples taken at intervals along the ground from plant communities (see Kershaw and Looney, 1985). The results from these samples can then be arranged along one or more axes, or community gradients, which may then link with obvious gradients in the environment. A simple example would be the transition from woodland to desert as one ascends a mountain, with its gradients of temperature, soil, moisture, exposure, and so forth. Given sufficient knowledge of an area, it is thus possible, by gradient analysis—the interpretation of gradients in both plant populations and abiotic factors— both to classify plant communities and to relate the occurrence of those communities to the physical environment (Fig. 3.11) (Whittaker, 1967).

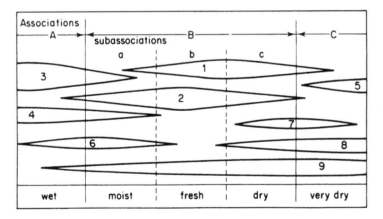

Fig. 3.11 The abundance and distribution of nine plant species along a moisture gradient. The horizontal length of each diamond indicates each species range, and the species relative abundance at any point along the gradient is proportional to the vertical height of the diamond at that point. Species 1 and 2 are character-species for association B, to which they are largely confined. Species 3–5 are character-species for the two adjacent communities. Species 6–8 are differential-species that distinguish the moist, or dry, subassociation from the "typical" (fresh) subassociation b. Species 9, with a wider distribution, may be a character-species for a larger grouping of plants. (From Westhoff and van der Maarel, 1973.)

With gradient analysis, plant species can be used as indicators of environmental conditions. This idea of indicator species has been taken over into ecotoxicology, but before discussing this we need to consider succession.

Succession

It is a fact of simple observation that communities do not remain the same for ever (Burrows, 1990). Not only do individual organisms die, but the abundance of different species changes. Moreover, these changes are not random—similar habitats tend to have similar sequences, and it is this sequence of change that is called succession. The time-scale for these changes may range from hours or days in the succession of populations associated with a carcass to, usually, centuries in a mature forest.

Whittaker (1975) describes the changes that occur in some lakes of northern temperate climates. A mat of vegetation, often moss, spreads out from the margins of the lake to cover the entire water surface, when the lake becomes a quaking bog. The dead layers of undecomposed vegetation from this covering mat gradually fill the entire lake basin, whilst the surface vegetation progresses via shrubs to trees to form a forest. A forest may present a fairly static picture, but it can be shown both from records and from transects that even such seemingly permanent communities have developed from earlier different assemblages of plants.

In more general terms, communities change as their environment changes, and these environmental changes can be caused by the community itself, as in the transition from bog to woodland, or be external as in the effect of fire or other disturbance such as pollution. Given an absence of external disturbances a succession ends with the climax community, that combination of species best suited to that particular climate and site.

These events can be described in terms of the community as a supra-organism, and Odum (1969) summarized succession as:

(1) An orderly process of community development that is reasonably directional and therefore predictable.
(2) The result of the community modifying its environment. The community controls the nature of the succession that it passes through, even though the physical and chemical environment determines the pattern and rate of change and often limits how far the succession can go. Development through succession is then analogous to development in an individual organism.
(3) A process that ends in a stabilized self-regulating, or homeostatic, ecosystem (climax community) in which there is maximum biomass per unit of available energy flow.

It is now less clear how frequently communities attain this stability (Putman, 1994). Certainly the climax community is relatively stable, with approximations to steady-state conditions for species populations and for cycling of nutrients, and with total respiration equalling gross primary production if there are no extraneous inputs or losses. But this stability is only relative, and depends on the time-scale, because it implies environmental stability. For example, sub-alpine forest near Vancouver has been undisturbed for more than 1500 years. It contains four dominant species, with no other common tree species. The pollen record suggests that all four species have flourished until recently, but now as trees die they are being replaced preferentially by only one of these four species, the Pacific silver fir (*Abies amabilis*). Lertzman (1992) suggests that the silver fir is being favoured by a warmer climate.

The idea of an indefinitely stable climax with no directional change is untenable because over long time-scales climate changes, but one can distinguish between communities that are changing rapidly and those that are changing very slowly. It is more realistic to say that relatively stable communities are the net result of many factors, such as soil, climate, biota and chance.

Some communities have cyclic changes. For example, ling or heather (*Calluna vulgaris*) dominates many Scottish heaths. As the plants age they become less vigorous, their branches are no longer dense enough to exclude lichens, and the lichen *Cladonia sylvatica* starts to appear on the heather plants. In due course the heather dies, and subsequently the lichen dies too, leaving a patch of bare ground, which is then invaded by another shrub, the bearberry (*Arctostaphylos uva-ursi*) (Watt, 1947). The bearberry plants are then invaded from the margins by young heather plants, which eventually oust the bearberry completely. This is clearly a cyclic process (Fig. 3.12) in which *Calluna vulgaris* is the dominant species: *Cladonia* and *Arctostaphylos* only become locally dominant where old *Calluna* plants have died. The whole

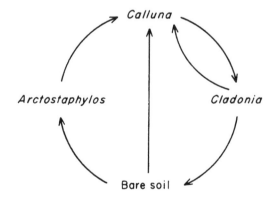

Fig. 3.12   The cyclic sequence of change between the chief species found in some heather communities (dwarf callunetum) on the Cairngorms, Scotland. (From Watt, 1947.)

climax community consists of a mosaic of these different stages: a "shifting-mosaic steady state" (Bormann and Likens, 1981). Many communities exhibit this pattern of cyclic change (Remmert, 1991), and often four phases can be recognized: pioneer, building, mature and degenerate (Krebs, 1994).

Odum (1969) also suggested that diversity increases during succession, with a more equal occurrence of, and more, species, but this is not invariably true. Good evidence one way or the other exists only for forest succession in temperate climates, where the generalization is not confirmed (Drury and Nisbet, 1973). Different groups of species within a community have different patterns of change in diversity during succession: the diversity of one group may decrease at the same time as the diversity of another group increases, and increased diversity is by no means an exclusive property of the later stages of plant succession (Whittaker, 1965). For example, the redwood forests on the coasts of California and Oregon are probably amongst the most productive climax forests to be found in temperate zones, but they have very low diversity.

However, doubts have long been expressed about this whole approach of community as supra-organism (see Whittaker, 1975) and coherent counter-proposals were advanced by Connell and Slatyer (1977). They argue that the mechanisms of succession have not been sufficiently elucidated, for several reasons. First, direct evidence is only available for the earliest stages of succession, when many of the species are short-lived and therefore amenable to experimentation. Later arrivals persist for much longer and the later stages of succession have therefore to be reconstructed from indirect evidence. Secondly, some possible mechanisms, such as the effects of grazing animals (Duffey *et al.*, 1974), have been ignored. This is in part at least because most studies have been restricted solely to plants. Plants are indeed the primary producers, they usually have the largest biomass, and they give the community its structural form or physiognomy. Hence competition between plants has been regarded as the principal determinant of the course of succession. Thirdly, mechanisms have not been defined clearly or stated as hypotheses that can be tested.

Connell and Slatyer (1977) suggested three possible ways in which plant succession might operate after a disturbance had produced a large empty site. Model I (the "facilitation" model) assumes that only certain species that come early in the succession are capable of colonizing the site. This accords with Clements' view of a community as a supra-organism. In contrast, the other two models both assume that any individual of any species that happens to arrive at the site is capable of colonizing it, although all models accept that certain species will tend to appear first because of their colonizing abilities. All models also suppose that the first colonists will so modify the site that it becomes unsuitable for those species that normally occur early in the succession. The three hypotheses then suggest three different ways in which subsequent species appear. Model I suggests that early occupants modify the environment so that it becomes more suitable for species that come later in the succession. Model II (the "tolerance" model) suggests that the sequence in which species appear depends solely on their speeds of dispersal and growth, so that the species already present have little effect on the subsequent recruitment of later species, whereas model III (the "inhibition" model) suggests that the species already present make the environment less suitable for subsequent recruitment of later species. The available evidence does suggest that all three models of succession can be observed in the sequence at one site (Southwood, 1996). However, the important point is that all three hypotheses rely not on the idea of a community as a supra-organism, but on succession as a process that relies on two factors: the probabilities that propagules of different species will be present, and the abilities of these propagules to survive, develop and reproduce themselves.

Drury and Nisbet (1973) add that colonizing ability depends on dispersal mechanisms and tolerance of physical stress, and that colonizing ability and growth rate tend to be inversely correlated with size at maturity and with longevity. A species place and role in succession is determined by three main groups of attributes: its method of persistence, its ability to establish and grow to maturity and the time needed to reach critical stages in its life history (Noble and Slatyer, 1980). These characteristics are suggested as a sufficient explanation for the broad features of succession.

Current evidence favours the idea that communities are not supra-organisms, and Krebs (1994) gives a good review of this topic. However, concerns about pollution have stimulated discussion of an extreme development of community as supra-organism, the Gaia hypothesis. Gaia was the Greek earth-goddess, and Lovelock and Margulis (1974) were struck by the planet's persistent physical and chemical disequilibria. They suggested that "... early after life began it acquired control of the planetary environment and that this homeostasis by and for the biosphere has persisted ever since". Without plants there would be very little atmospheric oxygen, and Fig. 3.13 illustrates the interdependence between the evolution of organisms and changes in the atmospheric content of oxygen and ozone. Life is defined as one of those phenomena in which entropy* decreases at the expense of free energy taken from the environment, mostly the sun's radiant energy. Gaia is singular and persistent, with a large decrease of entropy. Gaia is then a hollow sphere, bounded by space and by the inner parts of the planet that are as yet unaffected by surface processes. Lovelock (1995) develops these ideas into the Gaia theory and states explicitly that there is no teleological intent. The biota and their environment form a tightly coupled system, one supra-organism, from which has evolved the ability to self-regulate climate and chemistry. Self-regulation is seen in terms of selection—organisms that change the environment adversely become extinct. Life delays the planet's increase of entropy, and the biota and their environment have evolved together by feedback processes.

---

* The less the entropy within a system the greater the proportion of thermal energy that can be converted into mechanical work.

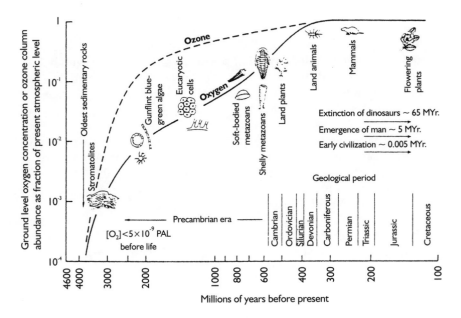

Fig. 3.13 Evolution of oxygen, ozone, and life on earth. In the absence of life, surface oxygen concentrations are unlikely to have exceeded ~ 5 × 10⁻⁹ of the present value. The build-up of oxygen to its present level is largely a result of photosynthesis. Early organisms would have found high oxygen concentrations toxic, but eucaryotic (nucleated) cells require at least several per cent of the present level for their respiration. Soft-bodied metazoans could have survived at similar oxygen levels, but the reduced surface oxygen uptake area available once the species had developed shells must mean that the concentration was approaching one-tenth of its current value about 570 Myr ago. Considerations such as these are used in drawing up the oxygen growth curve. Ozone concentrations are derived from a photochemical model, and ozone moved away from the earth's surface as the oxygen concentration increased, until now it forms a layer in the stratosphere (see Chapter 5). Life could not have become established on land until there was enough ozone to afford protection from solar ultraviolet radiation. (From Wayne, 1991.)

This theory undoubtedly has a strong appeal for some. Lovelock (1995) quotes one prominent environmental campaigner to the effect that this theory is too important, both as a symbol for environmentalists and as a way to unify disparate interests, to be discarded. The final defence of the theory appears to be that its critics lack independence of mind. Nevertheless it is difficult to see what it adds that is scientifically useful to the current view that organisms survive if they are adapted to their environment and that with time organisms also modify their environment. As Holland (1984) put it, "we live on an Earth that is the best of all possible worlds but only for those who have adapted to it". Lovelock (1995) argues that the notion of "adaptation" is tautologous because on the planetary scale life and environment are so tightly coupled, which highlights the essential point: are we discussing living organisms, or life as a metaphysical process? At the severely practical level organisms can evolve in very different environments. There is a distinct and rich invertebrate fauna associated with deep-sea hydrothermal vents, adapted to survive not only the high pressures and lack of oxygen but also high temperatures and chemical stress from exposure to toxic metal ions and hydrogen sulphide (see e.g. Vetter *et al.*, 1991; Dixon *et al.*, 1992).

✱

But we still have to face the problem: how do we assess or predict the effect of pollution on a community? Do we have to consider the community one population at a time, or is a more holistic approach possible? In practice, some of the currently favoured approaches rest on the assumption, often implicit rather than explicit, that communities are supra-organisms:

(1) The term indicator species, which is used in the classification of communities (p. 59), is also used in ecotoxicology, with a variety of meanings. Sometimes it indicates the idea that knowledge of one species within a community will indicate the well-being or biological health of the whole community. This seems a reasonable proposition if one accepts the traditional view of community as supra-organism, but I would suggest that it is, in fact, misleading. There is so far as I am aware no fundamental reason from community structure to suppose that any particular species within the community will give a better measure of impact from pollutants than will another. Pollutants will affect populations of particular species, and which species are first affected will depend on the relative degrees of exposure and susceptibility, and these are functions much more of the particular pollutant and of the individual species than of the community. Nor is there any obvious reason to suppose that, regardless of which species are initially affected by the pollutant, there are certain indicator populations that will inevitably react to the fact that pollutants have affected one or more other species within the community. An indicator species can only be used to assess the impact of pollution on a community if quite a lot is known about both the pollution and the community.
(2) The concept of biological or environmental health can also be misleading. One may properly refer to the health of individual organisms, but not so easily to the health of a community. A community may change markedly if affected by a pollutant, but it will just become a different community that is neither more nor less "healthy", just different. It may be a less desirable community, for economic, social, scientific or aesthetic reasons, but that is quite a different matter.
(3) Woodwell (1970) deduced from studies on ionizing radiation, eutrophication and persistent pesticides that "... the ecological effects of pollution correspond very closely to the general 'strategy of ecosystem development' outlined by Odum" (1969, referred to on p. 61). The precise

meaning of that is not quite clear, but it includes the idea that pollutants produce similar changes in many different ecosystems. In particular, pollutants simplify the structure of both plant and animal communities.

Regier and Cowell (1972) and Whittaker (1975) developed this theme, and suggested that although the effects vary with both the type of pollutant and the community, the effects may sometimes be described as retrogression—a reversal of the trends that characterize many successions. In such instances pollution will decrease species diversity, productivity, biomass and also the structural complexity of a plant community. This suggestion is intuitively attractive: community development entails the development and use of separate niches by different species. However, the evidence is not as strong as it might be. Results from the Brookhaven National Laboratory in New York provide the major support for these conclusions. An oak-pine forest at Brookhaven was irradiated for years from a ^{137}Cs source. Within 6 months a vegetation gradient developed towards the source, and this gradient became more pronounced during the next 7 years:

(1) Original oak-pine forest.
(2) An oak zone, where pine was eliminated.
(3) A shrub zone.
(4) A sedge zone.
(5) A central devastated zone, where only some mosses and lichens survived.

Woodwell (1970) summarized this gradient as a change in structure, with associated changes in primary production, respiration and nutrients. These changes are similar to those along gradients of increasingly severe conditions, such as increasing altitude on a mountain. Fire and other disturbances produce similar results, with the "generalist" species surviving most readily.

It is doubtful whether this is a very useful way to consider the effects of pollution, if only because Woodwell's own examples do not fit his description very well. Mosses and lichens are commonly associated with severe conditions, as in the Brookhaven experiment, and yet they appear to include some of the species most susceptible to sulphur dioxide (see Chapter 8). Eutrophication results not only from man's activities: it is also a natural late stage of development for many lakes (Harper, 1992), and can therefore hardly be described as an example of retrogression. For pesticides, Woodwell quoted the effects of the herbicide 2,4,5-T in Vietnam as an example of retrogression, because the forest was replaced by species of bamboo. It seems simpler to explain this particular instance as an example of selective toxicity: bamboos are grasses, and grasses in general are not easily killed by this herbicide. Other detailed criticisms are possible. I would not deny that the changes wrought by pollutants that affect many species within a community may sometimes resemble a process of retrogression, but would argue that examples of this type should not be taken to imply a generality.

In practice one cannot usually hope to know all of the impacts that a pollutant may exert on a community, and the choice of indicators for effects is a value judgement (Kelly and Harwell, 1989). One school of thought argues that ecological effects of pollutants cannot be interpreted solely by studies of specific populations, because this omits possible effects on "the subtle interactions among populations" and on "the structure and function of the ecosystem itself" (Levin and Kimball, 1984). This is a defensible proposition when dealing with soils, which are in practice treated as "black boxes": provided the input and output of energy and nutrients are unaffected, effects on individual species are taken to be of little concern, if only because such effects, if they occur, are unlikely to be noticed unless the net transfer of energy and nutrients is affected. One might ask though, why is the continued normal functioning of soils deemed to be important? The proposition would seem to contain the implied value judgement that soils matter because plants depend directly on soils for their survival (see Smith, 1981), and thus indirectly so do the associated animal populations. The proposition becomes more suspect when considering some other habitats. Thus the disappearance or severe reduction in populations of the peregrine falcon (*Falco peregrinus*), discussed in Chapter 9, had rather little discernible effect on ecosystem structure and functioning, but was nevertheless a matter of concern. Even within soils, a full scientific understanding of changes in nutrient and energy flow still requires knowledge of the populations of species involved.

I would conclude that, for ecotoxicology, the proper centre of emphasis is on populations of individual species. The effects of pollutants on populations within a community can be complex, and—apart from the obvious effects of reduction or elimination of populations—resurgence, population increase or introduction of rarer species, sublethal effects and genetic changes may all be part of the changes that occur.

4   Genetics of Populations

The previous two chapters suggest the population as the relevant ecological unit for ecotoxicology. So far we have considered population size and age structure, which depend on the rates of birth, death, immigration and emigration. We will now consider another important characteristic of populations, their genetic composition.

Individual members of any species are not all identical, and I take it as general knowledge that these variations depend in part on differences of inheritance and in part on differences in environment. The relative importance of nature and nurture has long been controversial in biology and, of course, it also spills over into politics. It is sufficient for our purpose to note that much of the variation between individuals is inherited from their parents.

It is also common knowledge that relatively few offspring of any species survive to reproduce. Charles Darwin calculated that even one pair of elephants, perhaps the slowest breeding animals of all, could produce at least 15 million offspring after 500 years. Obviously, no environment could sustain such rates of increase for long, and so Darwin concluded that natural selection must occur (Darwin, 1859): some individuals will have a higher probability of survival than others and, on average, such individuals will then leave more descendants than other less well adapted individuals. Nothing was known at Darwin's time about genetic variation and mechanisms of inheritance, and most of what was known about heredity related to domestic plants and animals, but it was sufficient to indicate that there are differences that can be inherited. It is, of course, axiomatic that natural selection cannot act on non-heritable features. Darwin therefore went on to suggest that, because the environment changes from place to place and from time to time, populations of one species will diverge in their characteristics until eventually new species evolve, because natural selection will favour the survival of some individuals rather than others.

We are not concerned too much in ecotoxicology with the evolution of new species. Futuyma (1986), Ridley (1993) and Avise (1994) give excellent introductions for those who are. We are concerned though with the role of pollutants in natural selection. It has been shown many times that pollutants can exert powerful selective forces, and we need therefore to understand something of the mechanisms of inheritance, and of how natural selection acts on populations.

Mendel (1865, with English translations by Bateson (1901), Bennett (1965) and Stern and Sherwood (1966)) advanced our understanding enormously when he published the results of his breeding experiments with the garden pea, *Pisum sativum*. The pea plant is particularly suitable for breeding experiments because the flower structure is such that the flower is normally self-pollinating: the female egg is usually fertilized by male pollen from the same flower. There is therefore little risk of accidental fertilization by pollen from another plant, although it is a simple matter to cross-pollinate artificially.

Much of the gist of Mendel's work can be illustrated by citing just one of his experiments, in which plants were either self-fertilized or cross-fertilized. To understand the results from these experiments, it is important to grasp that the nature of the seed depends on characters inherited with the gametes, the pollen and egg cells, which fuse to form the seed. Mendel selected pea plants that bred true, when self-pollinated, to produce seeds that were coloured either yellow or green. He then cross-pollinated plants that produced yellow seed with those that produced green seed and found that, regardless of which gamete came from which plant type (yellow or green seeded), the seeds produced by these cross-pollinated plants were always yellow. These were the so-called first filial generation, or F_1 hybrids. Plants grown from these F_1 seed, and self-pollinated, did not breed true, but produced 5474 yellow seeds and 1850 green seeds, a ratio of about 3:1. Other experiments with other characters confirmed and extended these observations, and one can draw several important conclusions (Fig. 4.1):

(1) Some characters, such as yellow seed, are dominant to other, recessive, characters such as green seed.

(2) Each individual plant carries pairs of hereditary factors (now called genes). If both genes for a character are the same, the plant is said to be homozygous for that character. If they are different, it is heterozygous.

(3) Only one of each pair of genes passes into each individual male or female gamete, so that when an egg is fertilized the embryo acquires a pair of genes, one from both of its parent plants, for each character.

(4) Which one of a pair of genes segregates into any particular gamete is determined by a random process.

(5) Inheritance is particulate, not blending; the genes are transmitted unchanged from generation to generation.

(6) One needs to distinguish clearly between the physical appearance of an organism (its phenotype) and its genetic constitution (its genotype): an individual may carry one recessive gene without any physical sign of its occurrence.

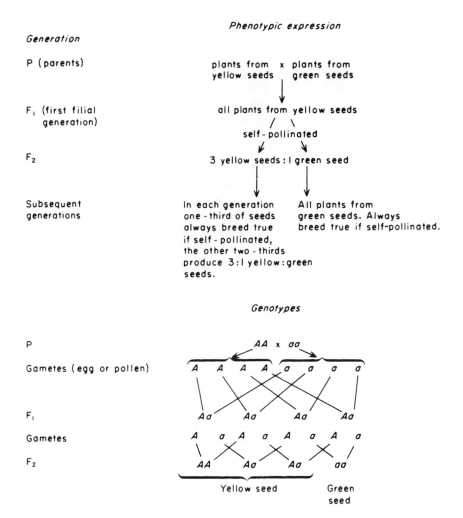

Fig. 4.1 Outline of some of Mendel's results obtained when breeding peas (*Pisum sativum*). *A* indicates the dominant gene for yellow seed, *a* the recessive gene for green seed.

However, genes do not always fall into this simple dominant/recessive pattern. Some may be incompletely dominant in the heterozygote, showing a transition stage between the phenotypes of the homozygous dominant and recessive conditions. Later workers also found that there are often more than two alternative forms (alleles) of a gene.

The significance of Mendel's work was not appreciated until about 1900. Then cytological techniques and breeding experiments during the next few decades established that the genes of plant and animal cells occur within the cell nuclei, where they are arranged at specific sites, or loci, along threads or chromosomes, whose principal chemical constituents are proteins and nucleic acids (Lewin, 1997). It was also shown that these chromosomes duplicated themselves at every cell division. Metabolic studies showed that some "errors of metabolism" in man result from the lack of single specific biochemical reactions, because the appropriate enzyme is missing, and whose production can be controlled by single genes. This knowledge developed into the idea that genes control the formation not only of enzymes, which are all proteins, but of proteins in general.*

Avery *et al.* (1944) then showed that the genetic material in a bacterium consists of the nucleic acid DNA (deoxyribonucleic acid), and in 1953 Watson and Crick first suggested the three-dimensional structure of DNA from which has developed all the subsequent work on the genetic code.

The essential features of this code, as generally understood at present, can be stated quite briefly, for the system is breathtakingly simple yet comprehensive. Genes are arranged along chromosomes, which in essence may be regarded as giant molecules of DNA. The DNA molecule consists of two intertwined helical chains of many nucleotides, with ten nucleotides in both chains for each complete turn of the helix (Watson, 1965). Each nucleotide also contains a purine or pyrimidine base, which is linked to the base in the corresponding nucleotide in the other helix (Fig. 4.2). It is these linkages between the two helixes that constitute the genetic code. There are only four possible bases: two purines (adenine (A) and guanine (G)), and two pyrimidines (thymine (T) and cytosine (C)). The bonds between bases on the two helices are specific hydrogen bonds between a purine base on the one chain and a pyrimidine base on the other chain. Adenine pairs with thymine, guanine pairs with cytosine, and the sequence of these bases on the DNA specifies the type of protein that it can form. All proteins consist of amino acids, there are only 20 amino acids in living organisms, and it takes a sequence of three successive bases on the DNA molecule to specify one amino acid. With four bases, there are obviously 64 possible sequences of bases, or codons, and most amino acids are therefore specified by more than one codon. In addition, three of the codons indicate "stop". This means that the genes, which contain all of an organism's inherited nature, contain instructions solely for assembling proteins, and that all of the inherited differences between individuals result from differences in sequences of amino acids.

One other aspect of the structure of DNA deserves brief mention. Obviously, only one strand of the double helix can contain the genetic message, because the other strand will have the complementary set of bases, which would give a very different message. So if we know the sequence of bases on one strand, we also know the corresponding sequence of bases on the other strand. This suggests, and it has been confirmed, that genes can

---

* When proteins, such as haemoglobin, contain more than one type of polypeptide the gene controlling the formation of each polypeptide is called a cistron.

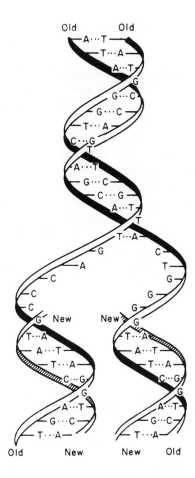

Fig. 4.2 Diagram to illustrate the double helix of DNA (deoxyribonucleic acid), with the two polynucleotide chains linked by complementary base-pairs (adenine (A) with thymine (T), and guanine (G) with cytosine (C)). Replication occurs when the two strands separate and both act as templates on which new complementary strands are formed. (From Watson, 1965.)

reproduce themselves by separation of the two strands of DNA, when both serve as templates upon which the other strand is duplicated. Occasionally, something goes wrong with the replication process, and one or more genes may be altered, lost or gained. Such spontaneous changes, or mutations, are rare, but some chemicals and forms of radiation can increase their frequency. Mutations are usually less favourable to the organism than the original gene, and are often sufficiently unfavourable to be lethal. Nevertheless, mutations in the reproductive cells are of crucial importance: when favourable they are the source of new genetic variation in subsequent generations.

This knowledge about gene structure and function modifies the Mendelian view of inheritance:

(1) The classification of genes as dominant or recessive needs to be reinterpreted. The "recessive" gene that produces no visible effect is presumably still producing protein, and we will consider this a little more later, when we discuss melanism.

(2) There is no known way in which the structure of DNA can be altered by information from the cell. This should finally put to rest the idea that acquired characteristics can be inherited.

(3) We now have an alternative, supplementary, way of regarding a population. Not only can it be considered as a collection of individual organisms, it can also be considered as a collection of genes, the gene pool, which is distributed amongst these individual organisms. Moreover, there are often several alternative forms, or alleles, of a gene that can occur at a given position, or locus, on a chromosome. It follows then that no one individual organism can contain all of the alleles that are available to a population. Thus in a very real sense the individual organism can be regarded as a subunit of the gene pool, not entirely autonomous.

For completeness one must add that most plant and animal cells also contain additional genes in organelles outside the nucleus, in mitochondria and, for plants, in plastids (including chloroplasts). These are inherited almost entirely from the female parent, and probably derive from cells of other organisms that have established a symbiotic (mutually beneficial) relationship with the "host" cell (Gray and Doolittle, 1982; Wallace, 1982).

We started by considering the effect of natural selection on a population with various phenotypes. We can now reassess this, and consider how natural selection affects the nature of the gene pool, the relative abundance of different alleles. We can use the same example, of pea plants that produce yellow or green seeds.

It should be intuitively obvious that although the proportions of genotypes and phenotypes may change from one generation to the next, unless there is differential selection for different genotypes, the frequencies of the two genes within two successive generations will remain unchanged. One may expect the proportions of two alleles in one generation to be reflected in the proportions of these alleles in the gametes, which then determine the proportions in the next generation. This does imply various assumptions, of course. In particular, for this example, it assumes there is random cross-pollination and that the population of pea plants is too large for chance effects to have a significant impact. This idea of constant proportions for two alleles, the Hardy–Weinberg law, was first enunciated by Hardy and Weinberg quite independently of each other, in 1908. This law also states, given the above assumptions that the frequencies of the different genotypes will also remain constant after the second generation, no matter what the frequencies were in the first generation.

The proof is simple and revealing. With the symbols of Fig. 4.1, let the proportions of the three genotypes $AA:Aa:aa$ in the parental generation be $P:2Q:R$. The F_1 generation will then contain these genotypes in the proportions $(P + Q)^2:2(P + Q)(Q + R):(Q + R)^2$ (Table 4.1 and Fig. 4.3), which we may also express as $P_1:2Q_1:R_1$. Now the frequency of the allele A in the parental generation may be denoted as p_0, when the frequency of a, q_0, must be $(1 - p_0)$. These allele frequencies can be expressed in terms of the genotype frequencies:

$$p_0 = \frac{2P + 2Q}{2} = P + Q$$

and

$$q_0 = \frac{2Q + 2R}{2} = Q + R$$

when $p_0 + q_0 = P + 2Q + R = 1$, as originally stated.

We can also express the genotype frequencies in the F_1 generation in the same terms (Table 4.1), and can therefore express the allele frequencies in the F_1 generation in terms of allele frequencies in the parental generation:

$$p_1 = \frac{2p_0^2 + 2p_0q_0}{2} = p_0(p_0 + q_0) = p_0$$

and similarly

$$q_1 = \frac{2q_0^2 + 2p_0q_0}{2} = q_0(q_0 + p_0) = q_0$$

This demonstrates that the frequency of alleles remains constant from generation to generation. We have also seen that genotype frequencies can be expressed in terms of the frequencies of alleles in the previous generation (bottom line of Table 4.1). It follows then that, after the parental generation, all subsequent generations will contain the same proportions of the three genotypes. The same principle applies to loci with multiple alleles (Weaver and Hedrick, 1995).

However, our most important assumption was that of no differential selection. Selection may act against individuals of certain phenotypes, and consequently there will be differential survival of different genotypes. This selection can occur at several stages: individuals may be removed from a population (by death, predation, disease, and so on) before they are ready to reproduce, individuals may survive but be unable to breed, breeding individuals may differ in the number of offspring they produce, or selection may

Table 4.1 The frequency of cross-pollination and the proportions of different genotypes in the offspring of pea plants with two alleles, A and a, for seed-colour[a]

Parental genotypes ♂ ♀	Frequency of cross-pollination	Frequency of genotypes in the F_1 generation		
		AA	Aa	aa
$AA \times AA$	$P \times P$	P^2		
$AA \times Aa$	$P \times 2Q$	PQ	PQ	
$AA \times aa$	$P \times R$		PR	
$Aa \times AA$	$2Q \times P$	PQ	PQ	
$Aa \times Aa$	$2Q \times 2Q$	Q^2	$2Q^2$	Q^2
$Aa \times aa$	$2Q \times R$		QR	QR
$aa \times AA$	$R \times P$		PR	
$aa \times Aa$	$R \times 2Q$		QR	QR
$aa \times aa$	$R \times R$			R^2
TOTAL FREQUENCIES:		$(P + Q)^2$	$2(P + Q)(Q + R)$	$(Q + R)^2$
Frequencies expressed in terms of allele frequencies in parents:		p_0^2	$2p_0 q_0$	q_0^2

[a] $P{:}2Q{:}R$ are the proportions of the three possible parental genotypes $AA{:}Aa{:}aa$.

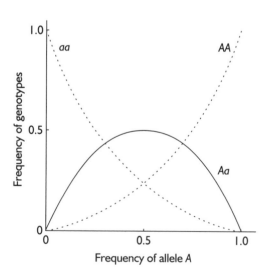

Fig. 4.3 The effect of allele frequencies on genotype frequencies in a population with two alleles, A and a, at one locus and which is not subject to differential selection.

act on the gametes themselves. Whatever the stage, the net effect on transmission of genes from one generation to the next is the same: some genes are
more likely than some others to be transmitted to the next generation.

Going back to Mendel's example of yellow and green seeds, let us suppose
that one generation of plants consisted of the genotypes AA, Aa and aa in the
proportions 25:50:25. Then p_0 and q_0 both equal 0.5. We have seen that, without
differential selection, p_1 and q_1 will also both equal 0.5, and we would also expect
P:$2Q$:R to remain at 25:50:25. Let us suppose though that one genotype is less
well adapted to survive and reproduce. Such a genotype is said to be less fit,
and one can measure the relative fitness (w) of different genotypes
(Strickberger, 1985). For algebraic convenience, the equivalent term s, the selection coefficient, is often used, where $s = 1 - w$. If the genotypes AA and Aa are
equally fit, and fitter than the genotype aa, then they will have a fitness of 1.0,
and a selection coefficient of zero. The fitness of the genotype aa is $1 - s$. Table
4.2 shows the effects on genotype frequencies after one generation has been
exposed to differential selection. The frequency of gene A has changed from
$(1 - q)$ to $(1 - q)/(1 - sq^2)$, and the frequency of gene a has changed from q to
$q (1 - sq)/(1 - sq^2)$. Thus the relative fitness for gene A is $1/(1 - sq^2)$, and for gene
a is $(1 - sq)/(1 - sq^2)$. In less precise terms, the relative frequency of gene A has
increased, whilst that of gene a has decreased. In more general terms, we can
say that if genes A and a are of equal fitness, their proportions will remain constant in the population, whereas if one is fitter than the other, its proportion in
the population will increase in successive generations.

One might extrapolate from this simple example to conclude that given
enough time a population will become genetically uniform, with only the
most successful of the alleles surviving, all less successful alleles having been
eliminated by natural selection. This is obviously not so in fact, but to give a
detailed rebuttal we must consider the nature of genetic variation in a little
more detail.

A useful way to start might be to consider the fate of a new mutation that
occurs in the gametes of one individual. Any offspring will necessarily be

Table 4.2 The effect of differential selection, for different genotypes, on the proportions of genotypes in the next generation

| | Genotypes | | | Totals |
	AA	Aa	aa	
Relative proportions, in equilibrium, before differential selection occurs	$(1 - q)^2$	$2q(1 - q)$	q^2	$= 1$
Frequencies	0.25	0.50	0.25	$= 1$
Fitness	1	1	$1 - s$	
Relative proportions after 1 generation exposed to differential selection	$(1 - q)^2$	$2q(1 - q)$	$q^2(1 - s) =$	$1 - sq^2$

heterozygous for this mutation. If it is recessive, then there will be no selection in the first generation for or against the mutation. Either way, dominant or recessive, whilst the number of individuals with this mutation is low, chance effects may well eliminate it. If we ignore that possibility, the fate of the mutation depends not only on its dominance, but also on its relative fitness. One extreme case would be a lethal dominant mutation, with zero fitness and a selection coefficient of one. This would disappear immediately. All other situations are rather less clear-cut. Harmful mutations are eventually eliminated, whereas advantageous mutations will eventually become widespread in the population. But these processes can take many generations, and theoretical calculations suggest that these may sometimes be numbered in tens of thousands of generations. The key determinants will be the degree of dominance, the magnitude of the selection coefficient and for disadvantageous genes the balance between their mutation rates and their selection pressures, and chance.

For well-adapted species in stable environments, most mutations are harmful and recessive, but one should not then deduce that a new mutation that has never arisen before is likely to be recessive. Dominance and recessiveness are not simple inflexible conditions. The degree of dominance exhibited by a gene is influenced by interactions with other genes at other positions on the chromosomes, and the current view is that most mutations have appeared many times before, and natural selection has influenced the gene complex to minimize the harm caused by the mutation, which becomes recessive. On reflection, it is not surprising that most mutations are harmful. If we accept that mutations arise as chance events, it simply indicates that organisms have already, by natural selection, become very well adapted to their environments. There is then relatively little scope for genetic improvement, unless the environment changes, when a new situation arises in which mutations are sometimes advantageous. It is important perhaps to stress here that the effects of genes do interact. For example, one might suppose that a mutation that enabled a bird to lay more eggs per clutch would be advantageous to that species. In fact, birds in temperate regions do already lay clutches that tend to produce the largest number of offspring (Godfray et al., 1991). Larger clutches may produce fewer surviving young for a range of reasons, such as insufficient food or more predation. This is an r-type strategy. In genetic terms, natural selection will favour any phenotype that enables parents to leave more offspring, and is correlated with the fact that catastrophes do occur from time to time (e.g. Fig. 2.11). In the tropics catastrophes are rarer; it is therefore perhaps important to devote relatively more effort to maximizing the carrying capacity of the environment—a K-type strategy. Possible strategies might include avoidance of predators, and reducing intraspecific competition at any given density. Thus relatively less effort will be devoted to rearing young, and clutch sizes do tend to be smaller (Cody, 1966). Whether the details of these interpretations are correct is perhaps arguable, but they do illustrate the essential point, that modification of one function implies the need to modify other functions too.

We did assume that the population was large enough to avoid significant chance events. It is unlikely that the proportions of alleles and therefore of genotypes in successive generations will always conform exactly to their theoretical proportions. This random genetic drift clearly becomes more important in smaller populations (Weaver and Hedrick, 1995), but its exact ecological significance is still a matter for dispute (Ford, 1975). It would probably be fair to say though that it is now attributed less importance than it used to have, because selection pressures are now thought to be more powerful. Examples of the powerful selection pressures that can be exerted by pollutants will be discussed later in this chapter. It used to be thought that selection coefficients were of the order of 0.01 (Fisher, 1930), but it has now become apparent that selection pressures are often much stronger and commonly exceed 0.25, which reinforces the idea that populations are well adapted to their environments. It must be added though that the value of most selection coefficients is still an open question, because characteristics tend to be studied whose frequencies are known to change quickly. Neutral mutations, which have no selective advantage or disadvantage, are another cause for argument. Some argue that they must be very common, others that they are uncommon events.

Here we approach the limits of present knowledge. There is much elegant mathematical theory about the consequences of different selection coefficients, mutation rates and population sizes. It is difficult to apply much of this theory to real populations because the rates of these processes are so difficult to estimate accurately. Endler (1986) gives a detailed appraisal of ideas and knowledge about natural selection.

Two alleles at one locus do often coexist indefinitely in a gene pool. This phenomenon is most easily studied with genetic polymorphisms, where a characteristic that is under genetic control occurs as two or more distinct types, not as a continuous range of variation: one example is the existence of distinct blood groups in human populations. Ford (1940) defined a genetic polymorphism as the occurrence of two or more discontinuous forms in the same locality in such proportions that the rarest form cannot be maintained merely by recurrent mutation. Polymorphisms may be transient, as when a favourable mutation is replacing an existing allele in a population, or balanced, when a departure in either direction from the optimum proportions is disadvantageous. In practice, some caution is needed. The occurrence of two or more forms in a species does not necessarily imply a genetic polymorphism, because various developmental and environmental effects can also permit the occurrence of distinct forms.

A locus may remain heterozygous because the relative fitness of different alleles changes from time to time or in different places. Environments do change with distance, and nearly every group of organisms has been found to form clines: genetic changes along geographic gradients (Endler, 1977). Nor is the environment constant within one restricted area: there is a mosaic of microhabitats, and the environment is dynamic, not static. All of these factors may act to continually alter the balance of advantage between two alleles, and so may enable both to survive.

A third possibility is for the heterozygote to be fitter than either of the homozygotes. There are several well-known instances where two alleles survive because of heterosis ("hybrid vigour"), when the heterozygote is the fittest genotype. The best-studied example is the occurrence of sickle-cell anaemia in man (see Allison, 1956). Oxygen is carried with the red cells of the blood by haemoglobin, a large protein molecule with well over a hundred known mutant forms. One of these mutants, haemoglobin-s, occurs in distorted blood cells, which are sickle- or spindle-shaped instead of disc-like. The gene is denoted Hs, in contrast to the normal allele Hn, and the homozygote $HsHs$ condition is potentially lethal, with most such individuals dying before adolescence. Heterozygotes ($HnHs$) are not anaemic and can lead normal lives, although their red blood cells may distort when there is little oxygen in the blood. In some parts of Africa up to 40% of the population is heterozygous for this gene, and it correlates with the occurrence of the most severe form of malaria, a disease caused by a protozoan parasite that reproduces within the red blood cells. People of the normal $HnHn$ genotype often die from this disease, but heterozygotes are much more resistant, because infected blood cells tend to collapse and so retard development of the parasite. Thus in these malarious parts of Africa the heterozygous individuals have the balance of selective advantage over both the normal genotypes who may die of malaria, commonly in childhood, and the anaemic ($HsHs$) genotypes who die of anaemia. By chance, the activities of Dutch slave traders have confirmed this account. They used to export natives from Ghana to Surinam, which used to have a high incidence of malaria, and to Curaçao, which did not. Descendants of these slaves in Surinam still have a high frequency of haemoglobin-s, but not the descendants in Curaçao (Allison, 1975). So here is an instance where selection pressures do not lead to the replacement by one allele of all other alleles.

Kimura (1968) proposed an alternative view. He argued from the differences between species in their haemoglobin molecules that random genetic drift is important in the evolution of species. This view was then extended to intraspecific variation: most mutations are so deleterious that they are quickly eliminated, but polymorphisms within species also result from mutations that are neutral, with no selective advantage or disadvantage, or near-neutral with a mean selection coefficient of 0.001 (Kimura and Ohta, 1971; Kimura, 1983, 1991). Polymorphisms are then seen as the product of neutral and near-neutral mutations and random genetic drift.

Many tests are needed, from the biochemical to selection tests on whole organisms to determine whether alleles are neutral (Powers *et al.*, 1993), but there are cogent arguments against this view (Gillespie, 1987), which assumes that the optimal phenotype for each locus remains constant for millions of years. Much of this argument is concerned primarily with the evolution of new species, but we are still left with the more quantitative question, how common are genetic polymorphisms?

This question was almost impossible to answer until we learnt something about the nature of the genetic code. Genes are not entities that control single characters, but segments of DNA molecules that specify proteins. The interactions of these proteins will, for example, determine the colour of a seed. This deeper understanding of the nature of genetic variation has given us a new way of studying that variation. Classically, one observes differences in phenotypic expression, characters such as seed colour or resistance to malaria, but much genetic variation may be expected to be less obvious than this. Unless and until we can detect these less obvious forms of genetic variation, we cannot determine how much genetic variation does exist in a population, and therefore how much scope there is for genetic adaptation to the environment. Hubby and Lewontin (1966) argued that any mutation in a gene should change, delete or add at least one amino acid in the corresponding protein. Enzymes, as proteins, are direct consequences of genetic activity, and it has been found that they can vary in the details of their structure. These are loosely called isozymes but more accurately, if produced by different alleles of the same gene, allozymes. In some instances these differences will alter the enzyme's net electrostatic charge, when the allozymes can be distinguished if exposed to a suitable electric current in a gel, a process called electrophoresis. If we accept that genes have multiple effects, we may expect that such cryptic diversity in the fine detail of enzyme structure will sometimes be correlated with more obvious features, and the study of allozymes has added a useful tool to the study of genetic variation in populations.

This approach has many advantages. Electrophoresis can be used with single macroscopic organisms (Hubby and Lewontin used single fruit-flies (*Drosophila*)) to distinguish phenotypes for specific enzymes, and breeding experiments with pure strains of the different phenotypes allow the genetics of the phenotypic differences to be analysed. Hubby and Lewontin found, in fact, that all of the enzymes that they examined were controlled by single genes, and Lewontin and Hubby (1966) first demonstrated that differences between individuals in the structure of their allozymes and other proteins are likely to be very widespread. They surveyed the degree of variation in allozymes for five natural populations of *Drosophila pseudoobscura*. They examined 18 loci (enzymes) and found from their samples of these five populations that on average each population, or gene pool, was polymorphic for 30% of these loci, with two or more alleles. They also calculated that, depending on the population, 8–15% of the loci in single individuals were heterozygous. There were three important biases in these estimates, all tending to lower the estimates of polymorphism:

(1) Not all amino-acid changes within proteins will be detectable by electrophoresis.
(2) They used cultures bred in the laboratory from the original wild populations, and this would itself reduce variation.

(3) Only small samples were taken from the wild populations, and inevitably some less frequent alleles would be missed.

Numerous studies of this type have now been published. Nevo *et al.* (1984) summarize the data for 1111 species where, depending on the taxonomic group, about 20–50% of the loci examined were polymorphic. The most useful measure is the heterozygosity (*H*), where for a single locus with *K* alleles, each with a frequency p_i,

$$H = 1 - \sum_{i=1}^{K} p_i^2$$

This measure is relatively independent of sample size, and, for a species that mates randomly, estimates the fraction of individuals that is heterozygous at that locus. The average heterozygosity ($\overline{H}$) for major taxonomic groups was about 0.04–0.15, although, because of the great variation between loci (Ward *et al.*, 1992), average values may be of limited biological interest. *Drosophila*, for example, is almost always homozygous for the malic enzyme while some esterases are almost always polymorphic (Gillespie, 1991). Moreover, electrophoretic data are probably not representative of all loci.

Electrophoresis only detects enzymes for which specific histochemical stains are available. Heterozygotes can be detected more directly, albeit less easily, be extracting DNA from the cell and isolating either short lengths of chromosomes with specific sequences of genes or even individual genes (Hoelzel and Dover, 1991; Hoelzel, 1992). These techniques confirm that there is an appreciable degree of variation in the sequences of DNA (genetic polymorphisms) within plant and animal populations, although not all of these variations alter sequences in the proteins produced by genes (phenotypic polymorphisms) (Lewin, 1997).

It is now time to consider the relevance of ecological genetics to pollution. Most current problems of pollution occur on a much shorter time-scale than that required for the evolution of new species, but we do need to give the lie to an opinion that has some adherents. It is sometimes suggested that it does not matter if pollution does eliminate some species, because this is the natural progress of evolution anyway. The critical difference between evolutionary change and that wrought by pollution is the speed: populations can disappear very rapidly from pollution, and if unchecked we would have a very impoverished fauna and flora. For practical purposes, if all populations of a species are eliminated, then that species is lost for ever. Sometimes individual populations appear to be irreplaceable too.

One of the very first recorded effects of pollution on wildlife, and perhaps the most striking evolutionary change ever to be actually witnessed, was the occurrence of melanism in moths. Melanic moths are darker than the norm, because more granules of melanin than usual are deposited in the cuticle and wing scales. The number of reported occurrences of melanism has risen steadily since about 1850. About 780 species of the larger moths occur in Britain, and samples of over 100 of these now commonly include melanic forms. Similar changes have been observed in both continental Europe and North America, commonly in association with industrial development, although in many species melanism occurred before the start of industrial development (Majerus, 1989).

In many instances these melanic forms have become predominant and spread very rapidly, and for a full understanding we need to understand the distribution, selective advantage and genetics of melanism. The first, and best-studied, example of melanism is *Biston betularia*, the peppered moth, which is particularly suitable for studies of this sort (Kettlewell, 1973). It is common, widespread, easy to capture, disperses well (which ensures adequate gene flow) and is easy to breed in the laboratory. The first melanic specimen was caught in 1848 at Manchester, and by 1895 98% of the moths in that area were melanic.

Before about 1850, melanic forms appear to have been maintained solely by mutation. Between 1848 and 1900, melanic forms were noticed for the first time in many widely separated places in Lancashire, the Midlands and eastern counties. The proportion of melanics increased rapidly in these areas, which were in general industrial zones and regions to the east of them. There is some variation in the degree of melanism, and three forms are recognized in *B. betularia*. The normal form, f. *typica*, is well adapted to what has usually been thought to be its normal daytime resting sites, commonly on exposed tree trunks, less commonly on walls and palings. When these surfaces are covered by lichens, f. *typica* is well-nigh invisible on them. It must be noted though that there has been little direct evidence of exactly where these moths do rest during the day (Clarke *et al.*, 1985). This form does show a considerable range of coloration, from heavily speckled individuals to pale, almost white specimens with fine, granular black markings, and breeding experiments suggest that genes at more than one locus are involved. The common melanic form, f. *carbonaria*, is controlled by a single dominant gene, and again there is a range of patterns, commonly all black, but sometimes with some light-coloured spots or patches. At least three other alleles can occur at the *carbonaria* locus, all dominant to f. *typica* and all recessive to f. *carbonaria*. These alleles produce a range of intermediate forms, known collectively as f. *insularia*, whose appearance overlaps with that of the other two forms.

Kettlewell went on to analyse the genetics in more detail. He mated heterozygous melanic moths from Birmingham with normal moths of a related Canadian species, *B. cognatoria*, taken from an area in Canada where melanics did not occur. The F_1 generation segregated as expected 50:50 into melanic and typical forms, but after repeating this twice more there was no longer a clear-cut segregation, but instead a complete range appeared of forms intermediate between f. *typica* and f. *carbonaria*. These results are

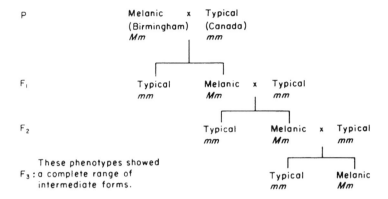

Fig. 4.4 Diagram to illustrate the genotypes of the crosses made by Kettlewell (1973) between melanic forms of *Biston betularia* taken from Birmingham, England, and typical forms of *B. cognatoria* taken from Canada. *M* represents the dominant gene for melanism.

surprising in terms of Mendelian genetics (Fig. 4.4), and indicate that the expression of the gene for melanism must be greatly influenced by the other genes present. If we ignore those genes that are linked to the gene for melanism by being on the same chromosome, 50% of the genotype in the F_1 melanic moths will be of Canadian origin. This will rise to 75% and then to 87.5% in the next two crosses. This explains the loss of complete melanism: the expression of the gene for melanism is controlled by the rest of the genotype.

A similar experiment in which heterozygous melanic moths from Birmingham were crossed with normal moths from Cornwall, again an area without melanics, had relatively little effect after three generations: the original melanic moths were all black, and those from the three filial generations were still melanic although the forewings were uniformly peppered with white scales. Thus they differed from the parent melanic moths from Birmingham, presumably because of the different gene complex in the Cornish population, but they also differed from the earliest melanic specimens taken in the middle of the nineteenth century, when such moths frequently had small wedges of white markings on both forewings and hindwings. This suggests that the earliest f. *carbonaria* may have been genetically different from present-day melanic individuals in Birmingham.

Kettlewell also took the darkest forms of the F_3 generation from crosses between melanic moths and the Canadian species, and bred them with f. *typica* from Birmingham and from Cornwall. In both instances the offspring fell closely into the two groups f. *typica* and f. *carbonaria*. Full dominance was achieved in one generation, which implies that the British populations contain genes that ensure the expression of dominance for melanism, and that these genes are also dominant but not linked with that for f. *carbonaria*.

The biological significance of melanism was a matter for debate for some decades (Ford, 1975). Kettlewell (1955, 1956) made the first significant observations. He both observed for several species of bird that the degree of predation by birds on resting moths depended on the degree to which they matched their background, and showed experimentally that the relative survival of the melanic and normal forms differed in rural and industrial areas. Known numbers of marked individuals of both forms were released and recaptured, and the difference in recapture rates of the two forms was found to be far greater than could be expected by chance (Table 4.3). If one takes either site alone, one could explain these results as a difference between the two forms in their life-span, migration or liability to recapture. Taking the results from the sites together one is forced to conclude that they indicate differential predation. The melanic form is better concealed at the urban site, and the typical form is better concealed at the rural site. The difference in predation rate, and hence of selection pressure, appears to be very pronounced, and Haldane (1924, 1956a) calculated that for the Manchester area, in the second half of the nineteenth century, the melanic forms must have had a 50% better chance of survival than the typical forms. This is, indeed, very high selection pressure. Even so, the melanic form does not always replace the typical form completely. Haldane, and others since, have suggested that the heterozygote may be fitter than both homozygotes for one or more factors other than predation by birds. This would indicate that there is a balance of selection forces between the different phenotypes, as in sickle-cell anaemia.

Kettlewell organized three surveys, extending from 1952 to 1970, of the frequency of melanism in Britain (Fig. 4.5). Since the allele for f. *insularia* is masked by that for f. *carbonaria*, the true frequency of f. *insularia* is probably higher than these figures suggest, but in general the high frequencies of melanic forms correlate with the industrial areas of Britain. Subsequent surveys by other workers confirm this correlation (Lees, 1981). High frequencies of f. *carbonaria* also occur throughout eastern England from north to south, even though far removed from industrial centres, and Kettlewell suggested that melanic forms occur where surfaces are blackened by soot deposits from smoke. These would, of course, be the industrial areas, but whether this is a

Table 4.3 The relative recoveries of marked individuals of two forms of the peppered moth, Biston betularia (f. typica and f. carbonaria) from two sites, one rural and one urban

Site	Form	No. released	No. recaptured	% recapture
Rural	*typica*	496	62	12.5
(Dorset)	*carbonaria*	473	30	6.3
Urban	*typica*	201	32	15.9
(Birmingham)	*carbonaria*	601	205	34.1

Data from Kettlewell (1955, 1956).

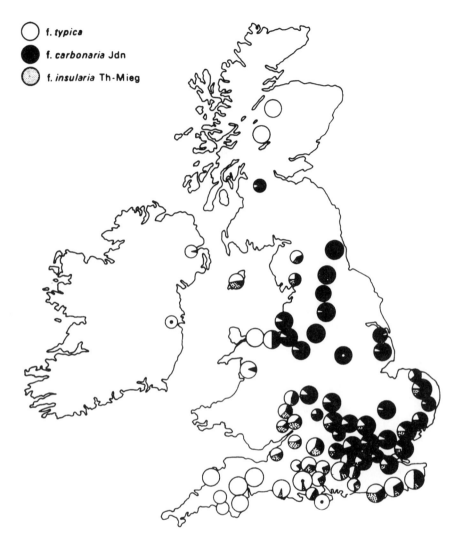

Fig. 4.5 The relative frequencies of the normal and two melanic forms of the peppered moth, *Biston betularia*, in Britain. The results are based on more than 30 000 records collected from 1952 to 1970 at 83 sites. (From Kettlewell, 1973.)

sufficient explanation for a high proportion of melanics in the rural areas of eastern England is more debatable, even though the predominant winds are south-westerly.

Bishop (1972) studied a cline from Liverpool for a distance of 50 km southwest into rural Wales. The frequency of f. *carbonaria* declined from 85–95% around Liverpool to 5–10% in Wales. He estimated survival, migration and reproductive rates, and the relative predation rates by birds on dead specimens of f. *typica* and f. *carbonaria* glued to tree trunks at seven sites along the cline. Although the data were admittedly incomplete, there appeared to be a far higher proportion of melanic moths in Wales than theory predicted. Later work on an extended transect from Manchester to Betws-y-Coed suggested that the concentration of sulphur dioxide is important, because it reduces the diversity of and cover by lichens on tree trunks (Fig. 4.6) (Bishop *et al.*, 1975). Sulphur dioxide gas travels further from a source than do smoke particles, and does appear to have affected the lichen flora in most of eastern England (Fig. 8.13). Melanic moths are then better camouflaged on tree trunks that have been denuded of lichens and or are blackened by soot.

As always, predictions depend on their assumptions. Recent work implies that the estimates of relative fitness for the different genotypes need to be revised. Mikkola (1984) argued that earlier experiments on the effect of substrate colour on differential predation (see Table 4.3) are probably misleading, because the moths used in these experiments settled on atypical backgrounds during the first day after their release. From observations on moths in experimental cages, Mikkola suggested that moths normally rest in less exposed situations on the underside of horizontal branches. Preliminary observations (Howlett and Majerus, 1987; Liebert and Brakefield, 1987) support this suggestion, and a pilot test (Table 4.4) confirmed, as expected, that predation on f. *carbonaria* is significantly higher in rural areas, and lower in urban areas, than it is on f. *typica*, and also that significantly fewer of the concealed moths are eaten than of the exposed ones. Moreover, most important, at both sites concealment reduced predation less for f. *typica* than it did for f. *carbonaria*. The authors point out that these results need to be confirmed, but these results do suggest that Bishop's estimates (1972) of predation rates may have overestimated the predation on f. *carbonaria* in rural Wales, and hence predicted the presence of fewer melanic moths than were actually observed.

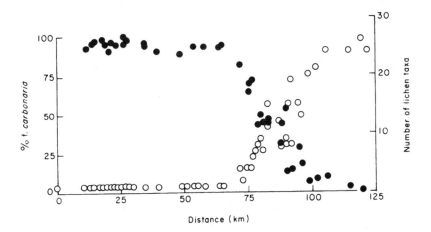

Fig. 4.6 The frequency of f. *carbonaria* (●) in samples of the peppered moth, *Biston betularia*, and the number of different lichens (taxa) on oak trees (*Quercus* spp.) (○), along a transect from the Manchester area to central Wales. Distances are measured from the most north-easterly site. (From Bishop *et al.*, 1975.)

Table 4.4   Differential predation on *Biston betularia* at two woodland sites. Numbers out of 50 specimens that were eaten after 72 hours' exposure

Site	Form	Position[a]	
		Exposed	Concealed
Rural	*typica*	16	13
(New Forest)	*carbonaria*	31	20
Urban	*typica*	29	25
(Stoke-on-Trent)	*carbonaria*	20	14

Data from Howlett and Majerus (1987).
[a] All specimens were glued onto tree trunks, either relatively exposed, or relatively concealed 5 cm below the junction of a branch with the trunk.

Migration rates, which were estimated from observation of adult males, may also need to be revised (Brakefield, 1987). When first instar larvae hatch they are suspended on silk threads and could therefore disperse long distances as part of the aerial plankton, which could explain the unexpectedly high incidence of f. *carbonaria* in eastern England.

The current mathematical model predicts the rate of decline in the frequency of f. *carbonaria* when pollution decreases by assuming a linear relationship between the mean winter concentration of sulphur dioxide and visual selection (Mani, 1990). This correlation does not of course imply that sulphur dioxide affects fitness directly. The *carbonaria* gene also confers a non-visual survival advantage, although heterosis is not invoked. The available field data for the frequency of f. *carbonaria* in England and Wales do not falsify the model's predictions (see Fig. 8.14 for one example). Lack of more experimental data prevents one from being more definite (Mani and Majerus, 1993).

Several points are worth emphasizing. Pollution in this instance is not having a direct effect on the moth populations, nor indeed on their predators, but an alteration to the habitat has altered greatly the relative fitness of different genotypes. The population response under such a strong selection pressure occurred in a few tens of generations, and the effect of the allele for melanism does depend greatly on the whole gene complex, but the precise nature of this selection is still uncertain. Melanism also illustrates the difficulty (discussed further in Chapter 9) of producing adequate proof, or disproof, of cause and effect when pollutants are thought to be causing major biological effects.

A second striking example, where pollution acts directly on the population, is metal resistance in plants. Soils sometimes contain, naturally, relatively high concentrations of one or more heavy metals, what the geochemist calls "anomalies". High concentrations can also result from mining activity. Either way, although many species of plant are unable to survive on such soils, because of metal toxicity, some species do survive and flourish. The first comparative study was described in a brief paper by Prát (1934), who found specimens of *Melandrium silvestre* growing on the waste tip from a copper mine at Piesky in Czechoslovakia. He collected seeds from both these plants and from other specimens growing in the Botanic Gardens of the University in Prague, and planted them in pots of garden soil. Other pots had in addition a range of copper carbonate concentrations, increasing in steps by a factor of two from 0.7 to 25%. Seeds from the Botanic Garden produced plants that grew poorly with 3% copper carbonate, and with higher levels did not develop beyond the cotyledon stage, and eventually died. In contrast, plants from seeds from the copper mine appeared healthy at all concentrations, although those with 25% copper carbonate were less vigorous than those with lower exposures.

There were no further studies of this type until the 1950s, when studies in both Germany and Great Britain showed that a considerable number of plant species can produce genotypes resistant to one or more heavy metals (Antonovics *et al.*, 1971; Baker, 1987), where resistance means a genetically-based decrease in response of a population to a pollutant as a result of exposure to that pollutant. It needs to be noted that botanists have preferred to use the term "tolerance" for the resistance of plants to metals, on the grounds that such plants do not exclude metal from their tissues, they tolerate it (Bradshaw, 1982). However, some resistant insects also "tolerate" insecticides in their tissues. In all known instances of plants occurring in both normal and contaminated soils, the resistant genotypes are uncommon on uncontaminated soils. It is difficult, however, to quantify the degree of a plant's resistance. Not only is it difficult to decide how exactly to measure the amount of metal in the soil, for it can occur in more than one form, but also the metals are often not distributed uniformly in the soil. Wilkins (1957, 1960) overcame this difficulty for the grass *Festuca ovina* by measuring the degree to which a standard concentration of heavy metal retarded root growth in water culture. Tolerance is then measured by the ratio of growth rates in normal and tolerant plants, but in a different environment from that in the field, and the ratio may vary with environment (Wilkins 1978; Macnair, 1989), although Baker and Walker (1989) conclude that the limitations of this method are minor compared to its speed and simplicity.

One of the best-studied plants is the grass *Agrostis tenuis*, which is common on the contaminated soils adjacent to mines for heavy metals. Populations of *A. tenuis* that are resistant to the heavy metals occur on contaminated soils, whereas adjacent populations on uncontaminated soils lack this resistance. Walley *et al.* (1974) compared seed from three strains of *A. tenuis*:

(1) A commercial susceptible strain.
(2) A population from Parys Mountain, Anglesey, Wales, which was growing on mine waste that was contaminated with high levels of copper, and which was resistant to copper.

(3) A population from Trelogan, Flintshire, Wales, which was growing on mine waste that was contaminated with high levels of zinc, and which was resistant to zinc.

As might be expected, the percentage germination and the growth of survivors on contaminated soils was in general better for plants derived from the appropriate resistant strain than from the susceptible strain (see Fig. 4.7 for the results with copper). There was also a considerable degree of variation in the degree of metal resistance between individual plants grown from seed taken from the susceptible strain. A small proportion (1–2%) survived a low degree of copper or zinc contamination and, for copper, a few individuals were almost as resistant as plants from the resistant population (Fig. 4.7b).

These results demonstrate that heritable variation in the degree of metal resistance exists in normal populations of A. tenuis, so that, if exposed to soils contaminated by copper or zinc, natural selection will increase the proportion of individuals that are resistant to the contamination (see also McNeilly and Bradshaw, 1968). In more quantitative terms, in contaminated soils, normal genotypes usually have a fitness of zero compared with their fitness in uncontaminated soils, and resistant genotypes have a fitness of 0.5–1.0, the exact value depending on the degree of soil toxicity (Bradshaw, 1976). Such figures suggest that a suitable plant population could adapt rapidly to metal contamination.

The speed with which resistance develops, when heritable variation occurs, was studied in a related grass species, Agrostis stolonifera (Wu et al., 1975). Grasslands of known, different, ages in the vicinity of a metal refinery at Prescot, Lancashire, were heavily contaminated with copper, so much so that young grassland, established 5 years earlier, had large areas of bare ground, but the oldest grasslands, 70 years old, had a complete cover of plants, dominated by A. stolonifera or A. tenuis, and with few if any other species present. Plants of A. stolonifera from uncontaminated sites were not resistant to copper, but plants from the contaminated sites showed an increasing frequency of resistant individuals, and an increasing degree of resistance in individuals, as the age of the grassland increased (Fig. 4.8).

We have already seen (Chapter 2) that immigration and emigration of individual organisms is one aspect of population size. A parallel process is gene flow, the transfer of genes, between populations. Pollen and seed may be expected to cross the boundary between adjacent resistant and susceptible populations of grass, so an appreciable degree of gene flow, movement of genes, may be expected to occur between the two populations near the boundary, and the degree of resistance will then depend on the balance between selection pressures and the rate of gene flow. McNeilly (1968) illustrated this balance very clearly in a copper mine at Drws y Coed in North Wales, where the contaminated area of about $300 \times 100 \text{ m}^2$ was surrounded by pasture that also contained A. tenuis. A notable feature was that the mine was on the floor of a U-shaped valley that ran in an east–west direction, with winds predominantly from the west.

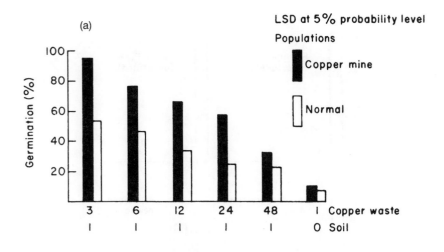

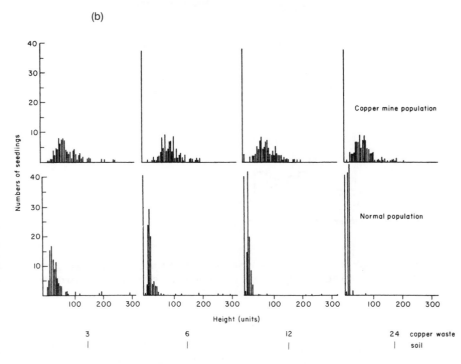

Fig. 4.7 The germination of seed, and growth of seedlings, of the grass *Agrostis tenuis* on a range of copper-waste/soil mixtures. One population of seed came from plants resistant to copper, on Parys Mountain, Anglesey. The other population came from a susceptible commercial strain of grass. (a) Germination, as a percentage of that on normal uncontaminated soil (LSD, least significant difference); (b) frequency distributions for the height of seedlings after 4 months' growth. (From Walley *et al.*, 1974.)

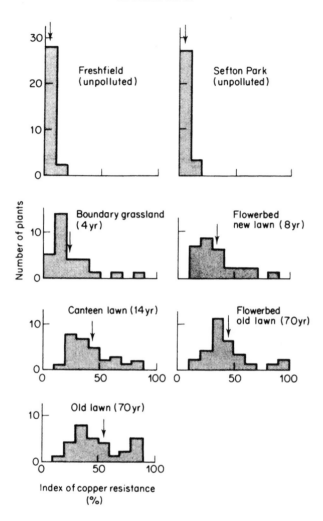

Fig. 4.8 The distribution of copper resistance in samples of the grass *Agrostis stolonifera* taken from seven populations of different ages around a copper refinery at Prescot, Lancashire. The index of copper resistance was given, under standardized conditions, by:

$$\frac{\text{mean length of the longest roots when grown in a solution with copper}}{\text{mean length of the longest roots when grown in a solution without copper}}$$

↓ indicates a mean value. (From Wu *et al.*, 1975.)

McNeilly studied plants on two transects, one downwind from the mine, the other across the valley floor, in effect upwind of the mine. At each site on a transect, seed was taken from 15 plants, and both the seed and the parent plants were tested for resistance (Fig. 4.9). On the upwind transect there was an abrupt transition, within a distance of 1 m, between resistant and suscep-

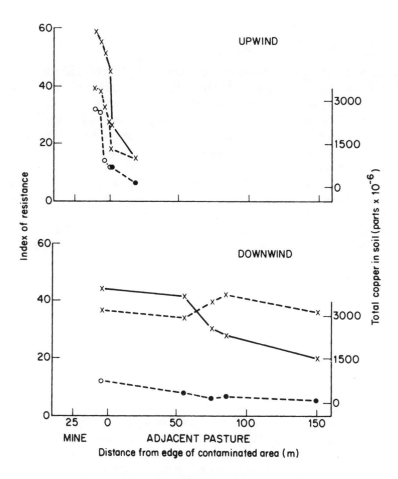

Fig. 4.9 The degree of copper resistance in seed and plants of the grass *Agrostis tenuis* on two transects across a copper mine at Drws y Coed, North Wales. The index of resistance is the ratio of root growth in solution with and without copper (for details see McNeilly and Bradshaw, 1968). (X—X) Resistance in adult plants; (X---X) resistance in seed from the same plants; (○---●) total copper concentrations in the soil, (○) a site where the paucity of species denotes copper toxicity; (●) a site where the vegetation is normal. (Data from McNeilly, 1968.)

tible plants, right on the boundary of the contaminated soil. Seed from resistant plants was less resistant to copper than the parent plants, which may indicate a significant input of genes from normal susceptible plants upwind of the mine. By contrast, resistance spread a relatively long way downwind beyond the region of contaminated soil in plants from the other transect, and seed were more resistant than their parents, again due presumably to a significant degree of gene flow, from resistant plants on the mine area. This would also explain high copper resistance in seed from plants downwind from the mine.

These data show clearly that there are strong selection pressures at work, with only plants that are resistant to copper on the contaminated soil, but rather less strong selection for normal genotypes on the uncontaminated soils. Experiments confirmed that only resistant plants could grow on contaminated soils, but both types of plant germinated and grew equally well on uncontaminated soils. However, when the resistant and normal genotypes were grown together on uncontaminated soils, the normal plants grew more successfully than the resistant genotypes: the normal genotypes were the more successful competitors, with a coefficient of selection against resistant plants on normal soils of 0.53.

This example has at least two striking similarities to melanism in *Biston betularia*. Selection pressures are very strong, and both genotypes are competitively superior in their appropriate environments. In contrast, the precise nature of the selection pressure for metal-resistant plants is rather more obvious than for melanism, selection acts much more quickly on the gene pool, and the resistant plants are not so markedly different phenotypically from normal plants of the same species. Indeed, metal resistance on contaminated sites is noticeable principally because other common species are absent, and many of the relatively obscure differences that have been described for metal-resistant plants are probably independent adaptations to other environmental gradients that are more or less correlated with the gradient of metal content in the soil (Antonovics *et al.*, 1971): much is still unknown about the mechanisms of metal tolerance (Verkleij and Schat, 1989; Jackson *et al.*, 1990; Peterson, 1993). So one is led to the question: how frequently does pollution alter the gene pool of exposed populations? How often is there both selection pressure and the appropriate variation, so that evolutionary change can occur? (Bradshaw and McNeilly, 1981)

Tolerant populations contain alleles for tolerance before they are exposed to metals (Wu, 1989), so one approach to this question is to screen populations for the degree of genetic variation between individuals in their resistance to pollutants. We have referred to the paucity of species on metal-contaminated soils. Gartside and McNeilly (1974) tested the distribution of copper resistance in seed from eight plant species thought at that time not to occur in copper-contaminated area, and also from *Agrostis tenuis*, which does. Ten thousand seeds of each species were tested for germination and growth. Three species (*Plantago lanceolata, Anthoxanthum odoratum* and *Trifolium repens*) had no survivors on contaminated soils, and four species (*Lolium perenne, Poa trivialis, Cynosurus cristatus* and *Arrhenatherum elatius*) had only slight resistance. The other two species, *Agrostis tenuis* and *Dactylis glomerata*, had 0.08% or 0.8% (both figures are stated in the text) of their individuals fully resistant to copper. Both of these species are, in fact, found naturally on contaminated soil (Bradshaw and McNeilly, 1981). From this and other similar work we may tentatively suggest that many species lack the appropriate alleles that would enable them to adapt genetically to high selection pressures from pollution. Even when alleles for tolerance do occur, they are not necessarily present in all populations of a species (Al-Hiyaly *et al.*, 1993).

The alternative approach is to study populations that have been exposed to pollutants in the field. We will consider first the reactions of plants to sulphur dioxide. Seed was taken from specimens of the annual weed *Geranium carolinianum* at six sites at various distances from a power plant in Georgia, USA, that had been emitting considerable quantities of sulphur dioxide for nearly 30 years (Taylor and Murdy, 1975). Short-term, high-level fumigation experiments (2080 µg sulphur dioxide/m³ for 12 h) showed that seed from plants with high exposures to sulphur dioxide produce plants with less visible leaf damage after fumigation. Subsequent breeding experiments showed that several genes must influence the degree of resistance (Taylor, 1978).

Similar work started in Great Britain, on a species of grass, when Bell and Clough (1973), who were working at Helmshore, about 24 km north of Manchester, on the resistance of ryegrass (*Lolium perenne*) to sulphur dioxide, discovered the indigenous population was more resistant to sulphur dioxide than was the cultivated variety, S23. Bell and Mudd (1976) then transplanted two indigenous populations of *L. perenne*, from Helmshore and from the coast near Blackpool where exposures to sulphur dioxide would be expected to be much lower. Not surprisingly, both populations grew best in their own environments: their genotypes were well adapted to their local environments, but at least part of the explanation may have been the lower resistance to sulphur dioxide of the coastal population.

Horsman *et al.* (1978) therefore made an intensive study to see whether populations of ryegrass are genetically adapted to withstand their local ambient concentrations of sulphur dioxide. They chose two sites on Merseyside in the Wirral peninsula—Newsham in the centre of Liverpool and West Kirby 16 km to the west-south-west—which had had six- to eight-fold differences in mean winter concentrations of atmospheric sulphur dioxide during the previous 20 years. The sites were both in parkland of similar aspect, altitude and soil, so that their relative proximity to each other minimized the risk of a confounding effect from genetic adaptation to factors other than sulphur dioxide. Thirty-six individually identified clones of *L. perenne* were taken from both sites. They were exposed to a low or high level of sulphur dioxide in a wind-tunnel for 8 weeks, when yield of dry matter, both living and total (living plus dead), was measured (Table 4.5). Yields from the two populations were comparable at the low exposure to sulphur dioxide of 35 µg/m³, but the higher exposure of 650 µg/m³ reduced yield significantly in the West Kirby clones. There was a smaller, and statistically insignificant, reduction in the Newsham clones. The high level of sulphur dioxide also increased the weight of dead shoots in the clone from West Kirby, which suggests that not only was there "invisible injury" but also a higher rate of senescence. Horsman *et al.* (1979a) confirmed and extended these observations, which suggest that the high levels of sulphur dioxide in the urban areas of Merseyside

Table 4.5 Yields of dry matter from clones of perennial ryegrass, *Lolium perenne*, taken from two sources[a]

Source of experimental clones	Mean winter concentrations of SO₂ (μg/m³) Around 1960	Present levels	Experimental concentrations of SO₂ (μg/m³) Yield of living dry matter (g) 35	650	Yield of total dry matter (g) 35	650
Newsham	600–700	≈200	0.91	0.82	1.18	1.18
West Kirby	≈80	30–40	0.90	0.64	1.19	1.00
At $P<0.05$, least significant difference between populations:			0.08		0.10	
At $P<0.05$, least significant difference between treatments:			0.14		0.19	

From Horsman *et al.* (1978).
[a] Plants were exposed in a wind-tunnel (air-flow 120 m/min) for 8 weeks to two levels of sulphur dioxide. Results are the means of two sequential experiments.

exerted a strong selective pressure until the early 1960s, and that these effects have persisted, although concentrations of sulphur dioxide have decreased markedly.

We have seen that pollution exerts selection pressures, so that if there is a suitable range of genetic variation, gene frequencies will change (Bradshaw and McNeilly, 1981). Pollution is usually a novel situation, gene pools have therefore not been selected in response to pollution, and so one might expect that populations will have a wide range of relevant genetic variability (Bradshaw, 1975, 1976). But not all species appear to contain the potential for genetic resistance (Bell et al., 1991).

There is evidence to suggest that L. perenne can develop this resistance quite rapidly. In 1975 trial plots of 1 m² were sown with seed of the cultivar S23 in Philips Park, in an industrial area of Manchester with a long history of damage to plants by air pollution. These plots were maintained as near monocultures by selective weeding and mowing, which also minimized reproduction by seed and gene flow from immigrant seed. In 1976 and in each year from 1978 to 1982 thirty tillers were selected randomly, grown in greenhouses, exposed to acute fumigation with sulphur dioxide (3000–6000 μg/m³ for 6 h) and the percentage of leaf necrosis assessed 3 days later. Results were compared with those for control plants grown from the original batch of seed (Wilson and Bell, 1985). By 1979, four years after sowing, the plants in Philips Park had become more resistant than the controls, but two years later, when ambient sulphur dioxide levels had decreased, there was again no significant difference (Fig. 4.10). The method by which plants were sampled meant that either some of the more susceptible genotypes had died or that susceptible genotypes grew less vigorously. By 1981, when sulphur dioxide levels had fallen, the difference in tolerance had been lost. Evaluation of the data is complicated by the variability between years in the controls' response to fumigation, and breeding experiments would be needed to confirm that the observed tolerance is genetic, not physiological adaptation. If one accepts that the tolerance is genetic, the presumption is that tolerant genotypes were selected in the first years of exposure, but that the less tolerant were able to survive for at least these few years and flourished again as sulphur dioxide levels decreased, when they were no longer at a competitive disadvantage.

Pesticides provide both the most voluminous data on the development of resistance in the field and the most detailed studies of mechanisms for resistance. Resistance was first detected in 1908 when a field population of the San José scale (Aspidiotus perniciosus), an insect pest of fruit trees, was found to be resistant to lime sulphur (Melander, 1914). By 1948, 40 years later, resistance to one or more insecticides had been reported in 14 insect species, but then the widespread use of the new synthetic insecticides accelerated the development of resistance, and by 1984 some populations of 447 species of insects and mites were known to be resistant to one or more insecticides (Committee on Strategies, 1986). Similar problems have also developed with herbicides

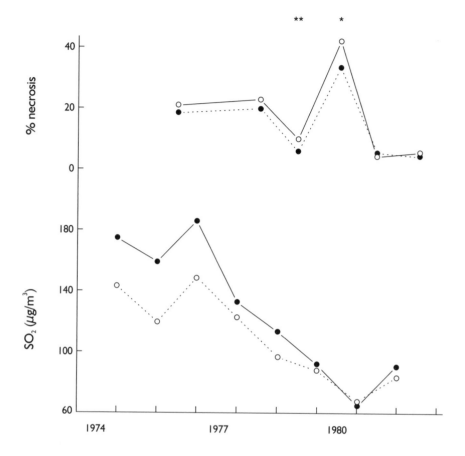

Fig. 4.10   Susceptibility of *Lolium perenne* in Philips Park, Manchester, to sulphur dioxide during the years 1976–1982. ●—●, ambient mean winter concentrations (October–March) of sulphur dioxide; ○- - -○, ambient mean annual concentrations. There were no records for August–October 1981, the months with usually the lowest concentrations, so the values given for that year are probably too high. Susceptibility was measured by exposure for 6 h to 3000–6000 µg sulphur dioxide/m³. ●- -●, necrosis in plants taken from trial plots in Philips Park, which were sown in 1975; ○—○, necrosis in control plants grown in an unpolluted atmosphere from the seed sown in 1975. * and ** indicate significance at the 0.05 and 0.01 probability levels respectively. (Data from Wilson and Bell, 1985.)

(LeBaron and Gressel, 1982; LeBaron, 1991), fungicides and rodenticides, and the same essential principles seem to apply to all of this diverse range of chemicals and species.

Initially, it was a matter for debate whether a field population could contain genes for resistance to a synthetic insecticide to which it could never have been exposed before. Certainly, the occurrence of insecticide resistance in exposed populations does not necessarily indicate a genetic mechanism: there is always the possibility that the individual organism becomes physiologically or biochemically adapted as a result of exposure. For example, exposure to insecticide may induce the formation of an enzyme that detoxifies the insecticide (Gil et al., 1968). However, it has now been established beyond doubt that many species of insect can develop genetically controlled resistance to insecticides (see Wood, 1981).

Only some species appear to have the appropriate genetic variability that enables resistance to develop. At one extreme, no species of tsetse fly (*Glossina* spp.) had evolved resistance to any insecticide, after 25 years of control by insecticides (Wood and Bishop, 1981). At the other extreme, the housefly *Musca domestica* had become resistant, in Denmark, to at least 17 different insecticides (Keiding, 1975). Conversely some chemicals appear to select for resistance more readily than do some other chemicals: weeds have developed resistance to triazines more quickly, in more species, than to phenoxy herbicides (LeBaron and Gressel, 1982).

There are many possible mechanisms of insecticide resistance (Georghiou and Saito, 1983; Mullin and Scott, 1992), different populations of the same species may develop different genetically controlled mechanisms for resistance to the same insecticide, and individual insects sometimes contain genes for more than one resistance mechanism. The various mechanisms can be classified in general terms as:

(1) Changes in structure or behaviour that reduce exposure (Roush and Daly, 1990).
(2) Reduced penetration rate of the insecticide into the organism (Oppenoorth, 1985).
(3) Increased detoxification rate (Soderlund and Bloomquist, 1990). There are many possible metabolic pathways, and it should be noted that metabolism of an insecticide does not necessarily mean detoxification. Some insecticides are relatively inactive until they have been metabolized. Resistant strains of at least two insect species produce enhanced amounts of detoxifying enzyme as a result of gene amplification: each cell contains many identical copies of the gene that controls production of the enzyme (Devonshire and Field, 1991).
(4) Perhaps increased excretion rates, although the evidence is inadequate at present.
(5) A change at the site of action. Thus changes in the structure of acetylcholinesterase can render this enzyme less susceptible to inhibition by organophosphorus and carbamate insecticides (Hama, 1983; Oppenoorth, 1985), and the nervous system can also become less sensitive to DDT, cyclodienes and the pyrethroids (Soderlund and Bloomquist, 1990). Similarly, it has been found in plants that triazine herbicides bind to specific sites in chloroplasts and thus inhibit photosynthesis (Gronwald, 1994). Resistant plants have different amino acids at these sites, thus limiting the attachment of triazine molecules.

A single population can develop resistance to more than one insecticide in two distinct ways. In multiple resistance, resistance is evolved independently after exposure to each particular insecticide. In cross-resistance, the development of resistance to one insecticide also confers resistance to one or more other insecticides, which may or may not be chemically related to the insecticide that stimulates the evolution of resistance within the population: cross-resistance evolves when one mechanism is suitable for more than one insecticide, and suitability is not correlated completely with chemical structure.

The likelihood that resistance will develop in a susceptible insect population when exposed repeatedly to insecticide is now of considerable practical importance, and, provided the appropriate genetic variation exists, can be influenced by many factors, which include:

(1) Dominant genes for resistance spread through a population more rapidly than those which are recessive.

(2) The shorter the interval between successive generations, the more rapidly is resistance likely to appear: an average of 10–15 generations is often sufficient for insect pests to develop resistance (Georghiou, 1980).

(3) The more mobile the species, the longer it takes for resistance genes to spread through the population. Influx of susceptible individuals and emigration of resistant individuals decrease the incidence of genes for resistance within the exposed population.

(4) Fecundity. The more offspring there are, the greater the chance of resistant individuals occurring. It may be no coincidence that herbicide-resistant plants are r-species (Hill, 1982).

It is worth reiterating the earlier statement (p. 77) that fitness is relative. Measures of resistance to insecticides, usually variations of the LD_{50} test, are not absolute, and can be markedly influenced by the environment. At the simplest level, the mode of exposure can influence insecticide toxicity. Busvine (1951) measured the resistance of two strains of housefly (*Musca domestica*) to DDT applied topically in one of two solvents. His results (Table 4.6) showed that the selective advantage conferred by any genes for resistance depends on the environment. This dependence of results on the particular environment makes laboratory studies on the selection of resistance inherently difficult: results cannot be trusted to predict the effect of selection in the field (Oppenoorth, 1985). This exemplifies a general difficulty for attempts to predict field effects of pollutants (see Chapter 7).

The genetics of resistance within single populations can be complicated (Wood, 1981; Roush and Daly, 1990). Resistance in field populations usually involves only one or two genes, but resistance is polygenic in laboratory populations, with each of many genes making a relatively small contribution to the total resistance (McKenzie and Batterham, 1994). Resistant strains taken from the field often respond to further selection in laboratory conditions, and attain much higher levels of resistance than occurred in the field population.

Table 4.6 Resistance of two strains of housefly (*Musca domestica*) to DDT applied topically in one of two solvents

Solvent	LD_{50} (μg/fly)		Ratio of LD_{50} values
	Susceptible	Resistant	(resistant/susceptible)
Mineral oil	0.44	7.2	15
Acetone	0.12	36	300

From Busvine (1951).

This difference probably results from different degrees of selection pressure. Field populations are, almost by definition as pests, both large and subject to a very high incidence of mortality from insecticides, which facilitates the spread of rare genes that impart a high degree of resistance. Such genes are often rare because they are deleterious when there is no exposure to insecticide. Conversely laboratory populations are small and exposure must be low enough to permit a viable population to survive and breed. Selection then occurs among many loci, each of which affects resistance to some extent. Similar principles apply to herbicide resistance (Maxwell and Mortimer, 1994).

It is not the purpose of this chapter to discuss in detail the genetic intricacies of resistance, but it is important to appreciate that the effects of pollutants can be modified by an organism's genetic constitution, and that pollutants can alter a population's gene pool. The interactions between pollutants and genes can be relevant both to understanding and to predicting effects, and as we shall see in Chapter 8, are potentially of great value for monitoring.

These last three chapters, on aspects of population dynamics, communities and gene pools that are of immediate relevance to ecotoxicology, all serve to emphasize the essential difference between toxicology and ecotoxicology. Toxicology is concerned with the effects of poisons on individual organisms, whereas ecotoxicology is concerned with effects on populations of individuals. The shift of emphasis has many implications, and three in particular should be noted now. First, the range of variables that affect population responses includes, but is greater than, the range that affects individual responses to pollutants. For ecotoxicology, increased attention needs to be paid to the influence of environmental conditions on the effect of pollutants. Secondly, sublethal effects on individuals may be as important as lethal effects. Thirdly, it is widely accepted in toxicology that different individuals of one species will not react in an identical manner to a toxin: the LD_{50} is based on this premise. Likewise, it is reasonable to suppose that different populations of one species may also not react in an identical manner to a pollutant. This takes us beyond the limits of our present knowledge, although some of the facts presented in this chapter on resistance to insecticides support the proposition. It is time now to consider the practical problems of biological effects, prediction and monitoring, within their ecological context.

5 | Effects on Habitats

Pollutants matter because of their effects on populations, and so, indirectly, on communities too, but pollutants act by their effects on individual organisms. These effects may be direct (see Chapter 6) or indirect, altering the habitat in some way. We have just discussed the pressures that select organisms for particular habitats (Chapter 4): any alteration to a habitat is likely to force change on the community. The question is not whether there will be change, but how much change?

Melanism is one example (see pp. 80–86). Another is eutrophication. Nitrogen and phosphorus are usually the critical nutrients that limit plant growth. A slight increase above natural levels may increase the number of species, but the water eventually becomes turbid and choked by luxuriant growth of a few species, frequently algae. This luxuriant growth can sharply reduce transmitted light through the water, and subsequent decomposition can reduce the water's oxygen content. Consequently many species may disappear. In brief, nutrient enrichment changes the balance of competitive advantage between species. Eutrophication can also occur naturally, but that is a very slow process.

The effects of habitat change are diverse, best assessed in terms of the environmental needs of individual populations, be it avoidance of predation, success against competitors, food or shelter requirements, and so on.

It used commonly to be believed that many if not all problems of pollution could be resolved by an adequate degree of dispersion and consequent dilution in the physical environment. This is now clearly untrue for pollutants that are released in sufficient quantities to alter habitats. Their effects can be widespread.

We will consider just one example in detail, the burning of fossil fuels (coal, oil and gas) with the release of carbon dioxide and sulphur dioxide. Carbon dioxide is long-lived, becomes almost uniformly dispersed in the atmosphere and is of global significance. Sulphur dioxide, with a shorter lifetime, is of regional importance. Both alter the habitat, but can also have direct biological effects. We will also consider briefly the effect of chlorofluorocarbons (CFCs) on both global warming and the ozone layer.

The "greenhouse effect"

"Global warming" evokes thoughts of increasing amounts of atmospheric carbon dioxide and consequent increases of mean global temperatures. There is no doubt that carbon dioxide levels are rising (Figs 5.1 and 5.5), but the consequences are still disputed. It should be realized at the outset that if the mean global temperature rises it would result from global changes in patterns of climate, and would not mean that global temperatures rise in unison. Of particular biological relevance, rainfall patterns would also change.

We are concerned with the effects on and interactions with organisms caused by the putative rise in temperature. However, first we need a brief review of the relevant physics of global temperatures and need also to consider some of the individual gases involved in global warming (see Houghton (1997) for a detailed introduction to global warming, and Wayne (1991) and Clarke (1992) for an account of atmospheric structure, transport and pollutants).

The Earth is warmed by radiation from the sun (range of wavelengths c. 180–4000 nm), and although temperatures close to the earth's surface do vary both daily and seasonally one can calculate a mean global temperature, which is stable to well within 1°C over decades. Direct geothermal heating of the atmosphere is negligible, and the earth must therefore return into space as much energy as it gets: part is simply reflected back, the rest is absorbed and then returned as thermal (infrared) radiation (Fig. 5.2), of much longer wavelength than the incident solar radiation.

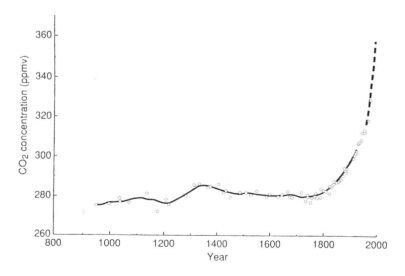

Fig. 5.1 Atmospheric concentration of carbon dioxide during the last 1000 years. O, samples taken from air bubbles trapped in Antarctic ice; -- indicates atmospheric samples taken at Mauna Loa, Hawaii (seasonal fluctuations not shown); the continuous curve indicates a 100-year running mean. (Adapted from Schimel *et al.*,

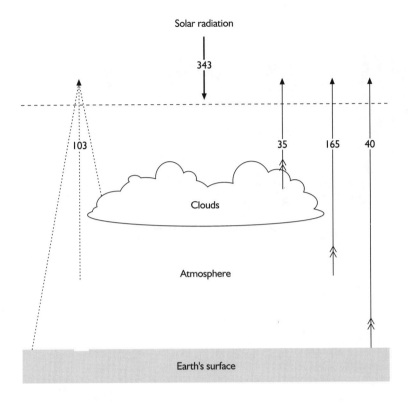

Fig. 5.2 Average radiation to and from the earth (watts/m²). →, thermal radiation from the earth's surface, atmosphere and clouds. ⤍, solar radiation reflected back into space. (Adapted from Houghton, 1997.)

About 85% of the atmospheric mass occurs in the troposphere, which extends on average to about 13 km above the earth's surface. Temperature and pressure decrease with altitude up to the tropopause, the boundary with the stratosphere, where the temperature gradient reverses because of stratospheric ozone, which absorbs some of the sun's radiant energy. The stratosphere, which reaches an altitude of about 50 km, contains most of the remaining 15% of the atmospheric mass, and about 90% of the ozone.

The mean temperature close to the earth's surface is about 15°C, some 20–30°C higher than physical calculations would suggest. The precise difference depends on what assumptions are made about the amount of ice present, and the amount of reflected solar radiation. This "greenhouse effect" is produced by gases in the atmosphere. Nitrogen and oxygen form about 99% by volume of the dry unpolluted atmosphere, and they neither absorb nor radiate thermal radiation, but some of the minor gases do (Table 5.1). These "greenhouse gases" each absorb radiation from only a relatively limited range of wavelengths, to different degrees, and at some wavelengths little of

Table 5.1 Radiative forcing—the effect on the planet's energy balance measured by the change in average net radiation at the tropopause—of both aerosols and greenhouse gases derived from man's activities, and of changes in two natural factors.

Greenhouse gases and other factors	Pre-industrial concentration (pre-1750) (ppmv/v)	Concentration in 1993 (ppmv/v)	Lifetime* in atmosphere (years)	Forcing (watts/m²)
Naturally occurring greenhouse gases				
Carbon dioxide	280	356	50–200[†]	1.56
Water vapour	variable	variable ($0–10^{4.3}$)		0.5
Methane	0.7	1.8	12–17[‡]	0.1
Nitrous oxide	0.275	0.311	120	0.2–0.6
Tropospheric ozone**		variable ($0–10^{3}$)	Continually formed and destroyed by photolysis	−0.1
Stratospheric ozone (depleted by CFCs)				
CFCs (see also previous factor)	zero	$10^{-3.3}$[§]	102[§]	0.3[¶]
Aerosols‖				
Sulphate aerosols			Days–months in troposphere	−0.25 to −0.9
Aerosols from burning of biomass				−0.05 to −0.6
Natural factors				
Change since 1850 in solar radiation				0.1–0.5
Mt. Pinatubo††				−4.0

Data from Houghton et al. (1995).
* Defined as the integrated loss divided by the total abundance.
† Cannot be defined precisely because different sinks have different rates of uptake.
‡ Includes effect of methane on the chemical processes that remove it.
§ Value for CFC-12, one of the most abundant CFCs.
¶ Value for halocarbons, which include CFCs.
‖ Indirect effect on cloud formation may have a similar additional impact. Large regional variation.
** Concentrations in the northern hemisphere may have doubled since pre-industrial times.
†† Example of a large volcanic eruption, which occurred in 1991. Effect lasts for a few years.

the earth's thermal radiation is absorbed by atmospheric gases (Brimblecombe, 1996). Most of the greenhouse gases occur in the troposphere, all except the CFCs occur naturally and apart from stratospheric ozone human activities have increased the amounts of all. Per molecule, their effects on global temperatures vary greatly (Table 5.2).

These greenhouse gases would elevate the earth's temperature simply by being present in the atmosphere. Like the glass in a greenhouse, they absorb some of the earth's thermal radiation and emit some of it back towards the earth, so reducing the net rate of radiation into space. However, this effect is minor, and it is the combination of three physical factors that permits the major effect. The rate at which matter absorbs and radiates thermal energy is proportional to T^4, where T is the absolute temperature (Garratt, 1992). Also both the atmospheric pressure and, usually, the temperature decrease with altitude up to the tropopause. Greenhouse gases in the air close to the earth's surface absorb some of the earth's thermal radiation. This warmer air rises by convection towards the tropopause, to be replaced by colder and denser air masses. The air expands as it rises, because of the lower pressure, and therefore cools (adiabatic cooling). The greenhouse gases then emit thermal radiation more slowly than they absorbed it, with the net result that the earth emits less thermal radiation into space and has a higher mean temperature than it would if the atmosphere consisted solely of nitrogen and oxygen.

The "natural greenhouse effect", from naturally occurring greenhouse gases and clouds, is generally accepted. The main debate centres on the "enhanced greenhouse effect", the extent to which greenhouse gases released by human activities are increasing the mean global temperature. The magnitude of this effect is indicated by the amount to which the earth's radiation budget has been perturbed. It is measured as the change in average net radiation (in W/m²) at the tropopause, and is called radiative forcing. By definition the incoming solar radiation is not a radiative forcing, but a change in the amount of solar radiation would be. A positive forcing tends to warm and a negative forcing tends to cool the earth's surface (Houghton et al., 1995).

Table 5.2 The relative effect of small increases in the current concentrations of greenhouse gases on their radiative forcing (defined in Table 5.1). Results expressed per molecule and per unit mass, relative to carbon dioxide

Greenhouse gas	Relative forcing	
	per unit mass	per molecule
Carbon dioxide	1	1
Methane	58	21
Nitrous oxide	206	206
CFCs { CFC-11	3970	12 400
CFC-12	5750	15 800

From Shine et al. (1995).

There can be little doubt that the atmospheric concentrations of green-house gases have increased since the start of the Industrial Revolution (Figs 5.1 and 5.3), but confounding factors, in particular aerosols and negative feedback mechanisms, counteract the tendency of increases in greenhouse gases to raise global temperatures.

Aerosols (usually defined as suspended particles with diameters within the range 10^{-3}–10 μm) enter the atmosphere from many sources, both natural and human, and can be particulate or gaseous in origin. Recent human activities have increased them, particularly as dust from disturbed soils (Li *et al.*, 1996), as sulphate formed from sulphur dioxide when fossil fuels are burnt, from the burning of biomass and as particulates from fuel combustion and industrial processes. Aerosols have a direct negative effect on temperature—they reflect solar radiation—and they can also have an indirect effect, principally by acting as nuclei for cloud droplets, so affecting the formation, lifetime and radiative properties of clouds. Dust particles also absorb thermal radiation (Lacis and Mishchenko, 1995). The overall effect is negative, so any steps to reduce sulphur dioxide emissions and problems with acid rain might then accelerate global warming, but estimates of magnitude of effect are highly uncertain. Moreover, aerosols are short-lived compared to the main greenhouse gases (Table 5.1) so any effects will be regional, not global, and cannot simply be offset against the effects of greenhouse gases (Jonas *et al.*, 1995). The direct and indirect effects of tropospheric aerosols are currently

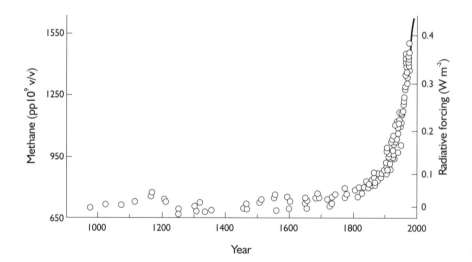

Fig. 5.3 Atmospheric concentration of methane during the last 1000 years. Samples taken from air bubbles trapped in Antarctic ice. The continuous line indicates atmospheric samples taken at Cape Grim, Tasmania. The right-hand axis indicates the degree of radiative forcing (defined in Table 5.1) caused by increases in concentration during the Industrial Revolution. (Adapted from Prather *et al.*, 1995.)

estimated to be of equal magnitude, with a total forcing effect of -1.0 W/m^2 (Houghton et al., 1996).

There are four important physical feedbacks, both positive and negative, that affect the magnitude and location of any temperature change:

(1) As temperature rises, more water evaporates, principally from the oceans. The extra water vapour, itself a greenhouse gas, in all probability also raises the temperature (McElroy, 1994).
(2) Changes in cloud distribution and content. Clouds reflect some solar radiation but they also absorb some of the earth's thermal radiation. The relative importance of these two opposing tendencies depends on cloud height and structure, but in general low clouds tend to lower temperatures, high clouds tend to raise them. Atmospheric temperature is much more sensitive to cloud changes than to proportionally similar changes in amounts of carbon dioxide (Table 5.3), and the future effect of clouds on temperature is a major uncertainty.
(3) The oceans and their currents have major effects on regional climatic changes. Not only do the oceans have a large heat capacity—the top 3 m of water has a greater capacity than the entire atmosphere—but unlike land surfaces, which have comparable specific heats, they can also store much of their heat for a relatively long time. They tend therefore to reduce local fluctuations of temperature. More importantly, ocean currents transport considerable amounts of heat from the equator to the polar regions (Woods, 1984). At high latitudes winds also contribute to this transfer. Without such transfer there would be a net gain of radiant energy at low latitudes and a net loss at high latitudes, with a balance at about 40°N for the northern hemisphere (Open University, 1989). This

Table 5.3 Estimates of the effects of changes in the amounts of carbon dioxide and of cloud on the current mean global temperature of 15°C

Greenhouse gases	Clouds	Change in temperature (°C)
As now	As now	0
As now	None	4
As now	As now + 3% high cloud	0.3
As now	As now + 3% low cloud	−1.0
Doubled CO_2 concentration, otherwise as now	As now (no additional cloud feedback)	1.2
Doubled CO_2 concentration, + best estimate of feedbacks	Cloud feedback included	2.5
None	None	−21
None	As now	−32

Adapted from Houghton (1997).

transfer affects regional climates. Conversely, climatic changes can also alter the course of these currents.

(4) Ice and snow reflect a large proportion of incident radiation. If temperatures rise enough for the ice to melt more radiation is absorbed, so tending to raise temperatures further.

Sources and sinks of greenhouse gases

The atmospheric lifetimes of most greenhouse gases except carbon dioxide are determined primarily by chemical reactions (see Prather *et al.*, 1995). We will consider only a few key reactions, many of which are initiated by photolysis with ultraviolet sunlight (UV).

Stratospheric ozone is formed naturally by UV radiation within the range 180–242 nm (Brimblecombe, 1996):

$$O_2 + hv^* \Rightarrow O(^3P) + O(^3P)$$
$$\text{photon} \quad \text{ground}$$
$$\text{state}$$

$$O_2 + O(^3P) \Rightarrow O_3$$

Ozone is highly reactive and is destroyed amongst other reactions by catalysis with chlorine:

$$O_3 + Cl \Rightarrow O_2 + ClO$$

and

$$O_3 + hv \Rightarrow O + O_2$$

$$\underline{ClO + O \Rightarrow O_2 + Cl}$$

$$2O_3 + hv \Rightarrow 3O_2$$

Each chlorine atom can therefore destroy many ozone molecules.

Molina and Rowland (1974) therefore suggested that CFCs could deplete the ozone layer, because, for example:

$$CFCl_3 + hv \Rightarrow CFCl_2 + Cl$$

The chemistry of chlorine formation in the stratosphere is not yet fully understood (Peter, 1994), but Farman *et al.* (1985) confirmed the existence of an ozone hole over Antarctica, and ozone depletion also occurs in mid-latitudes (Peter, 1994).

---

* The energy associated with one quantum of light (see Harrison and de Mora, 1996).

The depletion of stratospheric ozone lessens the screening of UV radiation from the earth's surface. This reduced absorption of solar radiation lowers stratospheric temperatures and so weakens the stratospheric temperature inversion. This in turn causes a slight cooling of the earth (Table 5.1), because there is less thermal radiation from the stratosphere to the Earth, and also affects tropospheric convection and circulation currents, so affecting weather and climate (Elsom, 1992).

Tropospheric ozone is unusual in the sense that emissions are negligible. Some transfers from the stratosphere, but the highest concentrations result as a secondary pollutant formed by photochemical reactions with nitrogen dioxide derived from vehicle exhaust gases (Watson et al., 1990; Clarke, 1992). It is the precursor for the hydroxyl radical, formed when ozone receives radiation below 310 nm:

$$O_3 + hv \Rightarrow O_2 + O(^1D)$$
$$\text{excited}$$
$$\text{state}$$

$$O(^1D) + H_2O \Rightarrow OH + OH$$

We have just seen that CFCs in the stratosphere increase the amount of solar radiation reaching the troposphere. CFCs thus increase the production of hydroxyl radicals, which are highly reactive, with a residence time of less than 1 s, and produce many reactive radicals and atoms. Hydroxyl radicals react with virtually all molecules that contain hydrogen, and with oxides such as CO, NO_2 and SO_2.

CFCs are destroyed principally by photolysis in the stratosphere, with atmospheric lifetimes of decades to centuries. They should now be being phased out from use, in accordance with increasingly stringent national and international regulations and agreements, most notably the Montreal Protocol of 1989 (with subsequent adjustments and amendments), because of their persistence and effect on the ozone layer (Elsom, 1992). Concentrations of CFCs are expected to decline during all of the twenty-first century.

Nitrous oxide arises predominantly from microbial activity in soils and oceans, with some addition from agricultural activities: amounts are uncertain, but production will increase if soils become warmer and wetter (Melillo et al., 1996).

Methane also comes principally from microbial action (Table 5.4), and the rate of production may be affected by changes in land use and the use of nitrogen fertilizers. Most is removed from the troposphere by hydroxyl radicals. There may also be positive feedbacks. If climatic change makes soils warmer and wetter they may produce methane faster, thus tending to raise global temperatures. In due course, if temperatures rose substantially, large amounts of methane could be released from high-latitude deep sediments where methane is hydrated under pressure.

Table 5.4 Estimated sources and sinks of atmospheric methane. Confidence limits have a 90% probability. The observed increase in atmospheric methane (Fig. 5.3) indicates that sources exceed sinks by 35–40 Mt/year, which compares well with the estimated value of 20 Mt/year

	Amount of CH_4 (Mt/year)	Total (Mt/year)
Natural sources		
Wetlands	115 (55–150) ⎫	
Termites	20 (10–50) ⎬	160 (110–210)
Oceans	10 (5–50)	
Other	15 (10–40) ⎭	
Anthropogenic sources		
Fossil fuels	100 (70–120)	100 (70–120)
Organisms		
Enteric fermentation	85 (65–100) ⎫	
Rice paddies	60 (20–100)	
Burning of biomass	40 (20–80) ⎬	275 (200–350)
Landfill	40 (20–70)	
Animal waste	25 (20–30)	
Domestic sewage	25 (15–80) ⎭	
All sources		535 (410–660)
Sinks		
Removal from atmosphere		
by tropospheric OH	445 (360–530) ⎫	
to stratosphere	40 (32–48) ⎬	515 (430–600)
to soils	30 (15–45) ⎭	

Data from Prather *et al.* (1995).

Carbon dioxide and the carbon cycle

Organic carbon compounds are essential for life, and most derive from photosynthesis (p. 46). Respiration (p. 47) returns carbon dioxide to the environment, and the atmospheric concentration of carbon dioxide is the product of biological, physical and anthropogenic processes (Fig. 5.4). The rate of photosynthesis increases each spring in temperate and high-latitude parts of the world as temperature and light intensity increase. This annual fluctuation causes corresponding annual oscillations in both hemispheres' carbon dioxide content (Fig. 5.5), but these oscillations are usually ignored in long-term data. Mixing between the hemispheres is relatively slow.

Antarctic ice-cores show that atmospheric levels of carbon dioxide have fluctuated between 190 and 270 ppm v/v during the last 220 000 years and correlate closely with atmospheric temperature (Jouzel *et al.*, 1993). Given the relative stability of atmospheric carbon dioxide levels for several thousand years before the Industrial Revolution (Fig. 5.6), it is assumed that until recently net fluxes between reservoirs were close to zero. Recent anthropogenic emissions, principally from the combustion of fossil fuels, although relatively minor, have dis-

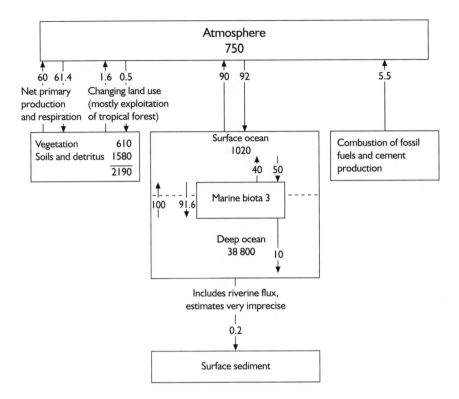

Fig. 5.4 Simplified diagram of the global carbon cycle. Values for reservoirs (Gt C) and fluxes (Gt C/year) are annual averages for 1980–1989, have considerable uncertainties, and many have significant yearly fluctuations. (Adapted from Schimel *et al.*, 1995.)

turbed that equilibrium, and fuel combustion, cement production and forest clearance currently contribute about 6.6 Gt of carbon per year as carbon dioxide to the atmosphere (Table 5.5). Nearly half, about 2.9 Gt C, is retained in the atmosphere, about 2 Gt C is absorbed by the oceans and the remaining 1.7 Gt C must therefore be absorbed by terrestrial ecosystems (for discussion of the evidence see Enting and Mansbridge, 1989; Tans *et al.*, 1990; Watson *et al.*, 1995). Biological feedbacks are involved in both the oceanic and terrestrial pathways.

Of the estimated 2 Gt C/year retained by the oceans, about 0.4 Gt C remains in the surface waters, and the rest transfers deeper in CO_2-laden water (Fig. 5.4) (Siegenthaler and Sarmiento, 1993). Most surface waters are well mixed, less dense and distinct from the deeper waters because of the pycnocline, a density gradient formed by vertical gradients of temperature (the thermocline—with surface waters warmed by the sun) and of salinity (the halocline—with surface waters diluted by rain and snow unless evaporation exceeds precipitation) (Angel and Rice, 1996). Most if not all photosynthesis occurs in these surface waters, because light intensity decreases with depth. Carbon dioxide

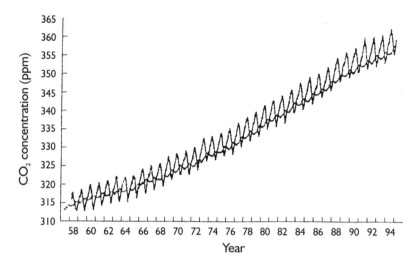

Fig. 5.5 Atmospheric concentration of carbon dioxide since 1958 at the Mauna Loa Observatory, Hawaii (oscillating line) and at the South Pole. The annual oscillations at Mauna Loa reflect the seasonal cycle in the rate of photosynthesis. (From Keeling *et al.*, 1995.)

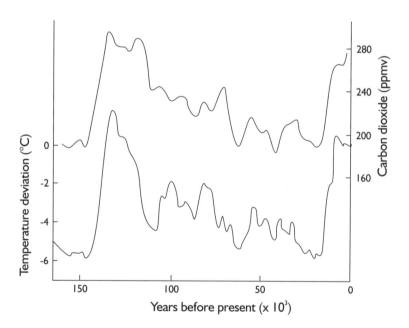

Fig. 5.6 Atmospheric concentration of carbon dioxide (upper line) and changes from the present mean temperature (lower line) in Antarctica during the last 150 000 years. Individual measurements of carbon dioxide concentration are average values for several hundred years. (Data from Barnola *et al.*, 1987, and Lorius *et al.*, 1990.) (Modified from Freedman, 1995.)

Table 5.5 Sources, storage and sinks for anthropogenic carbon dioxide: the average annual carbon budget for the decade 1980–1989. Confidence limits have a 90% probability

	Amount of C (Gt/year)	
	Released into atmosphere	Sequestered
Anthropogenic activity		
Burning of fossil fuels and cement production	5.5±0.5	
Tropical deforestation	1.6±1.0	
Regrowth of forests in mid- and high latitudes of the northern hemisphere		0.5±0.5
Storage in reservoirs		
Atmosphere		3.2±0.2
Oceans		2.0±0.8
Terrestrial sinks		1.4*±1.5
	7.1±1.1	7.1

Data from Schimel et al. (1995).
* Value obtained by balancing the budget.

exchanges freely between the atmosphere and ocean and therefore although most, in surface waters, occurs as HCO_3^- and CO_3^{2-} ions, with less than 1% as dissolved carbon dioxide, photosynthesis is usually limited by nutrient deficiency, not by lack of carbon. Nutrients are replenished each winter when the surface waters cool and so reduce the temperature and density gradients. Then mixing with deeper waters transfers nutrients upwards. This process is supplemented by recycling, atmospheric deposition and run-off of nutrients from land.

Although marine algae only contain about 3 Gt C, production may reach 50 Gt C/year (Fig. 5.4) (Melillo et al., 1996; see also p. 47). Some phytoplankton dies, some passes along the food chain, and both fates release particulate organic carbon (POC) and dissolved organic carbon (DOC—that which passes through a filter with 0.45 μm pore size) from disintegration and waste products. Some of this organic carbon is exported to deeper waters, the "biological pump", where most is eventually converted back by bacteria into inorganic carbon (remineralized) and returns to the surface waters. Some POC reaches the ocean bottom, where it may remain indefinitely, buried in the sediments (Longhurst and Harrison, 1989; Longhurst, 1991). A second route from surface waters is as inorganic carbon in calcium carbonate in the shells and exoskeletons of dead plants and animals, but most of this redissolves under pressure at a depth of several thousand metres. The biological pump thus reduces the total amount of carbon dioxide in surface waters and the atmosphere, increases the total carbon content of the deep oceans and sequesters some in sediments. Current understanding is that the biological pump continues to act as a natural background process that increases the total carbon content of the deep oceans but does not sequester any of the anthropogenic carbon (Siegenthaler and Sarmiento, 1993), although there is some dissent (Riebesell et al., 1993).

Increased atmospheric carbon dioxide could decrease the winter cooling of surface waters, with less mixing with deeper waters, less nutrient replenished and so less photosynthesis next spring. This would reduce the removal of atmospheric carbon dioxide by oceans, and less carbon would be pumped towards the ocean bottom. It is estimated that a doubling of atmospheric carbon dioxide may reduce the flux of POC to the ocean bottom by 10%, a significant positive feedback (Woods and Barkmann, 1993).

Such predictions must be tentative. The communities and production of marine ecosystems depend largely on the physical processes that control the supply of nutrients to the sunlit surface waters. Climatic change could alter the pattern of ocean currents and so alter nutrient availability. More directly, temperature change could alter physiological rates and community structure, and increased UV radiation could damage cells and affect production (see pp. 127, 199–200).

The terrestrial feedback is more direct, and probably negative. Enhanced levels of carbon dioxide increase the rate of photosynthesis and growth, which entails storage of carbon, in many though not all species of plant (Woodward, 1992): the "fertilization effect" (see pp. 117–118). Microbial respiration could also increase with rising temperature and if significant would reduce this effect.

Past changes in mean global temperature

The "enhanced greenhouse effect" is associated with the Industrial Revolution, which started in about 1750, so one may ask how stable were temperatures before 1750, and how have temperatures changed since that date? One must first distinguish between the Ice Ages, with extensive ice cover, and the interglacial periods such as now. The evidence suggests that during the whole of the Pleistocene period, which started about 1.6 million years ago, the amounts of glacial ice have fluctuated many times.

Temperatures from before recorded history can be deduced by analysis for isotopes of oxygen and hydrogen in ice cores (Duplessy, 1978). These show that temperatures in Antarctica have varied by about 7–8°C during the last 150 000 years (Fig. 5.6), with similar changes in Greenland. The primary cause of these temperature changes is variations in the earth's orbit round the sun, magnified by positive feedback from consequent changes in atmospheric carbon dioxide and methane and probably also of nitrous oxide: carbon dioxide and methane increased with rises in temperature or at least lagged by not more than 1000 years, and decreases of carbon dioxide lagged temperature drop at the onset of glaciation by several thousand years (Raynaud et al., 1993).

Periods of stable mean temperature lasted for 70–5000 years but the shifts from one mean temperature to another sometimes occurred within 10–20 years (GRIP members, 1993; Dansgaard *et al.*, 1993). Shifts in ocean circulation combined with changes in atmospheric carbon dioxide levels could account for such changes (McElroy, 1994), but our current predictive abilities for future changes of climate are inadequate. We do not really know why and when temperatures in interglacial periods are stable.

The last Ice Age ended about 8000 BC and since then temperatures have been relatively stable. Temperatures for the last 1000 years in a small region of England were estimated from documents, botanical data and weather records and show a range of about 1.5°C in mean temperature before the recent increase in greenhouse gases (Fig. 5.7). These estimates must have some error, but are essentially correct: vines grew as far north as Yorkshire during the warm period in AD 1100–1300, and the River Thames sometimes froze over in winter during the period 1400–1850. These fluctuations occurred before the recent increase in greenhouse gases, but do not necessarily indicate changes in global temperatures. They could well be normal climatic variations. Bradley and Jones (1992) summarize similar extensive indirect data on past climates for many parts of the world.

Reliable extensive temperature records exist only from the middle of the nineteenth century. They show considerable variation between years and that mean temperatures were steady during the periods 1860–1920 and 1940–1980 (Fig. 5.8). Particles emitted by volcanoes are one short-term cause of temperature variations: they reduce the amount of solar radiation reaching the earth's surface and so lower global temperatures, but only for a few years. Variability between years is more often explained by variations in the climate system, in particular by changes in the transport of heat by ocean currents.

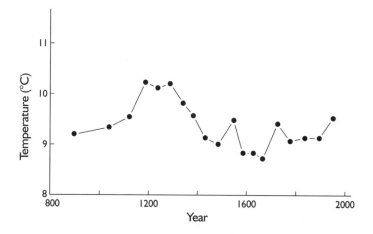

Fig. 5.7   Mean annual temperatures (for periods of 50 years from AD 1100–1950) in central England. Estimates based on botany (such as occurrence of vines), documents and meteorological records. (Data from Lamb, 1965.)

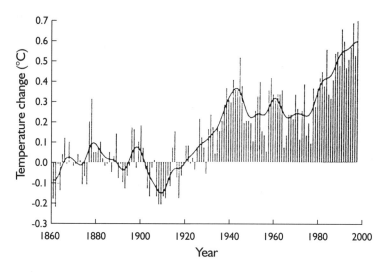

Fig. 5.8 Variations in the mean global annual temperature for each year from 1860. The last value is for January–September 1997. The curve indicates the running mean for each year, which is a weighted average based where possible on the range of values extending 10 years before and after the year in question. (From the Hadley Centre for Climate Prediction and Research.)

Not surprisingly, there are many opinions about the enhanced greenhouse effect. The Intergovernmental Panel on Climate Change (IPCC) has recently concluded that "the balance of evidence suggests a discernible human influence on global climate" (Houghton *et al.*, 1996), although some have doubts about details of the model of global climate. Critics argue that the climate model only fits the data because arbitrary values were assigned to the effect of aerosols, that the impact of some other factors is uncertain, that the correlation of temperature rise with amounts of greenhouse gases is poor and that important known atmospheric processes are not included in the models (see, e.g., Lindzen, 1994). Clearly many questions remain to be solved. Thus Andreae (1996) argues that observed temperature changes since the nineteenth century accord far better with model predictions when due allowance is made for the effects of dust from disturbed soils.

If there is no global warming then negative feedback must be counteracting the effect of enhanced levels of greenhouse gases. One may then expect at least regional climatic changes.

Biological effects of global warming

Quite apart from the direct effects of increasing levels of carbon dioxide, climate change has innumerable potential impacts on and interactions with ecosystems. Other human activities, which affect most communities, compound the difficulties of prediction.

Fourier suggested in 1827 that human activities might modify climate, but widespread concern only developed in the 1970s (Strain and Cure, 1985). There are now many speculations about possible effects (see Melillo *et al.*, 1990) and predictions can range from the most gloomy (US Council, 1980) to the distinctly optimistic (Simon and Kahn, 1984). We cannot yet realistically assess the likelihood and extent of most possible effects, but a few general points can be made.

Rising sea levels

Mean global temperatures are currently predicted to rise by a few degrees centigrade within the twenty-first century, which would affect coastal habitats. Sea levels have risen by an average of 1–2 mm/year for the last 100 years, because of thermal expansion of the oceans and the melting of mountain glaciers and of the margins of the Greenland ice sheet (Robin, 1986). Effects at individual sites have also been influenced by local vertical land movements and by regional differences of up to two-fold from the mean rise in sea level. These variations arise from changes in meteorological, oceanographic and tectonic conditions (Warrick, 1993). This impact of rising temperatures may be mitigated by increased accumulation of snow in Antarctica. Future rises of sea level will occur in part because of past anthropogenic emissions, and the most optimistic assumption of future demands for energy implies a further mean rise of about 18 cm by AD 2100. It appears unlikely on any probable future pattern of energy use that the rise would exceed 1 m. A rise in sea level would be significant for low-lying coastal areas liable to inundation, salt intrusion or erosion (Warrick and Oerlemans, 1990).

Effects of carbon dioxide on photosynthesis

It is well known that increased levels of carbon dioxide often increase the yield of agricultural crops, but the link between increased photosynthesis and growth in response to increased carbon dioxide is not simple.

Most plant photosynthesis occurs in the leaves, where carbon dioxide enters through stomata (pores) whose apertures vary with ambient conditions. Stomata have two conflicting functions: they tend to open wider to facilitate inward diffusion of carbon dioxide but to close to minimize water loss by the outward diffusion of water vapour (Hopkins, 1995). The first constant step in carbon fixation is always for carbon dioxide to form a 3-carbon sugar, with three known variations in the way this is achieved, each with its own ecological consequences. In C_3 plants carbon dioxide combines with an acceptor molecule, a 5-carbon sugar, to form immediately two molecules of 3-carbon sugar. C_4 plants first form a 4-carbon sugar, and are able to photosynthesize with lower internal concentrations of carbon dioxide. So when plants need to conserve water, by reducing their stomatal apertures, C_4

plants can photosynthesize more effectively than C_3 plants. Some plants typical of dry deserts have gone a stage further and are able to photosynthesize during the day despite closed stomata because carbon dioxide was taken in and stored as an organic acid during the previous night. These are the so-called CAM plants, short for crassulacean acid metabolism, because it was studied extensively in plants of the family *Crassulaceae*.

Enhanced levels of carbon dioxide increase the rate of photosynthesis more in C_3 plants than in C_4 plants (Woodward *et al.*, 1991; Allen and Amthor, 1995) but the consequent enhanced growth rate may decrease with time (Bazzaz, 1990), and sometimes increased photosynthesis simply alters the relative growth of different parts of a plant with little or no effect on total dry weight and carbon storage (Norby *et al.*, 1992). Respiration rate also determines production, and this rate can either increase or decrease with enhanced carbon dioxide.

Ecological effects

One might predict that rising levels of carbon dioxide would increase the productivity of plant communities. This has to be a very tentative prediction, because one is extrapolating from one level of organization to a higher more complex level, when new factors and interactions may yield unpredictable results (Weiner, 1996). One would also expect the competitive balance between different species to change, particularly between C_3 and C_4 species. Individual plants tend to use water more efficiently, because stomatal apertures usually decrease when carbon dioxide levels rise, which reduces water loss from transpiration, and also because of the increased rate of photosynthesis.

Most relevant data derive from short-term treatments in controlled environments. The effect of increased individual plant growth on plant survival and on population size is unknown, and species differ in their response to increased carbon dioxide, so effects on the degree and type of competition and interaction between species cannot be predicted, and the effect on plant communities is uncertain (Oechel and Strain, 1985). Factors such as light intensity, water, soil nutrients, salinity and temperature usually limit communities and affect the response of plants to enhanced carbon dioxide in different ways in different species (Idso and Idso, 1994; Koch and Mooney, 1996). Nitrogen is certainly an important variable, often the critical nutrient that limits terrestrial growth, although in many areas, especially in the tropics, phosphorus is also an important limiting factor. The effect of higher levels of carbon dioxide on net primary production increases with temperature, and although mathematical models do suggest availability of nitrogen to be the cause (McGuire *et al.*, 1992) we do not yet know whether a warmer climate would increase or decrease the amount of nitrogen available to plants (Davidson, 1995). Detection of effects of increasing carbon dioxide will be difficult and some will probably be unexpected. These effects need also to be assessed within the context of the concurrent climatic changes.

In principle one might expect three types of ecological effect. The gene pool of individual populations may adapt to rising levels of carbon dioxide (Thomas and Jasieński, 1996) and to changing climate. In geological time-scales, the concentration of carbon dioxide has probably tended to decline since pre-Cambrian times as oxygen levels have increased (Wayne, 1991), so species gene pools may already contain genes appropriate to rising levels of carbon dioxide. However, even if the population contains the relevant genetic variability the response time to climatic change is likely to be too slow for long-lived species with long generation times (Kingsolver, 1996). Community structure may then change as the changing environment changes the competitive advantage of individual species. Finally, populations may migrate.

The present distribution of different natural types of community, which range from forests to deserts, correlates with and is determined principally by regional differences of temperature and rainfall (Warrick *et al.*, 1986). Global warming could alter regional rainfall as well as temperature. The greater and more rapid the climatic change the greater the chance of local extinction of populations.

Human activities have restricted the range and available habitats of many species. They may then be less able to migrate to more suitable sites, even if the rate of climatic change is slow enough. Rare species with small ranges may be particularly vulnerable. Conversely, in some habitats human activities may be a major determinant of migration rates for plants, much higher than natural rates (Woodward, 1992).

Community changes lag climatic changes, so there may be complex transient effects on structure if communities cannot withstand the altering climates (Melillo *et al.*, 1996). The greater and more rapid the change the less the possibility that the original community will gradually be invaded and replaced by other species, and the greater the chance of a radical temporary change. Novel climates may produce novel plant communities (Webb, 1992). The associated animal species, many of which could migrate more quickly, may also be limited by the speed with which plant species can migrate.

A study on forests in the eastern USA suggests that global warming would yield higher temperatures and drier soils in that region. This could move the southern boundary of many tree species northwards by up to 1000 km during the twenty-first century, although the effect may be less severe because the elevated levels of carbon dioxide that cause the climate change will also enable trees to use water more efficiently (Smith and Tirpak, 1990). Similarly the northern boundary could move northwards by 600–700 km, but records from palaeoecology suggest that many tree species can only migrate naturally at less than 100 km per century. These conclusions imply several centuries during which these forest communities re-establish themselves, quite probably with different species and relative abundance. Any extensive forest die-back would also alter the terrestrial uptake and release of carbon.

The arctic tundra provides a detailed example of possible interactions. Tundra ecosystems are dominated by plants that can grow and reproduce at temperatures not much above freezing—typically mosses, lichens and dwarf shrubby plants. Dead plant roots form humus and peat, which decomposes so slowly that it accumulates. As it accumulates the lower layers freeze because the newer upper layers insulate the older material from the atmosphere, and this peat remains frozen indefinitely unless the climate becomes warmer. Tundra has therefore been a net sink for carbon dioxide in historic and recent geological times, but evidence from Alaska suggests that recent warming has increased the depth to which the soil thaws each summer, and the lower water table has enhanced soil drainage and aeration, thus increasing soil mineralization with a loss of carbon dioxide to the atmosphere. Concurrently nitrogen, often a limiting nutrient for plant growth, is released into the soil (Oechel *et al.*, 1993).

Although global warming would increase primary production, which would remove carbon dioxide from the atmosphere, arctic plants are inherently slow-growing, and carbon is probably being released faster by heterotrophic respiration as peat is broken down than enhanced plant growth is storing it (Oechel and Vourlitis, 1996). The longer-term result would depend on the balance between the rate of loss from mineralization and the rate and extent to which trees and shrubs migrate into the tundra and sequester carbon more rapidly than does the present vegetation. Plant migration would probably reverse the current decrease in carbon storage (Smith and Shugart, 1993).

In summary, we have only limited studies so far on the direct and indirect ecological effects of rising carbon dioxide, and these effects will vary with the ecosystem.

Sulphur dioxide

Sulphur, like carbon, is one of the elements essential for life. Too severe a deficiency causes death of both plants and animals, but excessive quantities can also be toxic. Sulphur is abundant in the earth's crust, with an average concentration in soils of 0.1%, and it cycles naturally between various parts of the environment (Fig. 5.9), although many of the estimates for the amounts and forms of sulphur involved are rather uncertain. This uncertainty arises in part from the paucity of data combined with high seasonal and regional variability (Bates *et al.*, 1992).

Natural releases of sulphur into the atmosphere come from three main sources (Table 5.6). Volcanic activity is episodic, and releases principally sulphur dioxide (Stoiber *et al.*, 1987). Sulphate particles are ejected from the seas' surface as spray from bursting bubbles, and amounts depend on the wind speed and the height at which measurements are taken. Most soon returns to the sea. Hydrogen sulphide from biological activity was thought to be the third main source, but we now know that marine and terrestrial organisms produce low concentrations of a range of volatile organic compounds, of

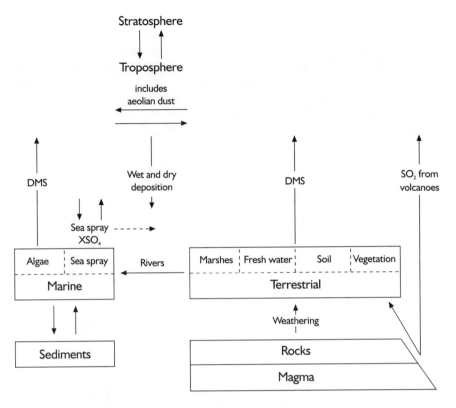

Fig. 5.9 Outline of the global sulphur cycle before man's activities had a significant impact. Only the principal form of sulphur is indicated for each route. DMS indicates dimethyl sulphide, XSO_4 indicates particulate sulphate compounds.

which dimethyl sulphide (DMS) appears to be the major component (Bates *et al.*, 1992). Most of these, like the inorganic compounds of sulphur, are soon oxidized to form sulphate aerosol particles, which can act as cloud condensation nuclei (Charlson *et al.*, 1987) and so affect cloud formation.

Man's activities have been releasing increasingly large and biologically significant quantities of sulphur (Fig. 5.10)—air pollution from the combustion

Table 5.6 Current estimates of global sulphur emissions into the atmosphere

	Oceanic	Terrestrial	Volcanic	Burning of biomass*	Other anthropogenic	Total Natural	Total Anthropogenic
Amount (Tg S/a)	15.4	0.4	9.3	2.2	76.8	25.1	78.9

Data from Bates *et al.* (1992).
* 95% from human activity.

of coal has been a local nuisance since the thirteenth century (Brimblecombe, 1987)—and most anthropogenic sulphur emitted into the atmosphere still comes from the combustion of coal and oil: an estimated 85% in 1976 (Cullis and Hirschler, 1980), with most of the remainder from the smelting of ores and refining of petroleum. Additional human inputs to the environment come via the rivers: the use of artificial fertilizers has increased the rate at which sulphur is leached from soils, and some sulphides are also released from mines into rivers (Brimblecombe *et al.*, 1989).

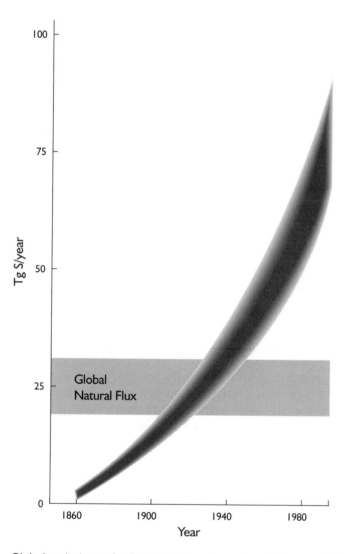

Fig. 5.10 Global emissions of sulphur into the atmosphere between 1860 and the late 1980s. Width of shading indicates the degrees of uncertainty. (From Penner *et al.*, 1994.)

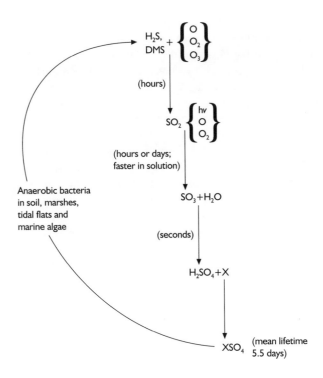

Fig. 5.11 Scheme of the transformations of sulphur in the lower atmosphere, with an indication of mean lifetimes (the reciprocal of the proportion removed in unit time). (From Kellogg *et al.*, 1972; Charlson *et al.*, 1987; Penner *et al.*, 1994.)

92% of the total sulphur emissions between latitudes 35° and 50°N are anthropogenic, with the major concentrations in north-west and central Europe and in north-eastern USA and Canada (Schneider, 1986), whereas they are only 0–70% of the total in latitudes between 80°S and 20°N (Bates *et al.*, 1992). Because sulphur has a relatively short atmospheric lifetime (Fig. 5.11) most returns to the earth's surface within about 3000 km of the source, so that any effects of pollution will be regional rather than global (Brown, 1982).

Sulphur contaminants can affect biota in diverse ways, and to some extent this diversity is linked with the diversity of forms and routes that sulphur can take. We must therefore consider first the return of sulphur compounds from the atmosphere to the earth's surface. In general terms, airborne substances can be deposited on the earth's surface by either wet deposition, in solution or in suspension in rain, snow and other forms of precipitation, dry deposition, as particles or gases, or occult deposition of mist, fog and cloud droplets (Fowler, 1984). The proportion of sulphur deposited wet increases with distance from the source. A further distinction must be made, for wet deposition, between rainout and washout. Both terms refer to processes that transfer material to droplets of water, which can happen either in clouds before they descend as raindrops (rain-out) or whilst they descend as raindrops (washout). Many mechanisms influence deposition. Deposits, wet or dry, may reach the earth's surface by the force of gravity, impaction, diffusion or turbulent transfer. Vegetation is therefore described as having a scavenging effect, when substances are filtered out of the air by the last three processes (Miller and Miller, 1980). Rain itself may reach the earth's surface under vegetation either as throughfall, or stemflow, and with both pathways rain washes filtered aerosols and gases off the vegetation and also contains leachates from the foliage (Miller, 1984). The degree of correlation between the rates at which sulphur is emitted and deposited decreases with time and distance from the source, because of variations in the rate at which sulphur dioxide is converted to sulphate and because wind speed and direction also vary (Goldmsith *et al.*, 1984).

The major effects of environmental contamination by sulphur include:

(1) The toxic effect of sulphur dioxide on plants. This is discussed later (see Chapters 6 and 8), but it is noteworthy that air concentrations at one site can vary appreciably (Lane and Bell, 1984). Devilla Forest, central Scotland, has industrial regions to the south and south-west, and

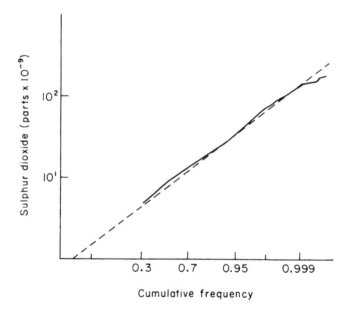

Cumulative frequency

Fig. 5.12 Variations in the concentration of sulphur dioxide in the air 1 m above the tree canopy at Devilla Forest, central Scotland, during 1978. The solid line indicates the cumulative frequency of individual analyses for sulphur dioxide that do not exceed specific concentrations; the dashed line indicates a log-normal distribution. (From Nicholson *et al.*, 1980.)

rural areas to the north, and concentrations have been found to range from below the limit of detection of about 5 parts $\times$ 10^{-9} to about 175 parts $\times$ 10^{-9} (<10–375 $\mu g/m^3$) (Fig. 5.12). For comparison, samples of air from uncontaminated sites, in Panama, Colorado, the Atlantic Ocean and the Antarctic, contained <1–4 μg sulphur dioxide/m^3, with probably similar concentrations of sulphate near the ground (Kellogg et al., 1972).

(2) The acidifying effects of sulphur dioxide on rain. Sulphur dioxide oxidizes to sulphuric acid (Fig. 5.11), but the effect on the acidity (pH) of the rain will depend on what other ions are also present (Howells, 1995). Sulphur dioxide has made rainfall appreciably more acid in and near some industrialized regions of the northern hemisphere (Smith, 1872; Gorham, 1958; Howells, 1995). The pH of rain can naturally be as low as 5.6, because of dissolved carbon dioxide, but acidic gases, in particular sulphur dioxide and nitrogen oxides, often cause a drop to about 4.0 in industrial regions. This "acid rain" can erode cuticular waxes on plant surfaces, leach nutrients from leaves and affect the occurrence of plant pathogens. The impact on well-buffered soils is probably negligible, but acid rain can affect the fauna and flora in regions such as the north-eastern USA, where soils are derived from granite and have little buffering capacity (Likens and Bormann, 1974). Water draining from poorly buffered soils into streams, rivers and lakes that contain poorly buffered water has also lowered the pH in those waters, with, in colder regions, episodes of relatively acid water occurring especially during the early stages of thaw at the end of the winter (Johannessen et al., 1980). Coincident with this acidification, there have been pronounced declines or losses of fish populations and of other species (Harvey, 1980; Dickson, 1986). Fish die from failure of their salt-regulating mechanisms, and although tests have shown that the pH of water has to drop below 4.6 before fish suffer anything more than a slight transient physiological effect, fish have disappeared from lakes with a pH value of about 5.0 (Muniz, 1984). The increased toxicity appears to be due in large part to aluminium leached from the soils by the acid rain (Fig. 5.13), although other factors may also be important (Howells, 1995).

(3) Effects on soils that are naturally deficient in sulphur. Contamination from sulphur dioxide can improve crop yields in such soils, until the critical exposure is reached at which sulphur dioxide starts to reduce yield (Cowling and Lockyer, 1976; Lockyer et al., 1976).

(4) In some regions an alternative or additional source of acidity comes from the mining of coal and other minerals, which may expose significant amounts of iron pyrite, FeS_2 (Glover, 1975). Pyrite oxidizes when exposed to air and water, and oxidation may be quickened by naturally occurring bacteria. One consequence is the release of sulphuric acid into waterways.

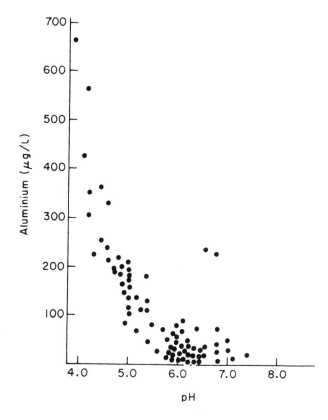

Fig. 5.13 The pH and concentration of total aluminium in the water of Swedish lakes during the summer and autumn of 1976. The two anomalous values are for lakes high in humic acids. (From Dickson, 1980.)

One may conclude that although it is important to know the rate at which a pollutant is released and the pathways it takes, the exposure of, say, a plant to sulphur will also depend on many other variables, both physical and chemical, and some of these will be affected by fine details of the plant's microhabitat. Attempts have been made to reduce damage by sulphur to ecosystems by calculating the "critical load": the deposition rate of sulphur below which harmful effects will not occur (Bull, 1991). Difficulties arise from the definition of "harm", changes with time and place in deposition rate, and the influence of other factors—including other pollutants—on the critical load. "Target loads", loads that are politically acceptable, may include a safety factor, when they are less than the critical load, or be greater when practical or economic factors are more important.

Effects of the greenhouse gas ozone

Tropospheric ozone is directly phytotoxic (Runeckles and Krupa, 1994), but stratospheric ozone has two indirect effects. Not only does its loss have a slightly cooling influence (p. 109), but its presence reduces the amount of damaging solar radiation that reaches the earth's surface. Solar UV-C, radiation of wavelength less than about 285 nm (definitions vary slightly; see Giese, 1978), can damage macromolecules such as proteins and nucleic acids, which are essential components of living cells. Major constituents of the atmosphere, especially oxygen, filter out radiation of less than 230 nm, but only ozone can attenuate longer wavelengths (Wayne, 1991). So destruction of the ozone layer by CFCs not only cools the earth slightly but also increases the incidence of lethal radiation on unicellular organisms and on the surface cells of multicellular plants and animals. Many present forms of life could not survive without the ozone layer (Fig. 3.13).

For all three of the pollutants we have considered—carbon dioxide, sulphur dioxide and CFCs—persistence is a key factor that determines the degree of their effect. We will see in the next two chapters that persistence is also of central importance for direct effects on individual organisms and for prediction of effects. To evaluate persistence we need to consider both the pollutant's mobility in the environment and its chemical stability. A toxic element such as mercury or cadmium is regarded as toxic and persistent regardless of its mobility, molecular form or stability. Toxic molecules may be persistent in two ways. They may be mobile in the environment but resistant to chemical change, and or they remain in one place for a long time.

6 | Effects on Individual Organisms

We have just considered pollutants where effects exerted indirectly on organisms through the habitat are the main or only concern. Many pollutants have only direct toxic effects, and here the approach of classical toxicology—the study of the poisoning effects of chemicals on individual animals—is highly relevant, although here too we will see that the environment needs to be taken into account.

Effects on individuals may range from rapid death through sublethal effects to no effects at all. Paracelsus observed over four centuries ago that "All things are poisons, for there is nothing without poisonous qualities. It is only the dose which makes a thing a poison." (Goldstein *et al.*, 1974). This statement highlights two important points. First, there is nothing intrinsically different about poisonous substances that separates them from all other elements and compounds. Secondly, whether or not a poison exerts its biological effects depends on the amount present. The most that one might be able to say about the difference between poisons and other substances is that a relatively small amount of a poison is usually sufficient to produce adverse effects. Paracelsus was, of course, talking about the effects of poisons on human beings, but he could equally well have been talking about pollutants and wildlife in general. Hence the relevance of toxicology, and the need for a quantitative approach, although Butler (1978) could still comment that most published work on pollution concentrates on effects and neglects the problem of estimating doses: "Without accurate knowledge of the dose there can be no quantitative assessment of the effects."

The most obvious effect of exposure to a pollutant is rapid death, but it is rarely instantaneous, and even cyanide takes at least some tens of seconds to kill a human being. Death is always preceded by lesser symptoms of malfunction, and a major concern in ecotoxicology is the frequency and extent to which plants and animals may survive the impact of pollution but function less effectively, or suffer what are commonly called sublethal effects.

The term "sublethal" is not quite so precise as might appear at first sight. Although it means an exposure that does not kill within a short time, it is reasonable to argue that any effect that impairs an organism's ability to respond to its environment is likely to shorten its life, but this is a question of probabilities, not of certainties. The dividing line, then, between a sublethal and a lethal effect, if the organism dies, is whether or not it was likely to survive in a less adverse environment. Clearly, there is no sharp distinction between favourable and adverse environments, and similarly we lack an absolutely clear-cut distinction between lethal and sublethal effects. The one point I would emphasize is that these effects occur, or do not occur, in individuals, not in populations. Perkins (1979) suggested that a sublethal exposure kills at most only a small proportion of a population, but the possibility that a sublethal exposure could cause a small proportion of individuals to die from acute toxicity seems self-contradictory.

The question of whether subtle sublethal effects can occur probably arose first with the effect of sulphur dioxide on plants. Acute but non-lethal injury shows most commonly as yellow patches on the leaves between the veins, although sometimes the damaged areas are brown (Mudd, 1975). The concentration of sulphur dioxide needed in the air to produce such symptoms decreases as the exposure increases from minutes to hours, and O'Gara concluded in 1922, from fumigation studies, that

$$(C - C_R)t = K$$

where C is the concentration of sulphur dioxide, C_R is the threshold concentration, t is the exposure time for damage to be initiated and K is a constant.

This equation embodies two important ideas. First, there is a threshold concentration (C_R) of sulphur dioxide below which damage does not occur. O'Gara deduced this threshold concentration to be 0.33 parts $\times$ 10^{-6} (943 μg sulphur dioxide/m^3) for his particular study, and it was presumed that at lower concentrations the plant could metabolize sulphur dioxide to non-toxic forms fast enough to prevent damage. There has been considerable debate about the concept of a threshold (Garsed, 1984). It can be argued for some causes of cancer that, in a statistical sense, there is no threshold, and that the incidence of cancer is proportional to the exposure. However, for many responses to many pollutants, there is a threshold level below which there is no effect, albeit that the precise level of this threshold will vary between individuals.

Secondly, the product of the amount by which this threshold concentration is exceeded, and the length of exposure needed to produce acute injury, is a constant. The value of C_R does depend on circumstances, but Cowling *et al.* (1973) summarized the available evidence for the species perennial ryegrass (*Lolium perenne*) as showing widespread acceptance of the view that prolonged exposure to concentrations of less than 430 μg sulphur dioxide/m^3 would not produce visible symptoms of damage. A prolonged argument had occurred during the intervening 50 years about the possibility of "invisible injury" from lesser exposures (see Mansfield and Freer-Smith, 1981). Stoklasa first suggested in 1923 that sulphur dioxide might reduce growth without visible symptoms of damage (see Ashenden and Mansfield, 1977). Subsequent experiments by several workers did not confirm this suggestion, and there existed two distinct schools of thought for five decades: that growth can or cannot be affected if there are no visible symptoms of damage. This is a particular illustration of a more general question of great importance in ecotoxicology: can pollutants affect organisms if there are no visible signs of damage?

For studies on the effects of air pollution on plants, this question was answered first for perennial ryegrass (*L. perenne*), of which various cultivars are used in agriculture. Bleasdale wrote a thesis in 1953, although not published as a scientific paper until 1973, in which he described some results with plants of the S23 cultivar grown in two greenhouses near Manchester. Air was passed continuously through the greenhouses. Ambient untreated air passed through one greenhouse, whilst air for the other was first "scrubbed" with water, which removed 98–100% of the sulphur dioxide. There were four exposure regimes, with plants exposed to ambient air for 0, 8, 16 or 24 h per day (Table 6.1). There were no abnormal lesions that could be ascribed to air pollution, but after 119 days there were differences in the amount of growth. The highest growth rates occurred with intermittent exposures to ambient air, and least growth of all occurred with plants exposed all of the time to ambient air. This is clear evidence that the air around Manchester contained factors that could affect plant growth. The concentration of sulphur dioxide varied, with a mean concentration of 47 μg sulphur dioxide/m^3 (Crittenden and Read, 1978), which was a very low concentration, below that which occurs in much of England (Fig. 8.13 and Table 8.3). These observations cannot be taken as direct proof that sulphur dioxide can cause "invisible injury", because air contains other contaminants besides sulphur dioxide, but Bell and Clough (1973) confirmed this effect by adding sulphur dioxide to air and getting a pronounced drop in growth without visible symptoms of injury (Table 6.2).

Table 6.1 Growth of 72 plants of S23 ryegrass (*Lolium perenne*) under each of four different types of exposure to polluted air near Manchester[a]

Daily exposure (h) to:		Number of tillers/ plant	Number of leaves/ plant	Dry weight (mg) of green shoot/ plant
Ambient air	"Scrubbed" air			
–	24	3.4±0.25	15.5±0.6	53±3
24	–	3.0±0.26	14.1±0.6	44±3
16 (16.30–08.30)	8	5.7±0.32	21.0±0.8	62±3
8 (08.30–16.30)	16	6.8±0.34	24.2±0.8	62±3

From Bleasdale (1973).
[a] Results are given as mean values ±SE.

Table 6.2   Yield of 144 plants of S23 ryegrass (*Lolium perenne*) after 26 weeks' growth from seed and with two levels of sulphur dioxide[a]

SO$_2$ exposure	Number of tillers	Number of living leaves	Dry weight (g) of leaves	Leaf area (cm²)
191 μg/m³				
(6.7 parts×10⁻⁸)	14.84±0.67	47.31±2.31	0.388±0.021	203.6±13.2
9 μg/m³				
(0.3 parts×10⁻⁸)	25.18±0.77	85.61±2.35	0.791±0.026	417.2±18.0
Decrease in productivity of plants exposed to high levels of SO$_2$ (%)	41.1	44.7	50.9	51.2

From Bell and Clough (1973).
[a] Results are shown as mean values ±SE: all differences between treatments are significant at the 0.001 probability level.

One must add though that the picture is not clear-cut. To take the simple practical application first, it would be rash to deduce from these two sets of experiments alone that the air quality over much of England is such that it will reduce the rate of growth in this cultivar. Many other variables affect response. One of the most obvious is wind speed, which will affect how quickly air is replaced around plants, and so affect the gradient of sulphur dioxide near plant surfaces. Wind speed may also affect the degree of stomatal opening, the principal route of entry for sulphur dioxide into plants. Plants of the S23 cultivar were grown in wind-tunnels with either clean air or air with about 315 μg sulphur dioxide/m³, when growth was unaffected by sulphur dioxide at a wind speed of 10 m/min, but was affected at 25 m/min (Ashenden and Mansfield, 1977).

It is then perhaps not surprising that there have been conflicting experimental results. Cowling and Koziol (1978) found no effect of 400 μg sulphur dioxide/m³ on plant yield, although some leaf injury was apparent. One possible explanation is that their plants were packed together much more densely, but many other environmental variables, such as light intensity, water stress, relative humidity, temperature and sulphur content of the soil may all influence results. We simply do not yet fully understand the situation: Bell *et al.* (1979) even got inexplicable variations between results from different experiments, but Unsworth and Mansfield (1980) emphasized that the way in which exposure is measured is of critical importance. One needs to measure the concentration of sulphur dioxide to which plants are actually exposed, at their leaf surfaces, rather than the concentration of sulphur dioxide in the incoming air.

To summarize, this work illustrates that the relationship between exposure and response can be influenced considerably by other variables (Guderian,

1977, 1985), that it can be difficult to devise meaningful measurements of exposure, and that sublethal exposures can affect organisms without obvious signs of injury.

Two other points must be noted. First, sublethal effects are not necessarily just preliminary stages in a sequence of events that leads to death. Thus, for the same genotypes (clones) of *Lolium perenne*, there is no correlation between the exposures to sulphur dioxide that produce acute and chronic injury (Fig. 6.1). This lack of correlation implies two different mechanisms of injury.

Secondly, it is usually assumed, implicitly, that sublethal exposure to a pollutant exerts either no effect, if below the threshold, or has an adverse effect whose severity increases with the dose. There is in fact evidence to suggest that there may sometimes be an intermediate degree of exposure that is beneficial to an organism, perhaps because it maintains the capacity for adaptive responses (Smyth, 1967; Laughlin *et al.*, 1981; Stebbing, 1982).

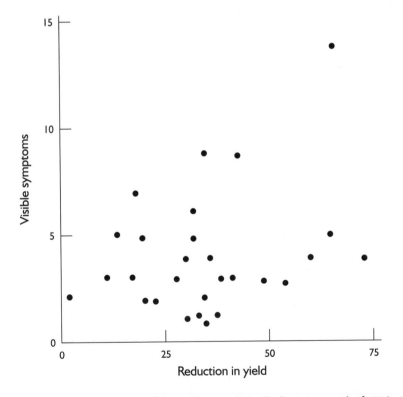

Fig. 6.1 Effects on 28 clones of perennial ryegrass (*Lolium perenne*) of acute and chronic exposure to sulphur dioxide (2600 μg SO_2/m^3 air for 2 weeks or 650 μg SO_2/m^3 air for 8 weeks). Acute effects measured as the percentage of tissue with visible symptoms, chronic effects as the percentage reduction in living dry matter when compared to controls exposed to 35 μg SO_2/m^3 air. (From Horsman *et al.*, 1979b.)

Different exposures to the same pollutant could conceivably have opposite effects, giving either a J-shaped or ∩-shaped response curve to increasing exposures. Certainly increasing doses of essential nutrients increase vigour until, for at least some nutrients, excess damages the organism, and Fig. 6.9 shows another example, although it may be debatable whether increased lysosomal activity is beneficial.

Schulz had already shown in 1877 and 1888 that low concentrations of some poisons stimulated the respiration and growth of cultures of yeast cells while successively higher concentrations reduced respiration until the lethal concentration was reached (Calabrese, 1994). A fungicidal compound extracted from *Thuja plicata* (western red cedar) had similar effects on the growth of some fungal species, and Southam and Ehrlich (1943) proposed the term hormesis for this stimulatory and presumed beneficial effect of sub-inhibitory concentrations of toxic substances on organisms. There is now considerable evidence for hormesis from a wide range of species exposed to a wide range of pollutants (Calabrese *et al.*, 1987), and effects may occur at all levels from subcellular to the whole organism, although not all inversions of response with increasing dose are beneficial (Furst, 1987).

It is uncertain how frequently such effects occur. Not only is there often a reluctance to accept the idea, in part at least because of the parallel with homeopathic medicine, but random variation in response between organisms can mask the true form of the response curve (Gaylor, 1994), as for example in Fig. 9.3. Also, experimental doses are usually chosen to elicit adverse effects and so tend to be too high to detect any opposite responses.

The diversity of known sublethal effects, for different species and different pollutants, is enormous, and many selective reviews have been written (e.g. Stickel, 1968, 1975; Moriarty, 1969; Sprague, 1971; Cooke, 1973; Jefferies, 1973, 1975; Mudd and Kozlowski, 1975; Corner, 1978; Davies, 1978; annually for freshwater organisms, e.g. Spehar *et al.*, 1979; some chapters in Treshow, 1984; Haynes, 1988; Bonga and Balm, 1989; Elzen, 1989; Thomas, 1989; Croft, 1990). The logical starting point is to consider how pollutants exert their effects. Strong solutions of a compound such as sulphuric acid will kill by chemical attack on all components of living organisms, but, apart from such unlikely disasters, pollutants usually act much more selectively. So far as our knowledge goes, pollutants, like drugs, act upon receptors, which are now envisaged as specific molecules, or parts of molecules, to which molecules of the pollutant, or of compounds derived from it, become attached. This combination of receptor with pollutant, the biochemical lesion (Peters, 1969), is the first step before any effects can become manifest (Ariëns *et al.*, 1976).

In fact, very little is known about the receptors for most pollutants, so we will take as an illustrative example one of the best-studied groups, the organophosphorus insecticides (Fig. 6.2) (Gallo and Lawyrk, 1991; Chambers, 1992). They were developed, of course, for their insecticidal activity, many are very toxic to vertebrates, and they can cause hazards to wildlife (e.g. Stanley and Bunyan, 1979; Hill, 1992). These compounds inhibit a whole range of esterase enzymes,

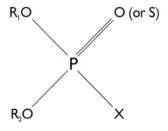

Fig. 6.2 General formula for organophosphorus insecticides. R is an alkyl group and X is the leaving group, which splits off from the molecule when hydrolysed.

and it is the consequences of inhibition of one of these enzymes, acetyl-cholinesterase (AChE), that produce the symptoms of acute poisoning (Wallace, 1992). Acetylcholine (ACh), the natural substrate for AChE, is one of the principal known transmitters of impulses across synapses between adjacent nerve endings, and across neuromuscular junctions. Nerve impulses stimulate the release of ACh, which transmits the stimulus across the gap to the adjacent nerve or muscle cell. The ACh is normally broken down rapidly, by hydrolysis, catalysed by the enzyme AChE (Fig. 6.3a). Inhibition of AChE means that ACh persists much longer, normal nerve functions are grossly disturbed, and a sufficiently severe disturbance ends in death (Eto, 1974). In brief, the lethal lesion disturbs impulse transmission across synapses and neuromuscular junctions, many physiological processes are disturbed in consequence, and death, in vertebrates at least, usually results from paralysis of the respiratory system (O'Brien, 1967).

The enzyme-mediated hydrolysis of ACh depends on the attack by a hydroxyl group attached to a serine molecule in the enzyme. Before hydrolysis can occur, individual molecules of ACh first bond to two sites on molecules of the enzyme AChE, one active or esteratic and one ionic (Wallace, 1992). The ionic bond forms first with the strong positive charge on the quaternary nitrogen atom in the choline moiety of ACh (Fig. 6.3a). Auxiliary binding forces form between two of the associated methyl groups and the enzyme (Gallo and Lawryk, 1991). The details are not yet clear, but the association of cationic (positively-charged) groups with this ionic site is thought to alter the enzyme's tertiary structure, or conformation, so that the carbonyl group in the acetate moiety of ACh aligns with and forms a covalent bond with the serine-hydroxyl group at the esteratic site with consequent release of choline. Hydrolysis then regenerates the acetylated enzyme, with release of acetic acid, and half-lives for recovery are measured in microseconds.

This serine-hydroxyl group in AChE is also the site of action for organophosphorus insecticides. In general terms, the organophosphorus molecule becomes attached to the serine group of the enzyme, in much the same way as does ACh, and Fig. 6.3b shows an example. Whether or not insecticides also form an ionic bond with AChE is uncertain (compare

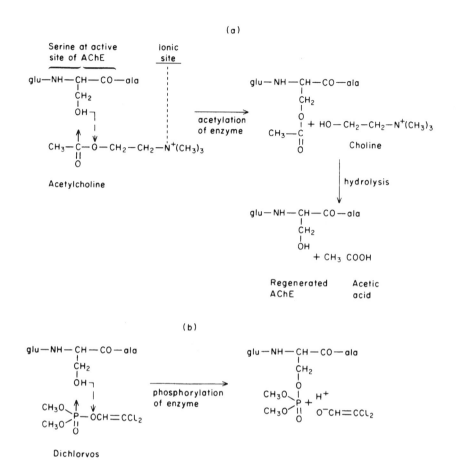

Fig. 6.3 Reactions between acetylcholinesterase (AChE) and two substrates: (a) acetylcholine; (b) an organophosphorus insecticide, dichlorvos (dimethyl 2-dichloro-vinyl phosphate).

Santone and Powis, 1991, with Wallace, 1992). Certainly though, in contrast to the reaction with ACh, this phosphorylation of the enzyme is practically irreversible, and spontaneous reactivation by breakage of the phosphorus –enzyme bond takes from hours to weeks, depending on the insecticide. Hence organophosphorus insecticides decrease the amount of AChE that is available, and so ACh accumulates at nerve endings. With some insecticides the ability of AChE to reactivate, either spontaneously or by chemical reagents, decreases with time. This ageing process results from the loss of an alkyl group from the phosphorus atom, although the detailed mechanism is unclear (Wilson *et al.*, 1992).

We have seen that the biochemical lesion that results in the acute toxicity of organophosphorus insecticides is in the serine group of AChE. However, the focus of interest is sometimes at more macroscopic levels of integration than the biochemical, when one might say that the site of action is in the nerve cord. On occasion, it is relevant to consider which is the critical organ, the organ within which the damage occurs from the biochemical lesion. This point was established in a detailed study on the effects of diazoxon, the active form of the organophosphorus insecticide diazinon, on the cockroach *Periplaneta americana*. Individual cockroaches were dosed on the ventral surface, at either the posterior end of the thorax or of the abdomen, with 6 or 8 μg diazoxon, a dose lethal to most individuals. The underlying ganglia of the nerve cord at these two sites (the metathoracic and sixth abdominal ganglia) were tested functionally with electrical stimuli and recording electrodes for the degree of interference with their normal conduction of nerve impulses. The ganglia were also examined histochemically for the degree of AChE inhibition. The results (Table 6.3) showed that:

(1) Behavioural symptoms of poisoning appeared sooner when cockroaches were dosed on the thorax.
(2) The functional activity of both ganglia was affected sooner when insecticide was applied locally, although regardless of where the insecticide was applied the metathoracic ganglion was affected before the abdominal ganglion.
(3) The functional activity of the metathoracic ganglion was severely affected in all cockroaches with severe symptoms of poisoning, bar one exception. In contrast, the abdominal ganglia regained their normal activity with time. Effects on activity were maximal at about 7½ h after dosing. In other words, the functional condition of the two ganglia was disturbed to different extents by the same dose, and the condition of the metathoracic ganglion gave a more reliable indication of the condition of the insect as a whole.
(4) The histochemical measurements of enzyme inhibition corresponded fairly well to the measurements of nerve function, except that the recovery of function in the abdominal ganglion was not accompanied by a recovery of enzyme activity.

These results illustrate in a simple fashion that the biological effects of a pollutant depend on the fine details of the exposure pattern. The surprising result was to find that activity of the sixth abdominal ganglion recovered in cockroaches that were still prostrate from diazinon poisoning, and that this recovery occurred although histochemical tests showed AChE to be still inhibited.

These discrepancies were resolved when it was discovered that there are several isozymes of AChE (Tripathi and O'Brien, 1973). That is to say, there are several closely related enzymes, similar but not identical in their chemical structure, which all hydrolyse ACh. The housefly (*Musca domestica*) has a very condensed central nervous system, with two principal ganglia, in the head and the thorax. Four isozymes of AChE occur in the head ganglion, and

Table 6.3 The progress of poisoning in cockroaches (*Periplaneta americana*) given a single dose of diazoxon on the surface of either the thorax (T) or the abdomen (A)[a]

Time since does applied (h)	Position at which dose applied	Dose applied (μg)	Condition of insect	Condition of nerve function		Index of AChE activity	
				T	A	T	A
0.5	T	6	S	B	N	3	5
	T	8	N	B	N	2.5	5
	A	6	N	N	N	5	5
	A	8	N	S	N	5	5
1	T	6	S	B	N	3	5
	T	8	S	B	N	2.5	3
	A	6	N	S	S	5	4.5
	A	8	N	S	N	5	2.5
2	T	6	B	B	N	3	5
	T	8	B	B	N	2.5	2.5
	A	6	S	B	B	3.5	3.5
	A	8	N	S	N	5	2.5
4	T	6	P	B	S	2	2
	T	8	P	B	N	2.5	2.5
	A	6	S	S	N	5	4
	A	8	P	B	B	2.5	2.5
7.5	T	6	P	B	S	3	3
	T	8	P	B	B	3	2.5
	A	6	P	S	S	3	2.5
	A	8	P	B	B	2.5	2
14	T	6	P	B	N	2.5	4.5
	T	8	P	B	N	2.5	4
	A	6	P	B	S	3.5	4
	A	8	P	B	S	2.5	1.5
24	T	6	P	B	N	3	4
	T	8	P	B	S	3	1.5
	A	6	P	B	N	3.5	3
	A	8	P	B	N	2	2
	A	8	P	B	N	2	1.5
	A	8	B } recovering	B	N	4	3.5
	A	8	N } recovering	N	N	3.5	3

From Burt *et al.* (1966).
[a] The condition of the whole insect, and of nerve function in the metathoracic and sixth abdominal ganglia, was assessed as normal (N), slightly affected (S), badly affected (B) or, for the whole insect, prostrate (P). The index of acetylcholinesterase (AChE) activity ranges from 1 (complete inhibition) to 5 (no inhibition).

another three, different, isozymes, in the thoracic ganglion. Groups of house-flies were given the LD_{50} dose of four organophosphorus insecticides, and the minimal activity was estimated for each of the seven isozymes (Table 6.4). It can be seen that isozyme 7 always lost virtually all of its activity, even in the survivors, so it cannot be the relevant site of action. Most of the other isozymes were inhibited to different degrees by the different insecticides, but isozyme 5, in the thorax, had a constant degree of inhibition, of just over 80%. These data suggest therefore that this is the isozyme whose inhibition is critical for life, and that the thoracic ganglion is the critical organ.

The fact that organophosphorus insecticides can inhibit a wide range of esterases prompts the question, can these compounds produce more than one biologically significant lesion? There is in fact much work to show that some, but not all, organophosphorus insecticides can produce, in birds and mammals, a second lesion quite unrelated to the inhibition of AChE (Johnson, 1975a, 1981; Baron, 1981). Symptoms do not appear until 1–2 weeks after exposure, when the hind limbs, and in severe cases the fore limbs too, become paralysed. Originally, this condition was referred to as demyelination, because the myelin sheath that surrounds most nerves degenerates. It is now considered that the primary damage is not to the myelin sheath but to the nerve axon itself, and the condition is more correctly called delayed neuro-toxicity, or organophosphorus ester-induced delayed neurotoxicity (OPIDN). The preferred explanation is that the primary lesion occurs in an esterase, neurotoxic esterase or, to avoid ambiguity, neuropathy target esterase (NTE), which belongs to the same group of enzymes as does AChE. Symptoms of poisoning appear, after a delay of 1–2 weeks, if more than about 80% of this enzyme is phosphorylated and so inhibited by an organophosphorus insecti-cide. Delayed neurotoxicity develops independently of acute toxic effects and for different compounds the severity of acute toxicity and of delayed neuro-toxicity tend to be inversely correlated (Windebank, 1987). Although this

Table 6.4 The minimal percentage activity of seven isozymes of the enzyme acetylcholinesterase in ganglia of the housefly (*Musca domestica*) after exposure to the LD_{50} dose of four insecticides

Body region	Isozyme	Activities following treatment with:				Range of minimal activities
		Malaoxon	Paraoxon	Diazinon	Dichlorvos	
Head	1	18	45	54	39	36
Head	2	42	53	28	72	44
Head	3	51	78	32	57	46
Head	4	28	67	67	70	42
Thorax	5	15	18	22	21	7
Thorax	6	5	38	1	16	37
Thorax	7	1	1	1	1	0

From Tripathi and O'Brien (1973).

delayed neurotoxicity can only occur if NTE is inhibited, inhibition by itself is insufficient. Neurotoxicity only occurs if the inhibitor can be aged, which needs at least one of the R-groups to be attached to the phosphorus atom by a labile bond, such as R–O–P, that can be detached by hydrolysis (Fig. 6.4). Ageing does not occur if the R-groups are attached through stable carbon bonds (Richardson, 1992).

Although NTE has no known function, experiments show that dosing with an inhibitor that does not cause delayed neurotoxicity and then with one that does prevents the development of neurotoxicity (Johnson, 1990). This suggests that despite the lack of known function inhibition of NTE is the primary lesion for delayed neurotoxicity, but that aged NTE then acts on another target before overt symptoms can occur. However, some recent work shows that when the order of dosing is reversed, with exposure first to a neuropathic organophosphorus compound, the subsequent exposure to an inhibitor that does not cause delayed neurotoxicity on its own can initiate (for a sub-threshold dose of the neuropathic compound) or increase the symptoms of delayed neurotoxicity (Richardson, 1992). Ageing of inhibited NTE may then not be a complete explanation for delayed neurotoxicity, but further progress probably depends on first determining the structure and function of NTE. It also becomes conceivable that exposure to more than one organophosphorus insecticide could lead to unexpected adverse effects.

Apart from delayed neurotoxicity, other effects of organophosphorus insecticides are less well understood. Johnson (1975b) makes the familiar, and convincing, argument for the specific instance of delayed neurotoxicity, that the considerable understanding we now possess of the primary lesion, structure – activity relationships, prediction of toxicity, and of possible preventive and therapeutic measures, stem to a large degree from fundamental studies whose relevance to these practical problems could not be foretold. I would extend the argument to declare that subtle sublethal effects can debilitate, and that therefore the more we know about the primary lesions, and related topics, the better our ability to predict the ecological consequences of pollutants. We have to accept though that pollutants may have more than one mode of action, different pollutants may interact in their effects and impurities may be important (e.g. phosphorothiolates in organophosphorus insecticides

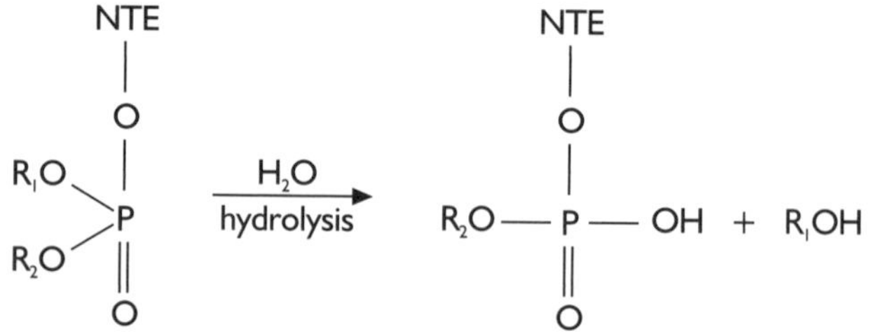

Fig. 6.4 The ageing of neurotoxic esterase (NTE) that has been inhibited by an organophosphorus insecticide.

(Thompson, 1992)). The ecological relevance of sublethal effects increases the further below a lethal exposure that they occur.

Sublethal exposures to organophosphorus insecticides can have subtle but ecologically relevant effects. Starlings (*Sturnus vulgaris*) given a single sublethal dose of dicrotophos, sufficient to inhibit 50% of the brain cholinesterase but not to kill any individuals, looked after their young less well than did the control birds, which would probably reduce nestling survival in natural conditions (Grue *et al.*, 1982). Further studies on starlings exposed to chlorfenvinphos showed that as the degree of inhibition of brain AChE increased from 10 to 40% different aspects of behaviour were affected and that the response also depended on environmental conditions (Hart, 1993). Experiments on standard laboratory animals confirm that these compounds can affect behaviour, possibly irreversibly (Annau, 1992).

This argument for the importance of sublethal effects may seem contrary to one of the predominant themes of this book, that effects on individuals only matter if they affect the population, but two examples may help to show how detailed knowledge of sublethal effects can improve our awareness of likely ecological effects.

We have already seen that effects by a pollutant on individual organisms do not necessarily imply ecological effects. The alternative situation, of no biological effects by a contaminant, cannot possibly have any direct ecological effects, although we need to add the rider that there may still be significant effects on the environment. This leaves the difficult intermediate area of uncertainty, where the question is to decide whether or not a particular exposure to a contaminant affects individual organisms in any way, and if so, whether they are debilitated, or less viable. This is the question that much routine toxicity testing is designed to answer, but without some knowledge of the biochemical lesions, sites of action and critical organs one can do little more than apply a standard battery of screening tests, such as tend to be applied in screening tests for regulatory agencies. In the short term, and in the practical world with our present degree of knowledge, it is often difficult to envisage anything better, but one has to accept that potentially important effects may be overlooked.

Our first example, the deliberate release of biocides from ships' hulls to prevent biofouling, illustrates an irreversible dose-dependent sublethal effect that is as important as are lethal effects (see p. 36). Marine ships rapidly acquire a surface film of bacteria and algae below the water-line, on which larger algae and invertebrates then develop (Evans, 1988). These growths markedly increase frictional drag, fuel consumption and, on steel hulls, corrosion. They are also expensive to remove. For control, hulls usually have surface coatings that incorporate one or more biocides, of which tributyl tin (TBT) is a recent example, introduced in the mid-1960s. Adequate control rarely lasted more than a year or so until the early 1970s, when triorganotins (usually referred to as organotins), commonly TBT, were polymerized with unsaturated monomers to form a surface film of methacrylate triorganotin copolymer on the hull with a high proportion of organotin (Fig. 6.5). Hydrolysis releases organotin at a relatively uniform rate to give a longer effective lifetime. With this "self-polishing copolymer (SPC) system" surface roughness and frictional resistance decrease as the polymer is eroded.

It was thought that once released the organotin would be diluted and degraded fast enough to minimize toxicity to non-fouling organisms. However, samples of unfiltered water from harbours and marinas in Canada, England and France exceeded levels known experimentally to be toxic in seawater, which implied the risk of adverse effects on marine organisms in coastal habitats (Cleary and Stebbing, 1985; Bryan and Gibbs, 1991). One of the most clearcut and sensitive of these effects from TBT is imposex, the irreversible development of male characters in female molluscs from other than normal biological causes such as injury and parasitism (Smith, 1981; Bryan et al., 1986). The common dog-whelk, *Nucella lapillus*, is particularly susceptible (Gibbs et al., 1988). Exposure of juvenile females, which are more sensitive than adults, to < 1 ng Sn (=2.5 ng TBT)/l of seawater initiates development of a penis and the associated duct, the vas deferens, although breeding is unaffected. Higher exposures of about 5 ng Sn/l or more induce greater development of both the penis and vas deferens. The vas deferens then blocks the oviduct with consequent

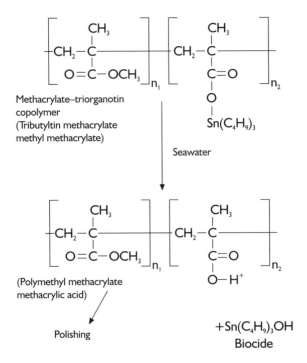

Fig. 6.5 Structure and hydrolysis of methacrylate–triorganotin copolymer. (From Evans, 1988.)

sterility because egg capsules cannot be released. Exposure to 10 ng Sn/l or more starts to suppress egg production in the ovary, and exposures above 25 ng Sn/l transform the ovary into a testis with seminiferous tubules and spermatozoa. The mode of action is not yet clear. Female *N. lapillus* exposed to TBT have higher tissue concentrations of testosterone, a hormone known to encourage male characteristics in vertebrates, but did not affect the amount of an oestrogen, which encourages female characteristics (Spooner *et al.*, 1991). Although the authors suggest that TBT inhibits the enzyme that converts testosterone into an oestrogen, thus initiating imposex, it could be that induction of imposex by TBT secondarily increases testosterone in females.

The effect on populations is more obvious. The dog-whelk lacks a planktonic mobile larval stage, and the adults are relatively immobile. Populations in habitats sufficiently polluted with TBT therefore become dominated by adult males, lack juveniles and eventually disappear. Sterilization of the females has effectively exterminated the dog-whelk from much of its European range, from Norway to Portugal (Gibbs *et al.*, 1991), during the last two decades. Although other species are less sensitive, many have been affected world-wide by exposure to TBT (Bryan and Gibbs, 1991).

A few local populations of *N. lapillus* have survived. The dog-whelk has disappeared from the southern side of the Thames estuary between Whistable and Ramsgate except for a local population at Dumpton Gap. As expected, this population contained females with at least some degree of imposex, although some showed no symptoms, and evidence from the adults suggested that only about 10% of the females with imposex were sterile. Unlike most affected populations, about one-quarter of the females with imposex (i.e. with a vas deferens) lacked a penis, as did a small proportion of the males (Table 6.5). Offspring from adults transferred to the laboratory, when exposed to TBT, showed similar responses, which suggests a genetic difference, perhaps with similarities to sickle-cell anaemia (p. 78): the presumed genetic disadvantage under normal conditions enables the population to survive exposure to TBT because a

Table 6.5 The incidence of standard and abnormal symptoms from exposure to tributyl tin in two populations of dog-whelk (*Nucella lapillus*). Samples taken during 1989–1990 from Dumpton Gap, on the Kent coast, SE England, and from the nearest surviving population, at Oldstairs Bay

Population	Females			Males	
	No symptoms	Imposex, with no penis	Imposex, with penis	Normal	No penis
Dumpton Gap	42	54	191	273	19
Oldstairs Bay*	0	0	35	56	0

Data from Gibbs (1993).
* This population was only just surviving: most of the females had not quite reached the stage where the oviduct is blocked.

proportion of females is little or not affected and can breed with the unaffected males. A somewhat similar situation occurs on the coasts near Brest in
north-western France (Huet *et al.*, 1996), and possibly elsewhere.

Effects on communities are more subtle. Moderately exposed rocky shores in
Britain have a mosaic of patches dominated temporarily by either seaweed
(*Fucus* spp.), which is grazed by limpets (*Patella vulgata*), or barnacles
(*Chthamalus* spp.). *Fucus* establishes itself most easily in crevices and on rough
surfaces, which may be either rock or rock covered by barnacles (Hawkins,
1981). High densities of barnacles reduce limpet grazing efficiency, which
enables *Fucus* to re-establish itself and smother the patches of barnacles until
the seaweed is again grazed by limpets and barnacles can re-establish themselves. The dog-whelk is an important predator of barnacles, so TBT might be
expected to affect communities on rocky shores (Hawkins *et al.*, 1994). Removal
of dog-whelks from experimental areas increased the lifetime of clumps of *F.
vesiculosis*, because the barnacles to which they were attached were not being
eaten, and increased the density of barnacles when dominant, which thus
reduced limpet grazing efficiency and so increased the incidence of clumps of
Fucus. So on two counts TBT might be expected to increase the relative abundance of *Fucus*, although there is not yet any clear observational evidence.

Our second example shows how knowledge of mode of action can sometimes help studies on sublethal effects. Like the organophosphorus insecticides, the organochlorine insecticide *p,p'*-DDT affects the nervous system,
but the biochemical lesion occurs in the axonic membrane of the nerve fibre,
not at the synapses (Narahashi, 1971; Bloomquist, 1996). The nervous system
has two functions: the conduction of nerve impulses, and neurosecretion,
which helps to integrate the hormonal and nervous sytems (Randall *et al.*,
1997). Behaviour and hormonal systems might then be the first places in
which to look for sublethal effects (Moriarty, 1971), and there have been
some relevant experiments on the effects of DDT on the thyroid gland of
birds, and on the thinning of shells from birds' eggs.

Jefferies and French (1971) dosed pigeons (*Columba livia*) with *p,p'*-DDT,
which was dissolved in olive oil within gelatin capsules and force-fed to the
birds daily. Apart from controls, individual birds received 3, 6, 9, 18, or 36 mg
DDT/kg of body weight per day for 8 weeks. Two of the birds on the top dose
rate died just before the end of the experiment, when all birds were killed
and their thyroid glands examined (Table 6.6). All doses enlarged the thyroid
to a similar degree, about double the weight of the thyroid in control birds,
and there were also effects at the cellular level. Histologically, the thyroid
contains numerous sub-spherical follicles enclosed within a capsule of con-
nective tissue, and each follicle consists of a single layer of secretory cells,
which surround a central lumen that is filled by a gelatinous material, or
"colloid". This colloid contains the thyroid hormones that the follicular cells
secrete. The size and number of the follicular cells, and the amount of colloid,
vary with the degree of thyroid activity (Jefferies, 1975). The control birds in
this particular experiment contained normal amounts of colloid material in
their thyroid glands, but the follicular cells of the dosed birds were so

Table 6.6 Effects on the thyroid gland of pigeons (*Columba livia*) when fed *p,p'*-DDT for 8 weeks[a]

Dose rate (mg DDT/ kg body weight/day)	Fresh wt thyroid gland (mg)	Area of colloid (mm²)/area of 0.3 mm²
0	45±3 (19)	0.039±0.007 (9)
3	89±26 (6)	0.009±0.008 (3)
6	70±18 (6)	0.005±0.004 (3)
9	78±15 (10)	0.005±0.003 (7)
18	95±14 (10)	0.010±0.006 (4)
36	85±9 (10)	0.003±0.002 (7)

From Jefferies and French (1971).
[a] Results are given as mean values ±SE; numbers examined are given in parentheses.

enlarged (hypertrophied) that they occluded the follicles, and they were also more numerous (hyperplastic) (Table 6.6 and Fig. 6.6).

Detailed interpretation of these changes is not easy. In particular, feedback mechanisms will tend to rectify departures from normal amounts of hormone circulating within the body, so that information on the degree of thyroid activity and of amounts of hormone stored within the gland do not tell us directly whether the amount of hormone available to receptor sites has altered. However, field observations and subsequent laboratory experiments showed that DDT and its metabolite (or conversion product) DDE were implicated in a widespread occurrence of abnormal eggshells and decline of breeding success in several bird species (see Moriarty, 1975a, and Newton, 1979, for general accounts). This prompted Haegele and Hudson (1973, 1977) to dose ring doves (*Streptopelia risoria*) with *p,p'*-DDE, and they found many aspects of breeding behaviour and egg physiology to be affected (Table 6.7).

Table 6.7 Effects on the breeding success of ring doves (*Streptopelia risoria*) when fed a diet containing 40 ppm of *p,p'*-DDE for 126 days

	Controls	Dosed[a]
Number of separate pairs of doves	12	12
Adult mortality	0	0
Number of times individual pairs nested	49	33
Mean number of days to renesting	6.5	16.9
Number of eggs laid	98	57
Mean number of eggs per clutch	2	1.7
Mean shell thickness of eggs laid in the 2nd–5th nestings (nm)	132	123
Mean mortality of young (%)	31	64
Number of young surviving for 21 days	35	10

From Haegele and Hudson (1973).
[a] All differences significant, at least, at the 0.01 probability level.

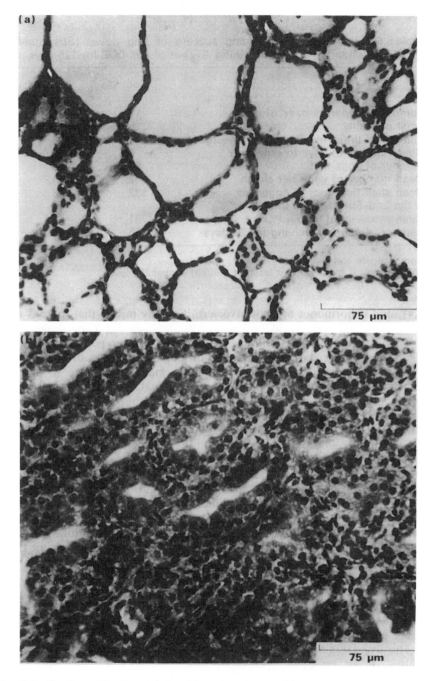

Fig. 6.6   Section of the thyroid gland from the pigeon, *Columba livia*. (a) Gland from normal pigeon, in quiescent state; (b) gland from a pigeon dosed with 36 mg *p,p'*-DDT/kg body weight/day for 6 weeks. The follicles are smaller, with more follicular cells and less "colloid". (From Jefferies, 1975.)

There was also some evidence that the same dose-level affects the amount of luteinizing hormone in the blood plasma (Richie and Peterle, 1979). Exposure of the same species to a mixture of organochlorine compounds, including DDE, affected the levels in blood plasma of several hormones, including thyroxine from the thyroid gland (McArthur et al., 1983).

We started this discussion of organochlorines and thyroids to see what relevance detailed knowledge of modes of action has for predicting effects on populations. We still do not fully understand the implications of hyperplastic thyroids for population dynamics beyond the vague general statement that such individuals must be more stressed. Disturbance of the thyroid could conceivably be the origin of the widespread occurrence in birds of eggs with thinner shells than normal, first discovered by Ratcliffe (1967, 1970) (Fig. 9.1), although there is no proof so far of a connection between effects on the thyroid and on eggshells. However, this is unimportant for present purposes: there is good evidence to show that DDT and DDE cause birds to lay

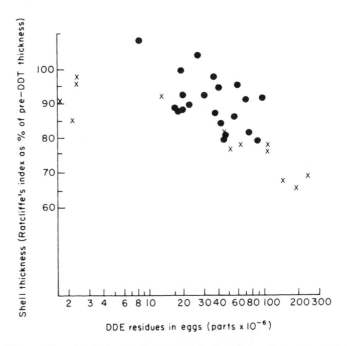

Fig. 6.7   The relationship between the concentration (based on dry weight) of *p,p'*-DDE in eggs of the American kestrel (*Falco sparverius*) and eggshell thickness (expressed as a percentage of the thickness of shells in eggs laid before DDT came into use). The nature of this relationship is similar for eggs collected from wild birds nesting in Ithaca, New York, during 1970 (●), and for eggs obtained in 1971 from captive birds fed food contaminated with *p,p'*-DDE (×). Note however, that there is a significant difference (P<0.01) in the absolute magnitude of the thinning associated with any particular concentration of DDE, discussed further on pp. 258, 259. (From Lincer, 1975.)

eggs with thin shells (e.g. Fig. 6.7), and if hormones are not involved, it simply means that there is an additional site of action.

This work on thyroids and eggshells illustrates four points of general significance. First, without detailed knowledge of the sites of action and the biological consequences, we rely very much on chance to detect effects that are significant, not just for individual organisms but also for whole populations. Hormonal disturbance and thin eggshells are both likely to affect a population's ability to survive, and neither is likely to be detected by routine toxicity tests.

Secondly, pigeons were fed DDT, but the brain and the liver contained higher concentrations of the metabolite DDE than of the original compound (Fig. 6.8), and subsequent experiments with pigeons dosed with p,p'-DDE had similar effects on the thyroid (Jefferies and French, 1972). The biological activity of pollutants often depends not only, or not at all, on the form that is ingested, but on one or more of the metabolites. Frequently, of course, metabolism is a process of detoxication, but not always. It then becomes difficult to determine how much, if any, of the effects of DDT on the thyroid are due to the direct activity of DDT and how much are due to its metabolite. There is little direct evidence about the mode of action of DDE. It is usually far less

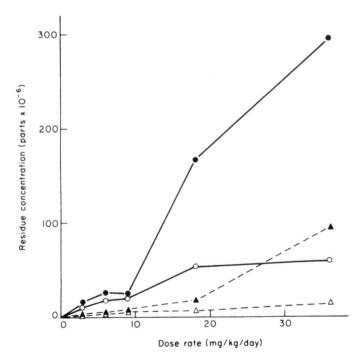

Fig. 6.8   The concentrations of p,p'-DDT and of p,p'-DDE in the liver and brain of pigeons (*Columba livia*) dosed orally for 56 days. (● — ●) DDE in the liver; (○ — ○) DDT in the liver; (▲--▲) DDE in the brain; (△--△) DDT in the brain. (Data from Jefferies and French, 1971.)

toxic than DDT, although the converse appears to be true for the pigeon (Jefferies and French, 1972), and physiological tests show DDE to have little effect on nerve axons (Narahashi, 1971).

Thirdly, DDT and its metabolite DDE are not the only possible causes of an enlarged, hyperplastic thyroid. Iodine deficiency, and goitrogens—those compounds that interfere with the control or synthesis of thyroid hormones—will produce the same result. So can environmental changes. The principal function of the thyroid in adult animals is to maintain a normal oxidative metabolism. Prolonged cold will cause the thyroid to release more hormone to stimulate heat production, and in extreme conditions the thyroid can enlarge and become hyperplastic. One may then reasonably expect that if environment affects thyroid activity, it would also influence the thyroid's response to DDT or DDE. There is some inconclusive evidence to suggest that the thyroid is less affected by DDT at high than at low temperatures (Jefferies, 1975).

Fourthly, it is worth commenting on the fact that eggshell thickness is usually regulated within definite controlled limits (e.g. Fig. 9.1). Claude Bernard made his celebrated remark "La fixité du milieu intérieur est la condition de la vie libre" over a century ago. In more prosaic terms, organisms tend to maintain constant many aspects of their internal environment, such as body temperature or thickness of shell deposited around eggs, and failure to do so can lead to malfunction and eventually to death. In 1929, Cannon coined the term homeostasis for this type of stability, which relies on feedback control by nervous and hormonal mechanisms for the maintenance of biochemical and physiological processes within definite controlled limits. Environmental factors that tend to disturb such homeostatic mechanisms are termed stressors; these can include pollutants, but the organism is said to be adapted and not stressed so long as the stressors do not upset the control mechanisms. This is a somewhat strict definition of stress, which is often used in a slightly wider sense, for the effect of stressors on processes such as growth, which are inherently much more variable. An organism may then be described as stressed, though possibly quite well adapted, if its growth rate is reduced, be it by pollution, lack of food, or any other cause. The idea of stress is imprecise—it can cover a diverse range of effects, caused by many qualitatively different substances and conditions—but it has stimulated some alternative ways of studying sublethal effects.

I have argued that the logical approach is to discover the primary lesions and to explore the ramifications, in terms of very specific physiological and biochemical processes, that ensue from these lesions. Not only is this knowledge important for the prediction of ecological effects, it may also enable one to select a physiological or biochemical function that gives an index of the degree of effect. For a simple practical example, inhibition of AChE or of other esterase enzymes is used to detect deaths from organophosphorus insecticides in wildlife (e.g. Weiss, 1961; Bunyan et al., 1968; Bunyan, 1973).

However, the four examples, of sulphur dioxide with plants, of organophosphorus and organochlorine insecticides with birds, mammals and insects, and TBT with molluscs, suggest that it is not easy to ascertain the primary lesion and its ramifications, and is not an approach which we can hope to undertake for all potential pollutants. At the opposite extreme, one can apply a standard battery of tests, but this is likely to miss sublethal effects that could affect the performance of whole populations.

The alternative approach is to concentrate not on the pollutant and its detailed mode of action but on the health of the individual and how well it is functioning, by measuring aspects of biochemical and physiological activity (Bayne *et al.*, 1979, 1985; McIntyre and Pearce, 1980; Calow, 1993a). Advantages of such measurements include rapid responses to pollutants and—as sublethal effects—they are sometimes very sensitive. However, response to a specific pollutant can be greatly influenced by many aspects of the environment as well as by other pollutants (Bayne, 1985).

Lysosomes, intracellular particles first described by de Duve in 1955 (see de Duve, 1963), are one possible indicator of well-being in animals. Lysosomes contain hydrolases (acid hydrolytic enzymes), whose function is to digest foreign and endogenous substances within the cell. Potentially, therefore, these enzymes could destroy the cell, but in health they are retained, inactive, within the lysosomes by a lipoprotein membrane. Many types of stress can alter the membrane structure, which then becomes more permeable and permits substrates to enter lysosomes and hydrolases to move out into the cytoplasm. Bitensky (1963) developed a simple histochemical index for the degree of lysosomal activity, in which the permeability, or "fragility", of the lysosomal membrane is measured by the time needed for substrate to enter the lysosome across the membrane. Many stressors have been shown to increase fragility. For example, in laboratory rats and guinea-pigs that were subjected to high temperatures, prolonged swimming, gravitational or emotional stress, all treatments increased the membrane permeability of lysosomes in the brain (Gabrielescu, 1970). Chemicals too can have similar effects. Moore and Stebbing (1976) exposed the marine coelenterate *Campanularia flexuosa* to low levels of inorganic copper, cadmium and mercury and concluded that lysosomes were affected by lower exposures than were needed to affect growth (Fig. 6.9).

Growth is another measure of stress, at a higher level of integration than lysosomal activity. There are many possible measures of stress, at various levels of integration between a biochemical lesion and a change of population structure, but growth is an intrinsically attractive measure.

Not only can growth be influenced by biochemical lesions of many different sorts, which may alter behaviour or induce metabolic or physiological incapacities, but growth, or the lack of it, which can affect maturity, senescence and reproductive potential, is also of obvious relevance to both birth and death rates, two of the key parameters in population structure. Growth studies of animals in the laboratory usually measure amount of food consumed and change

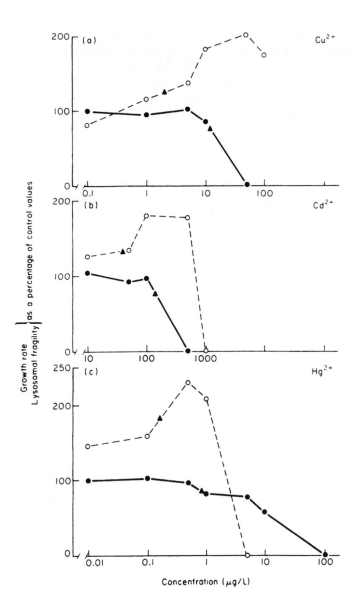

Fig. 6.9 The effect of low concentrations (μg/1) of (a) cupric, (b) cadmium and (c) mercuric chloride in sea water on the growth and lysosomal fragility of•the hydroid *Campanularia flexuosa*. Growth rate (●—●) and the amount of lysosomal activity (○--○) are expressed as a percentage of that in the controls. (▲) Threshold concentration, defined as that concentration of metal at which the distance between the fitted concentration–response curves and the control value is three times the standard error of the residual variation. (From Moore and Stebbing, 1976.)

of body weight, but Warren and Davis (1967), who were concerned specifically with fish, argued that although growth obviously does depend on food consumption, other variables affect this relationship, that we can only obtain a complete understanding of growth if we relate it to the effect of a range of environmental factors, and that this can be done by measuring the energy content of the food consumed and the uses to which this energy is put. The "scope for growth", defined as the difference between the energy content of the food consumed and all energy uses and losses apart from growth, can be expressed as:

$$Q_g = Q_c - Q_w - Q_r$$

where Q_c is the energy value of the food consumed; Q_w is the energy value of waste products in faeces and in urine, and lost through the gills and skin; Q_r is the energy utilized or released (respiration); and Q_g is the energy value of body materials (growth and gametes), the "scope for growth". All four of these terms are influenced by environmental variables, such as temperature, oxygen concentration, food availability and quality, such as adequate intake of vitamins and minerals. Elliott (1994) gives a detailed example of this approach for the brown trout (*Salmo trutta*).

Warren and Liss (1977) applied this approach to pollution studies, and argued that assessment of the impact of a pollutant on growth, or on any other measure of performance, requires performance to be measured in more than one environment. Exposure to pollutant is considered as an additional variable, whose effect on growth can then be assessed under a range of controlled environmental conditions. This is, of course, the standard approach in agricultural field trials, where Fisher's work impelled experimental design and the analysis and interpretation of data towards a high level of sophistication.

The mussel, especially *Mytilus edulis*, has been used extensively for studies on scope for growth. Simple measures of growth, in both laboratory and field studies, can of course be used to assess the effect of both the environment and any pollutant it contains. In bivalves like the mussel though, growth is not easy to measure: a large part of Q_g often goes to form gametes, which are released from the body, and the link between growth of shell and of soft tissues is not close. It is also difficult in the field to measure how much food is available, so it becomes difficult to distinguish effects of food from those of pollutant (Widdows and Donkin, 1992). Measurement of scope for growth has several advantages. Measurement is rapid, albeit sophisticated, it can be applied to both field and laboratory conditions and it can be related to amounts of pollutant in the body (Fig. 6.10) (Widdows, 1993). However, consistent results do require careful control of test conditions, and Grant and Cranford (1991) discuss the assumptions that are made when scope for growth is equated to actual growth.

The need to take account of variables other than the pollutant must be emphasized. Much of the published literature on sublethal effects is of limited

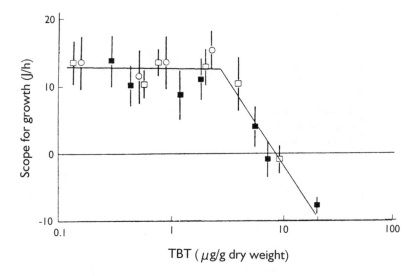

TBT (μg/g dry weight)

Fig. 6.10 Scope for growth (joules/hour, expressed for a standard dry weight) in mussels (*Mytilus edulis*) after 4 days' exposure to different concentrations of tributyl tin (TBT) in the ambient seawater. Initial concentrations ranged from 0.01 to 5 μg TBT/l seawater, replenished four times on both the second and third days. Ambient concentrations were therefore variable, and the scope for growth is therefore related to tissue concentrations of TBT. Each point is the mean±95% confidence limits for 16 mussels. □, ■ and ○ indicate results from three different experiments. (From Widdows and Page, 1993.)

value simply because effects are described for one set of conditions, often ill defined, with one type of exposure, and with no indication of the influence of other environmental variables. These are the same sort of criticisms that apply to the LD$_{50}$. One classic, oft-quoted, example that demonstrates the influence of environment is for the effect of DDT on birds and mammals. The highest concentrations of DDT occur in the fat reserves. Starvation mobilizes these reserves, when DDT concentrations sometimes rise in the blood and in some other tissues (Findlay and deFreitas, 1971; Hayes, 1975). This presumably explains the fact that starved animals will sometimes die of DDT poisoning whereas fed ones appear unharmed (e.g. Dale *et al.*, 1962; Bernard, 1963). This is more than a curiosity of the laboratory, for starvation must be a common experience for wild animals. Interactions between pollutants (e.g. Fig. 6.11) are just one form of this phenomenon.

This approach has been applied to some aspects of population survival in marine habitats (Newell, 1979), rather less with pollutants, but Fig. 6.12 illustrates one example. It is sometimes suggested that an organism already stressed by its environment is more likely to be harmed by exposure to a pollutant than is a relatively unstressed individual. Unless one can measure different types of stress in the same units, this proposition is of doubtful utility.

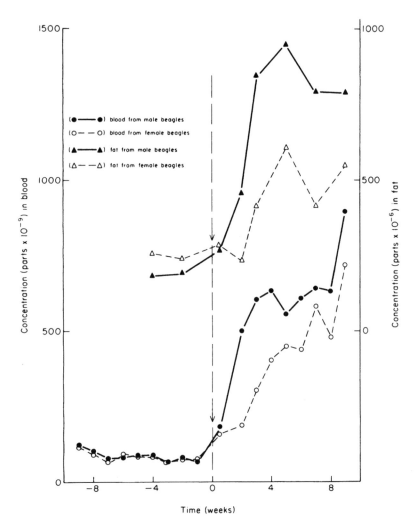

Fig. 6.11   The effect of aldrin on the concentrations of *p,p'*-DDT in beagles. Dogs were fed 12 mg *p,p'*-DDT/kg body weight on 5 days/week for 10 months, when they were fed, in addition to DDT, 0.3 mg aldrin/kg body weight on 5 days/week. Each concentration of DDT in blood or fat is the mean from four dogs. ↓ indicates the time when aldrin was first included in the diet. Concentrations of DDT in control dogs, which did not ingest aldrin, remained relatively constant during this period. (Data from Deichmann *et al.*, 1971.)

Much of ecotoxicology considers chemicals one at a time, although in practice of course contaminants rarely if ever occur in the environment without any other contaminants. This is particularly true of effluents and leachates. Does exposure to mixtures of chemicals produce greater, or lesser, effects than would be anticipated from exposures to the individual compounds?

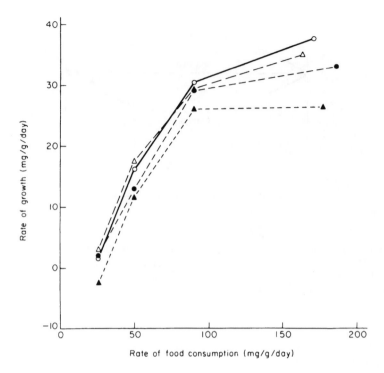

Fig. 6.12   Food consumption and growth of juvenile chinook salmon (*Oncorhynchus tshawytscha*) with different exposures to effluent from a pulp and paper mill. Such effluents will act at least in part by their tendency to deoxygenate the water as organic compounds are oxidized. There may also be direct effects on the fish, from a range of unknown compounds. (O—O) Control; (△--△) exposure to 0.25% v/v of effluent; (●--●) exposure to 1.0% v/v of effluent; (▲—▲) exposure to 1.5% v/v of effluent. (From Warren and Liss, 1977; original data from Tokar, 1968—reference not given.)

Acute toxicities of mixtures of compounds in aquatic habitats are sometimes assessed on the assumption that the toxicities of the individual compounds can be expressed in terms of a standard "toxic unit", when the concentration of each compound is expressed as a proportion of that required to produce a standard response such as the 48-h LC_{50} (Brown, 1968; Sprague, 1970). These "toxic units" are taken to be additive (e.g. Fig. 7.5).

In fact of course mixtures may be more or less than additive in their biological effects (Fig. 6.13) (e.g. Busvine, 1971; Runeckles, 1984), but it is important to distinguish between what might be called the biological and the statistical approach to this question. Published work on air pollutants has tended to adopt the statistical approach. In the simplest situation, experimental plants are allocated to four exposures: air, air with a specified level of either of two pollutants, or the two pollutants together. If the response to the combined pollutants is significantly greater, or less, than the sum of the responses to the

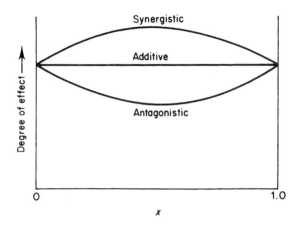

Fig. 6.13 Scheme to show three possible types of response to a mixture of two chemicals when *x* is the proportion of the first chemical and (1.0 − *x*) is the proportion of the second chemical present in the mixture. Proportions are based on fixed amounts of the two chemicals. (Adapted from Loewe, 1928.)

individual pollutants, there is an interaction between the pollutants (e.g. Table 6.8). This statistically significant interaction indicates that the degree of effect by either pollutant depends on the presence or absence of the other pollutant. Sometimes it is then deduced that a positive interaction indicates synergism, that the pollutants affect or potentiate each other's action (Tingey and Reinert, 1975). This is necessarily true only if the plants' response to either pollutant alone is linearly related to the degree of exposure. Similar arguments apply to the converse situation, for a negative interaction, of antagonism.

Synergism is more readily detected by asking whether the effect of exposure to a mixture of two pollutants is significantly greater than the effect of exposure to double the amount of either pollutant alone. Then one may logically infer that the pollutants are influencing each other's biological activity. The relationship is sometimes one-sided: the efficacy of some insecticides can be increased by a synergist, a compound which has little or no effect by itself.

Table 6.8 Interaction between sulphur dioxide and hydrogen chloride in their effect on the rate of photosynthesis, measured as a percentage of the control value, in spinach plants (*Spinacia oleracea*) after 53 hours' exposure

	Exposure to SO_2 (mg/m³ air)	
	0	0.9
Exposure to HCl (mg/m³ air)		
0	100	89
1.35	82	50

Adapted from Guderian (1977).

As we shall see in the next chapter, our ability to predict the effects of chemicals is rudimentary, so it is inevitable that our ability to predict the effects of mixtures is very poorly developed. As a result of these difficulties the extensive literature on the effects of mixtures of chemicals tends to emphasize theoretical models and methodological pitfalls (see Vouk *et al.*, 1987; Christensen and Chen, 1989). Most of this work relates to effects on individual organisms (see Yang, 1994b). Further complications, and even more uncertainty, prevail when the effects of mixtures on populations and communities are discussed.

*

We referred at the very start of this chapter to the need for precise quantitative assessment of the amount of a pollutant. Exposure is normally measured as the concentration of pollutant in the ambient air, soil, food or water, with some indication of the duration of exposure as well, but is of limited value, in part because the rate of intake depends not only on the amount in the environment but also on the efficiency of intake (Norstrom *et al.*, 1976; Opperhuizen and Schrap, 1987). Moreover, it is not easy to measure exposure accurately. Jameson and Walters (1984) discuss the difficulties of measuring exposure from food.

Comparative studies on the effects of air pollutants on plants have been hampered by the lack of data on amounts taken in (Runeckles, 1974), and a long-standing discrepancy between experimental results obtained by different authors for sulphur dioxide has been ascribed to differences in the ways of measuring exposures (Unsworth and Mansfield, 1980), although other factors may also be important (Cowling *et al.*, 1981). Finally, we need more detailed understanding of pollutant intake and distribution if we are to relate degree of exposure to degree of effect, a fact implied by the cockroach and housefly data (Tables 6.3 and 6.4). Here, too, may lie the explanation for much of the interaction between exposure and other environmental variables that determines the biological response: intake, metabolism and/or excretion of pollutant may change with environment to yield different amounts of pollutant retained within the body (Table 6.9).

Many of the ideas used for the interpretation of amounts of pollutant within organisms derive from pharmacology. Dose measures the amount of pollutant which an organism receives, commonly for animals by ingestion, injection or topical application, and is measured in units of mass, while dosages express doses as rates or ratios, such as mg/kg body weight/day (e.g. Fig. 6.8). Frequently only the exposure is known (e.g. Table 6.2).

The biologically important measure is the amount of pollutant at the site of action, and it is assumed for drugs and for some pollutants that there is a direct though not necessarily linear relationship between the degree of effect and the amount present at this site (Hodgson *et al.*, 1991). For other sites this relationship may be hysteresial (Colburn and Eldon, 1994). However, the effect of many pollutants is irreversible and the response may also be slow. The effective target dose that determines the extent of the lesion is then best defined as the integral over time at the site of action of the amount of available pollutant (Foulkes, 1989).

Table 6.9 The retention of three organochlorine compounds, labelled with ^{14}C, by groups of ten mosquito fish (*Gambusia affinis*), after 84 h of exposure in static tests with waters of two different salinities, which were both within the range of tolerance for this species

Compound	Amount in water at beginning (ng/l)	Salinity of water (%)	Weight of fish (mg)[a]	% of ^{14}C within fish[a]	Probability (P) that the observed difference was due to chance
p,p'-DDT	41	0.15	88±29	37.1±6.7	0.001
		15.0	97±35	21.6±4.3	
p,p'-DDE	32	0.15	151±36	51.0±8.6	0.015
		15.0	149±39	42.3±6.1	
p,p'-DDD	55	0.15	89±15	38.5±6.7	0.060
		15.0	97±23	32.8±4.9	

From Murphy (1970).
[a] Results are given as mean values ±SD.

Many studies have been made of doses and their subsequent distribution within the organism, but although an enormous amount of data has been published on results from chemical analyses of plants and animals, much is of doubtful scientific value. Although the details of chemical analyses for pollutants can be complicated and difficult, it is too easy to collect many biological specimens and to have them analysed for a range of pollutants. The precise questions to be answered need to be clearly formulated at the beginning (Holden, 1975).

Some pollutants, such as ozone, are so transitory that it is virtually impossible to detect them within tissues—one can only detect the effects. For most others, too, it is in practice impossible to determine amounts present at the site of action. However, for more persistent pollutants, the amounts present within the body can give some indication of both past exposure and the likelihood of biological effects. Gydesen (1984) reviews the types of mathematical model that have been used to describe the distribution of pollutants within organisms and ecosystems.

Data from laboratory experiments have usually been analysed by compartmental models (Robinson and Roberts, 1968; Task Group on Metal Accumulation, 1973; Moriarty, 1975b; Friberg *et al.*, 1979), which have been developed extensively in animal physiology and pharmacology (Atkins, 1969; Godfrey, 1983). A compartment is defined, strictly, as a mass of pollutant that has uniform kinetics of transformation and transport, and whose kinetics are different from those of all other compartments. A whole animal is envisaged as consisting entirely of compartments, which are linked in a mammillary system: the peripheral compartments are all linked to the central compartment, but not with each other (Fig. 6.14). A similar approach is beginning to develop in plant studies, too (Fig. 6.15). Equations derived from

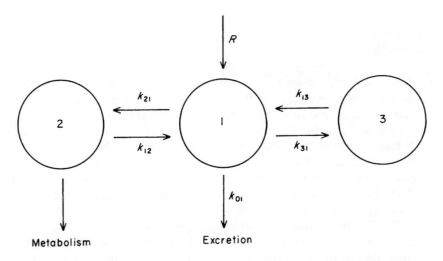

Fig. 6.14 A three-compartment model for the distribution of a pollutant within a vertebrate. Pollutant is absorbed into the blood (compartment 1) at a rate R, most metabolism occurs in the liver (compartment 2), and rates of transfer between compartments are indicated by the rate constants (k).

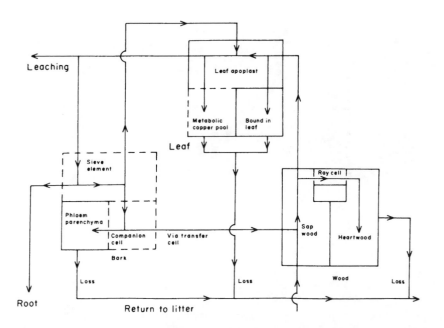

Fig. 6.15 A compartmental model for the distribution of copper in plants. (From Lepp, 1979.)

these models for the amount of pollutant within a compartment, during or after exposure, consist of a series of exponential terms (Figs 6.16 and 6.17). Several important assumptions are commonly made with these models:

(1) The central compartment, compartment 1, always represents the blood, haemolymph, extracellular fluid or transport system. This is the only compartment that receives pollutant from, or returns it to, the exterior. This makes reasonable physiological sense. A fish, for example, is likely to take up (at a rate, R) most of any pollutant into its bloodstream either through its gills or from food in its gut. Excretion, via the kidney (indicated by k_{01}), occurs from the blood. Compartment 2 might represent the liver, the principal site of metabolism. Other tissues of the body may be represented by additional compartments.

On the other hand, from what is known about the distribution of pollutants within organisms, compartments would be much more realistic if they represented cellular components. For example, most of the organochlorine insecticide in mammalian blood is likely to be associated with haemoglobin in the blood cells and with soluble proteins in the serum, much of the cadmium and of some other heavy metals in mammalian liver and kidney is associated with the protein metallothionein, and the kinetics of pollutants associated with different cellular components will be different.

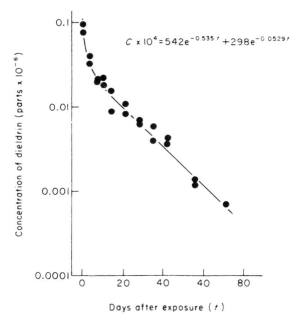

$$C \times 10^4 = 542e^{-0.535\,t} + 298e^{-0.0529\,t}$$

Fig. 6.16   Decrease in the concentration (C) of dieldrin in rats' blood during the first 71 days after exposure. Data fitted to an equation with two exponential terms. (From Robinson *et al.*, 1969.)

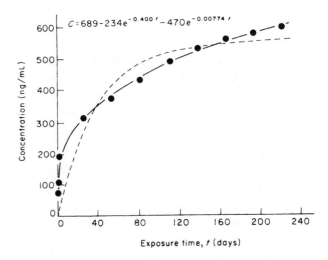

Fig. 6.17 Increase in the concentration (*C*) of dieldrin in sheeps' blood while ingesting 2 mg dieldrin/kg body weight/day. (—) Line derived from equation with two exponential terms; (– –) line derived from equation with one exponential term. (Data from Davison, 1970.) (From Moriarty, 1975b.)

The practical justification, at the present stage of development, for these simple compartments is two-fold. First, they have proved a very useful way of expressing experimental results (Figs 6.16 and 6.17), and secondly, discrete tissues and organs are relatively easy to analyse chemically.

(2) The model assumes that the rate at which pollutant leaves any compartment is directly proportional to the amount of pollutant in that compartment. The rate constants (fraction of pollutant transferred per unit of time) are indicated by *k*, with suitable subscripts.

(3) Even if there is reason to suppose that there are many compartments and there are analytical data with which to test this supposition, in practice, the residual variation in even very good data often makes it difficult to estimate parameters on retention or loss for a model with more than two or three compartments, and, at least for drug studies, absorption, metabolism and excretion are often represented adequately by two compartments (Riegelman *et al.*, 1968). Marcus (1982) points out though that accurate estimation of parameters requires simultaneous measurements of changes in pollutant content in all compartments, which entailed, in his particular example, metal content in nine tissues and organs, subdivided into thirteen compartments.

(4) The model assumes that the rate constants (see Fig. 6.14) are indeed constant in value. Stochastic models assume that the values of *k* fluctuate, although the mathematical solution then becomes more complex and one needs to know, or assume, the nature of these fluctuations (Godfrey,

1983). This is not just an academic point. Given one compartment only, where the value of k_{01} fluctuates randomly, a stochastic analysis can result in equations with more than one exponential term (Matis and Tolley, 1979), whereas the compartmental model would yield only one exponential term (see Moriarty, 1975b). Tiwari (1979) concludes that compartmental models may be appropriate when the parameters have "relatively small" variances. We know very little at present about the stability of rate constants.

Even when compartmental models are appropriate, individual animals may have different rate constants, and ideally one needs therefore to measure changes with time of pollutant burden for individual organisms. In practice of course animals usually have to be killed before their content of pollutant can be measured, and the data then bias the estimates of the mean values of the exponential constants for the experimental population (Marcus, 1983).

(5) The model defines compartments in terms of mass of pollutant, when the volume of tissue, in which that mass occurs, is not directly relevant. Physiological and pharmacological experiments usually have short time-scales, measured commonly in hours or days, when the volumes of tissues that contain the different compartments may not vary significantly, but pollution studies commonly last for weeks or months, when volumes may and often do change appreciably, commonly from growth or reproduction. Analytical results for pollution studies are usually expressed as concentrations of pollutant, not mass: a change in concentration may indicate movement of pollutant into or out of that compartment, but it may also indicate decrease or increase in size of the relevant tissue, which can then have repercussions on concentrations in other tissues.

The advantages of mass over concentration were illustrated by some data on eggs of the herring gull (*Larus argentatus*) (Peakall and Gilman, 1979). Between the 7th and 28th days of incubation, the lipid content of the eggs decreased from 9.0 to 4.1%, the proportion of water stayed almost constant at about 70%, and the total wet weight decreased by 18.7%. The mass of a totally persistent pollutant would, by definition, remain constant during incubation, but the concentration of pollutant would increase, by 23% if based on wet weight, 18% for dry weight, and 171% for lipid weight.

There are at least two reasons why results are expressed as concentrations. It can facilitate comparisons of results from individuals of different sizes, but this is only valid if the body burden of pollutant is directly proportional to body weight. Often body burden is not directly proportional, when it is often more appropriate to use analysis of covariance for comparisons of different groups of animals (Moriarty *et al.*, 1984). Secondly, concentration can also be a measure of degree of biological effect, when a variety of units may be used: fresh weight, dry weight, lipid or protein content. It is a matter of judgement and opinion which unit is the most appropriate.

Quite apart from normal growth, there may well be other, often seasonal, changes of volume. The highest concentrations of the lipophilic organochlorine insecticides occur in the fat reserves. Breeding activity (Deichmann *et al.*, 1972; Anderson and Hickey, 1976) and hibernation (Jefferies, 1972) can reduce these reserves, and so alter tissue concentrations and distribution. Concentrations of copper fluctuate seasonally in most tissues of the marine gastropod *Busycon canaliculatum* (Betzer and Pilson, 1974), and similar fluctuations can occur with liver protein and heavy metals in the starling, *Sturnus vulgaris* (Osborn, 1979). Phillips (1980) reviews, for aquatic organisms, the whole topic of seasonal fluctuations in tissue concentrations of organochlorine compounds and trace metals (see also Phillips and Rainbow, 1993). The evidence shows that increase in mass of pollutant, and decrease of volume in which that mass is contained, both increase the rate at which pollutant tends to leave that compartment. For this reason the use of concentration rather than mass is perhaps to be preferred, but unless changes or constancy of compartmental volumes are also known, it then becomes impossible to deduce the rate at which concentration in one compartment will rise as a result of intake from another compartment. In instances where there are appreciable changes in compartmental volume, it then becomes a question of whether these equations have any theoretical basis, but are justified simply by their ability to fit a line to the data. If all that is required is a good fit of the data to a line, other equations will probably serve just as well (Fig. 6.18).

Compartmental models do have relatively limited applications when biological effects are only related to steady-state concentrations. Pharmacologists have long recognized the need to integrate such studies on dose and distribution within the organism (pharmacokinetics) with studies on dose and degree of effect (pharmacodynamics) (Benet, 1994). One promising approach for drug studies has been to include a theoretical effect compartment (Fig. 6.19a): effects are assumed to be directly proportional to the amount of drug in this compartment. The rate constant k_{el} is set at a much smaller value than any of the other rate constants, so that the mass transferred to the effect compartment is negligible and cannot materially influence the concentration in other compartments (Sheiner *et al.*, 1979). Return of drug from the effect compartment can therefore also be ignored, and the rate constant k_{oe} therefore determines the temporal relationship between the degree of effect and the amount of drug in the central compartment (Fig. 6.20). Sometimes it is more appropriate to link the effect compartment to one of the peripheral compartments (Fig. 6.19b) (Colburn and Eldon, 1994). Sublethal effects from pollutants are often irreversible, but when the degree of response depends on the degree of exposure this could still prove to be a useful approach for pollution studies.

*

Laboratory studies have demonstrated several points of general significance. Figure 6.16 shows as an example the loss of dieldrin from blood (after cessation of intake), described by the equation for a model with two compartments. The

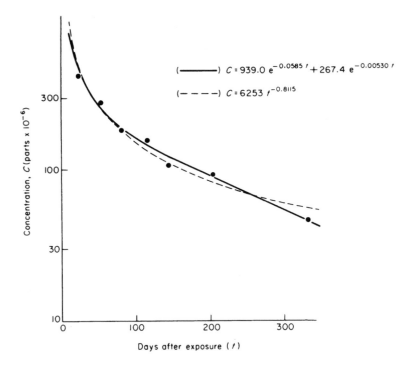

Fig. 6.18 Decrease in the concentration of p,p'-DDT in steers' omental fat after exposure. Two lines, derived from the compartmental model (—) and from a power function (– –), have been fitted to the data. (Data from McCully et al., 1966.)

two exponential constants (0.535 and 0.0529) are derived, theoretically, from the three rate constants for the two compartments. In practice, of course, they are estimated directly by fitting an equation to the data with a least squares procedure. In practical terms these data show that, for the first 71 days after exposure, the dieldrin in rats' blood could be considered in two parts. The part that is lost more rapidly has a "half-life" of 1.3 days ($\log_e 2/0.535$), whereas the part lost more slowly has a "half-life" of 13.1 days. The exponential coefficients indicate that, when exposure stopped, the rapid component was almost twice as large as the slow component (542:298, or about 9:5). The higher this ratio, the more rapidly most of the dieldrin disappears. It should be noted that, unless the simplest one-compartment model is used, it is meaningless to talk of a pollutant's half-life in organisms—the proportion lost in unit time is not constant.

Similar equations can be derived for the amounts of pollutant found within a compartment during chronic exposure (Fig. 6.17). This example shows clearly that the equation for a one-compartment model gives a poor fit to the data for amounts of dieldrin in the blood, whereas a two-compartment model gives a good fit. The compartmental model assumes that the rate constants for transfer of pollutant between compartments have the same values during and

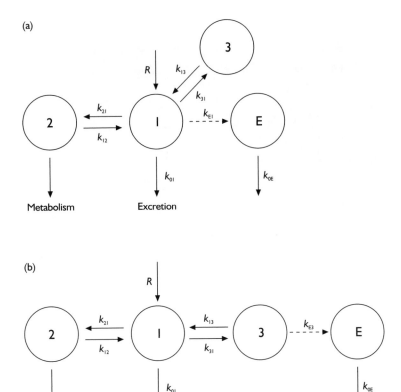

Fig. 6.19 A three-compartment model (see Fig. 6.14) with an additional hypothetical effect compartment (E) to which pollutant is transferred from either (a) the central compartment or (b) a peripheral compartment.

after exposure. There is, at present, little experimental evidence either way on this point, although, if true, there are probably two compartments for dieldrin in the blood (Moriarty, 1984). If blood is taken to comprise only one compartment, then excretion from the blood occurs about 300 times faster during than after exposure. The equation in Fig. 6.17 implies that, if exposure continues for long enough, then a steady state will be reached, in which the amounts of pollutant within each compartment stay constant. At this stage the rate of intake equals the rate of loss. There is considerable evidence to support this, although in some of the few experiments where exposure has continued for long enough, the steady state, apparently reached, does not continue indefinitely and amounts have increased again (Moriarty, 1974). This may result from some change in metabolism, due possibly to the presence of the pollutant or to some false assumption in the model, but at present we have no information on possible explanations. In theory, also, the amount of pollutant

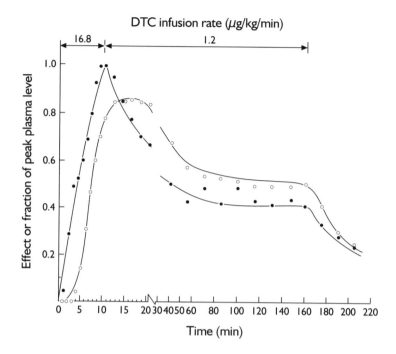

Fig. 6.20   Charges in concentration of *d*-tubocurarine (●) in the plasma (central compartment) and of degree of paralysis (○) in a human patient during and after intravenous infusion of the drug. Rates of infusion were 16.8 μg/kg body weight/min for the first 10 min and 1.2 μg/kg body weight/min for the next 150 min. (From Sheiner *et al.*, 1979.)

in a compartment, once the steady state has been reached, should be proportional to the degree of exposure. This does seem to be true as a first approximation in some instances, although the proportion retained can decrease as exposure increases (Fig. 6.21).

The purpose of laboratory experiments on exposure, residues and effects must be to help explain or predict field situations. We will discuss this in more detail in the next two chapters, but two points can usefully be emphasized now.

Species can differ appreciably in the way they take in, accumulate, distribute, and get rid of pollutants. Species differences are, of course, the major difficulty in estimating the toxicity of chemicals to man (Menzel, 1979). Table 6.10 indicates how animals differ in their retention of methyl mercury—the presentation is over-simple, without reference to specific tissues or changes with time in the rate of loss, but it does indicate the great differences that can occur.

It should also be anticipated, for genetic and other reasons, that there may be distinct differences between groups of individuals within populations, in

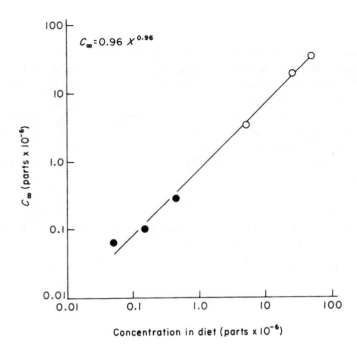

Fig. 6.21 Linear regression for the steady-state concentration (C_∞) of p,p'-DDT in eggs of white leghorn hens on the concentration of p,p'-DDT in the diet (X). Both scales are logarithmic. (●) Values calculated from data of Cummings et al. (1966); (○) values calculated from data of Cecil et al. (1972). (From Moriarty, 1975b.)

Table 6.10 The "half-life" for methyl mercury in a range of animals, not all identified to species

Animal	Approximate "half-life" (days)
Mouse	8
Rat	16
Squirrel monkey	65
Man	70
Seal	500
Poultry	25
Molluscs	700
Crab	400
Pike	700
Flounder	1000
Eel	1000

From Clarkson (1972).

the way they distribute and get rid of pollutants. In 1971 and 1972, thousands of people in Iraq suffered from poisoning by methyl mercury after eating treated cereal seed. Al-Shahristani and Shihab (1974) estimated the speed with which individuals lost methyl mercury from the body by analysing successive 1 cm lengths of hair (Fig. 6.22a), from which "half-lives" were estimated. Forty-eight individuals were examined and, as might be expected, there was considerable variation between individuals in their "half-lives" for methyl mercury. The significant point was that the distribution was bimodal (Fig. 6.22b), with 42 individuals within the range 37–93 days, five others within the range 117–120 days, and one value of 189 days. Al-Shahristani and Shihab suggested that these last six individuals may have had livers that excreted methyl mercury more slowly. This is plausible, and possible explanations include either a malfunction or a genetic difference. Whatever the explanation, other things being equal, such individuals are more at risk from poisoning by methyl mercury than are the rest of the population.

It is possible to argue that, for practical purposes, all we need is to analyse one tissue or organ for the pollutant of interest, and that the theory of compartmental models is irrelevant to the practical problems of pollution. I would suggest two counter-arguments. There is the general argument for models, that they help us to marshal the data and to formulate our ideas. More specifically, by estimating rate constants or parameters derived from them, it should enable us to measure the differences between species and individuals in the way they take in, accumulate, distribute, and get rid of

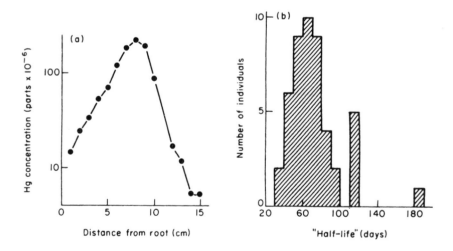

Fig. 6.22 Methyl mercury concentrations in hairs of people who ingested methyl mercury in Iraq, 1971–1972. (a) The variation in concentration along one hair, after exposure. The "half-life" can be calculated if one knows when exposure started, and assumes that the hair has grown at a uniform rate since then. (b) The frequency-histogram for the "half-life" of methyl mercury in 48 individuals. (From Al-Shahristani and Shihab, 1974.)

pollutants. It should also help us assess the more intractable problems of interactions between pollutants and the effects of other variables. Some recent work illustrates these points.

Loss of lipophilic compounds such as dieldrin commonly depends on pre-liminary conversion, by enzymes (mixed function oxidases) in the liver, to more water-soluble compounds, which are then rapidly excreted (Walker, 1975). One might then expect that the speed of these enzyme reactions would influence the persistence of such compounds within the body, and there is some experimental evidence to support this suggestion for dieldrin (Chipman and Walker, 1979). Walker (1978) estimated the activity of one group of these enzymes, microsomal mono-oxygenases, from the published literature, and found that enzyme activity correlated with two features: the taxonomic group (fish, bird or mammal) to which the animal belonged (see also Nebert and Gonzales, 1987), and the body size—the heavier the animal, the smaller the mass of organic compound that is metabolized per unit of time and of body weight. Subsequent work by Ronis and Walker (1989) sup-ports this relationship (Fig. 6.23). One would then expect the "half-life" to increase with body size. Such does appear to be the case (Fig. 6.24), although this apparent relationship must be regarded as tentative: Walker discusses the possible errors involved in his estimates of enzyme activity, we have

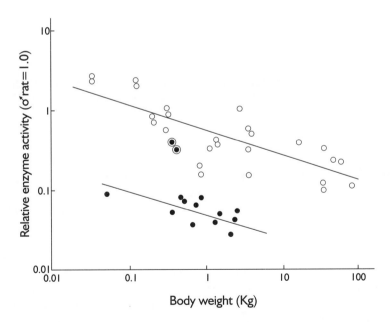

Fig. 6.23 Linear regressions of relative mono-oxygenase activity in liver on body weight. Data are for mammals (○) and fish-eating birds (●). Enzyme activity expressed as a rate per unit of body weight. The values for male and female puffins (◉) (*Fratercula arctica*) were excluded from the regression for birds. (Adapted from Walker, 1992; original data from Ronis and Walker, 1989.)

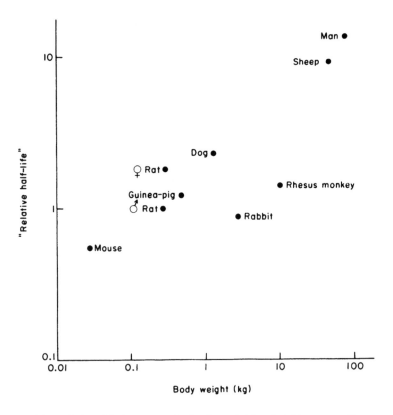

Fig. 6.24 The relationship between the body weights of eight mammalian species, and the mean "half-lives" of up to four drugs in plasma (or occasionally in body fat) compared with the "half-life" in the male rat, which was assigned a "half-life" of one. (Data from Walker, 1978.)

already discussed the oversimplication of "half-life", and the whole hypothesis rests on a limited range of compounds and species.

There are, not surprisingly, many other variables, such as the rates of blood flow to, and uptake of substrate at, the site of metabolism, that can also affect the correlation between *in vitro* metabolic activity and the rate of loss *in vivo* (Walker, 1981). All the same, one can usefully develop these ideas a little further. If the compartmental model is a reasonable representation of reality, then animals that lose pollutants relatively quickly after exposure, would also reach a steady state during exposure relatively quickly. Conversely, heavier species would take longer to reach a steady state. This may explain the seemingly anomalous situation with man, who shows no signs of reaching a steady state after 2 years' exposure to either dieldrin or DDT (Moriarty, 1975c). This is in marked contrast to other, lighter, species for which we have data, which approach steady-state concentrations much quicker (Moriarty, 1975b).

In summary, the effects of pollutants on populations are mediated via their effects, direct or indirect, on individuals, and the likelihood of these effects depends on the dose. Sublethal effects can be unravelled from knowledge of the mode of action. Alternatively, emphasis in the study of sublethal effects can be placed on the health of the individual organism. With both approaches, the effect of other environmental variables needs to be given much more prominence than heretofore, and this could profitably be linked with studies on amounts of pollutant within organisms. It is from this basis that we have to consider how best to predict and to monitor the ecological effects of potential pollutants.

7 Prediction of Ecological Effects

One has to accept that, at the present stage of knowledge, it is impossible always to predict with complete confidence the ecological effects of a new chemical. Indeed, it is an open question whether predictions ever could attain complete certainty. There lies a large range of uncertainty between those compounds that are likely to be safe or to be unsafe. To summarize some of the salient points in earlier chapters, there are many reasons for this:

(1) Often more than one compound is involved. Either the original product is a mixture of compounds, or there are biologically active impurities, or metabolites or breakdown products have biological effects. Often there is also the possibility that two compounds may interact in their biological effects.

(2) Distribution of a chemical in the environment, which affects exposure, is not uniform, some pollutants may occur in more than one form, and it can be difficult to determine the precise details of exposure.

(3) The relationships among exposure, amount of pollutant within organisms, and effects on the individual organism are complex.

(4) Species, and different groups within species, react differently to the same exposure, for both genetic and environmental reasons.

(5) The consequences of interactions between individuals within a population, and between species within a community, are also complex, and in many instances are little understood.

The need to predict ecological effects has intensified with the introduction of increasingly widespread and stringent rules and regulations for the control of pollution. One should always remember that both scientific and value judgements are involved, and Cairns et al. (1978) distinguished carefully between the risk, defined as the scientific judgement of the probability of harm, and the safety, defined as the value judgement of the acceptability of the risk, when evaluating a new chemical. Hazard may then be defined as the situation in which harm could occur, although these and related terms are sometimes used with other, conflicting, meanings (see Whyte and Burton, 1980; Royal Society, 1983).

The aim with any new chemical must be to continue questions and tests until the answer becomes that there obviously is or is not likely to be a significant effect, but it is conceivable that for some compounds no amount of testing will produce an unambiguous prediction. For example, some chemicals that experience suggests can be used without harm would probably not now be permitted as new chemicals. Conversely, there have been unexpected ecological effects with some chemicals. The decision, then, either to ban the chemical or to continue, possibly with careful monitoring or concurrent testing, will rest on the precise circumstances. It is important to realize though that routine testing is inadequate: as a Royal Society Study Group (1978) put it, toxicology "... is swamped by routine tests of limited value and governed by regulations rather than rational thought".

Prediction of biological effects can involve three questions:

(1) What is the relationship between exposure and the amount of pollutant within organisms? It is sometimes useful to distinguish, for animals, between intake of pollutant from food and directly from the abiotic environment, and three terms are commonly, but not universally, used to make these distinctions: bioconcentration, the increase of pollutant concentration from water when passing directly into aquatic species; bioaccumulation, which has a similar meaning, but indicates the combined intake from food as well as from water; biomagnification, which indicates the increase in concentration of pollutant in animal tissue in successive members of a food chain.

(2) What are the biological effects of these amounts of pollutant on individual organisms? Death is still the most commonly used measure of effect. When sublethal effects are considered their significance for population dynamics is often assessed by their effects on longevity and reproduction (e.g. Waldichuk, 1979). In one sense this must be valid, but the emphasis is too simple for the design of relevant laboratory tests. Thus melanism affects the survival of *Biston betularia* in the field, but results of laboratory tests for longevity are unlikely to differ for the melanic and normal forms. However, if one assumes that a population is as well adapted as it can be to its environment, then any sublethal effect that alters an organism's responses to its environment may be expected to reduce its survival ability, and so will be detrimental to the population. By this measure enzyme induction (the production of enzyme in response to the occurrence of substrate) is, within normal physiological limits, not a deleterious effect of a pollutant, but a non-adaptive change of behaviour is.

(3) What are the effects on ecosystems? It is necessary to remember that effects may be mediated not only by direct toxic effects on individuals, but also indirectly by altering the abiotic environment. Hynes (1960)

discusses several examples for fresh water, such as deoxygenation and the addition of suspended solids.

Korte (1977) emphasized that the risk of pollution depends not only on a chemical's toxicity, but also on its production, use and dispersion. He therefore suggested six questions whose answers, taken together, would enable us to assess possible risks of pollution from new synthetic compounds:

(1) Estimated amount to be produced?
(2) Intended uses?
    The answers to these first two questions indicate the amount that would be released into the environment.
(3) Tendency to disperse in the environment?
(4) Persistence ⎱ in both biotic and abiotic conditions?
(5) Conversion ⎰
(6) Ecotoxicological consequences: effects on individual organisms and hence on ecosystems?

In brief, we are concerned with the amounts of a pollutant, its physicochemical characteristics and consequent fate in the environment, and its ecological effects.

Korte *et al.* (1978) then developed the idea of a "profile analysis", in which the properties of the new compound are compared with the properties of a known similar compound for which more environmental data are available. This comparison is not exhaustive, but gives a preliminary assessment—commonly called an evaluation—to determine which compounds should first be subject to more detailed and expensive investigation. It is based on a series of standard tests, with no attempt at ecological simulation, which has the advantage that measurements relate to relatively well-understood systems, even if these are artificial and much simpler than real ecosystems, and yet attempts also to utilize the available ecotoxicological knowledge. The tests give answers to the first five questions only. Answers to the sixth, most difficult, question rest on the comparison between the two compounds. The crucial questions, then, are: how well understood is the ecotoxicology of the known compound, and how similar is it to the new compound? An important feature of this approach is the sequential nature of the questions. The amount of information required is determined as the questions are answered. For example, a chemical of low production, no tendency to disperse and rapid breakdown to known safe products is unlikely to need any further consideration.

Profile analysis rests on comparisons with similar compounds, and supposes that compounds with similar physicochemical properties are also alike in their ecological properties. The assumption when screening new compounds is that those with the same mode of action have similar molecular properties and that the strength of their biological activities can be related quantitatively to their molecular properties (Nirmalakhandan and Speece, 1988). Conversely, compounds with similar molecular properties also have similar biological activities. There is an unavoidable tension between this need for tests that are reproducible, sensitive, precise and as simple as possible and the need for ecological relevance. Calow (1993b) suggests that sometimes reproducibility should take precedence over relevance: otherwise the scientific and legal credibility of regulatory tests will be undermined. As we will see, this could then lead, at least in theory, to decisions based on irrelevant but accurate precise data.

Verhaar *et al.* (1992) used chemical structure to group many, but not all, organic compounds into four classes, which also differ in their biological activity:

(1) Inert. Non-ionized compounds with no specific mode of action that exert a narcotic effect, the so-called "baseline" or minimum toxicity. No other compounds are thought to be less toxic.
(2) Less inert. Not reactive under normal physiological conditions, but slightly more toxic than class 1.
(3) Reactive. No specific mode of action, but react chemically with organisms by, for example, an aldehyde group.
(4) Specific mode of action.

For a given degree of biological effect, concentrations of class 2 compounds are said to be 5–10 times lower than for class 1, and for classes 3 and 4 concentrations are 10–10^4 times lower.

Early studies on one group of class 1 compounds, the general anaesthetics and hypnotics, found that the partition coefficient correlated with biological activity. These compounds are usually thought to act not by virtue of their particular structure and interaction with a receptor, but by being present within cells in sufficiently high concentrations (Albert, 1985; Elliott and Haydon, 1986), when they occlude space within the cells and disrupt normal metabolic sequences (Crisp *et al.*, 1967). Lipnick (1989) reviews early studies, which started with Cros in France in 1863, and the lipid theory of narcosis, proposed independently by Meyer and by Overton in 1899 and 1901, declared that the depressant effect of these compounds increases with the value of the partition coefficient between a lipid and water. Ferguson (1939) made results more consistent by expressing amounts of narcotic not as molar concentrations but by an index of thermodynamic potential, and suggested that "chemical constitution" accounted for the residual variation between the efficacy of different physical toxicants.

Hansch *et al.* (1968) then showed for several structurally unrelated series of hypnotics that

$$\log(1/C) = -k(\log P)^2 + k' \log P + k''$$

and that when

$$\frac{\mathrm{dlog}(1/C)}{\mathrm{dlog}P} = 0, \text{ then } \log P \simeq 2$$

where C is the molar concentration of applied drug that produces a standard biological response, P is the n-octanol:water partition coefficient (often designated as K_{ow}), and the constants are estimated by a least-squares procedure (Fig. 7.1). In other words, the more readily a hypnotic leaves water for lipids, the greater its depressant activity, except for very hydrophobic compounds, whose activity decreases again. Very hydrophobic compounds are usually said to become trapped in external lipids and are therefore less able to enter cells, although this behaviour can also be interpreted as a slower rate of transfer across membranes, and therefore more time for loss from metabolism and excretion (Dearden and Townend, 1977). Additional terms have to be added to this equation for other drugs (Houston and Wood, 1980). Partition coefficients are then seen as the key determinant of the degree of effect from a given absorbed dose, and this use initiated an influential approach to the prediction of both amounts within and effects on organisms from exposure to pollutants (Connell, 1994).

Amounts within organisms

Several attempts have been made to apply a modification of this approach to organic pollutants, particularly pesticides: the partition coefficient is taken to

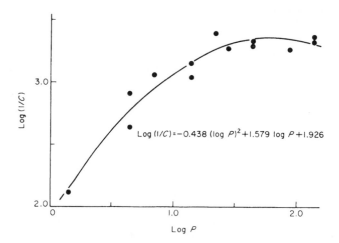

Fig. 7.1 The quadratic regression of log (1/C) on the logarithm of the n-octanol: water partition coefficient (P) for a series of barbiturates. C is the molar concentration (moles of free barbituric acid administered per unit weight of animal) injected into white mice for a standard biological response: that dose at which 50% of the mice were anaesthetized (AD_{50}), as indicated by loss of the righting reflex. (Original data from Cope and Hancock, 1939a,b; regression values from Hansch et al., 1968.)

be the principal determinant of how much of an applied dose is retained within an animal. Hamelink *et al.* (1971) concluded, from a series of experiments with DDT applied to ponds, that the concentrations of DDT and related metabolites found in the biota did not depend on transfer along a food chain. They also found that, for invertebrates in particular, there was a linear correlation between concentrations in the biota and those in water. They therefore suggested that accumulation of DDT and its metabolites depended on the phenomena of physical chemistry: the partition coefficient of the residue between water and body fat. Equilibrium would not be attained instantly, so concentrations in the biota would depend on the species, the exposure time and, most notably, on the fat content. Field data do often support the idea that fat content determines, to a considerable extent, the concentration of organic pollutants (e.g. Table 7.1), although goldfish (*Carassius auratus*) exposed in the laboratory to DDT or dieldrin showed no correlation between tissue concentrations of DDT or dieldrin and the lipid content of these tissues (Grzenda *et al.*, 1970, 1971). Moreover, concentrations of DDT (based on wet weights) and of PCBs (based on lipid weights) have sometimes been found to decrease with increasing lipid content (Earnest and Benville, 1971; Vreeland, 1974). One can plausibly argue that the partition coefficient indicates the potential of a compound for bioaccumulation, but factors such as metabolism and lack of transport across membranes may reduce this

Table 7.1 Logarithmic mean concentrations for four chlorinated hydrocarbons in three tissues from seven specimens of *Trigla lucerna* (yellow gurnard) taken from the English Channel (50° 26'N, 00° 03'E)[a]

Tissue	Contaminant			
	PCB	*p,p'*-DDE	*p,p'*-DDD	*p,p'*-DDT
Fresh weight				
Visceral adipose tissue	0.54±0.06	−0.25+0.07	−1.04±0.05	−0.40+0.06
Liver	−0.11±0.11	−1.06±0.11	−1.48±0.11[b]	−1.10±0.14[b]
Skeletal muscle	−1.28±0.06	−2.41±0.09	−2.85±0.08	−2.25±0.06
Difference between smallest and largest mean values	66-fold	145-fold	65-fold	71-fold
Lipid weight				
Visceral adipose tissue	0.64±0.06	−0.16±0.06	−0.94±0.05	−0.31±0.07
Liver	0.49±0.08	−0.47±0.10	−0.86±0.07[b]	−0.49±0.09[b]
Skeletal muscle	1.00±0.05	−0.14±0.06	−0.53±0.07	0.03±0.08
Difference between smallest and largest mean values	3.2-fold	2.1-fold	2.5-fold	3.3-fold

Values calculated from data by Ernst *et al.* (1976). From Moriarty (1977).
[a] Concentrations (μg/g±SE) are given for both fresh weight and lipid weight of tissue.
[b] Values based on six specimens.

potential. Different species can handle the same pollutant in significantly different ways and so retain different amounts.

Difficulties of comparison that arise from non-equilibrium conditions are
easily avoided by considering amounts acquired when a steady state has
been established. Neely *et al.* (1974) estimated, from steady-state concentrations, the bioconcentration of eight organic chemicals in muscle of rainbow
trout (*Salmo gairdneri*). These estimates involved some dubious assumptions,
but for the sake of argument one can accept that they estimated the ratio of
the steady-state concentrations of each chemical in the muscle to that in the
water. These were not a homologous series of chemicals, yet they found
there was a linear relationship between the logarithm of the bioconcentration and the logarithm of the *n*-octanol : water partition coefficient (Fig. 7.2).
Several subsequent studies have extended and confirmed this relationship
for other compounds and species (Davies and Dobbs, 1984; Samiullah,
1990a). Like hypnotics, a quadratic regression is more appropriate for very
hydrophobic ("superhydrophobic") compounds—those with a partition
coefficient greater than about 10^6 (Sugiura *et al.*, 1978). Various reasons have
been suggested for this deviation from a linear relationship (Connell, 1994).

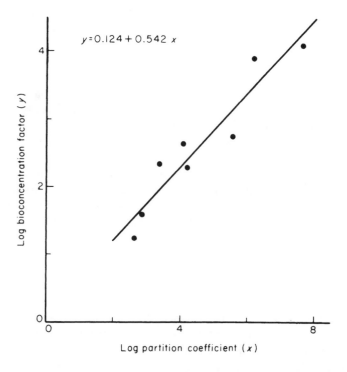

Fig. 7.2   The linear regression of the logarithm of the bioconcentration factor (*y*) on
the logarithm of the *n*-octanol: water partition coefficient (*x*) of eight organic compounds taken up from water into the muscle of rainbow trout (*Salmo gairdneri*).
(From Neely *et al.*, 1974.)

Clearly this approach gives only a first approximation (e.g. Table 7.4 and Ernst (1985)). Many modifications and refinements have been published (e.g. Galassi and Migliavacca, 1986) but, although partition coefficients can account for much of the variation between chemicals in their intake by animals, other factors can also be important. Those compounds that are metabolized may yield aberrantly low bioconcentration factors. Excluding such compounds, Zaroogian et al. (1985) concluded that partition coefficients predicted bioconcentration with an error of at least one order of magnitude for up to 30% of chemicals. Partition coefficients cannot take account of some factors. For example:

(1) No account is taken of differences between species in their rates of intake, or in distribution between compartments within the body, metabolism or excretion, unless one is prepared to ascribe any such differences to differences in lipid content. Within species, size may also be an important variable (Gobas and MacKay, 1987).

(2) No account is taken of size of dose: intake of organic pollutants by aquatic organisms is taken to be analogous to partitioning between two immiscible liquids. This is valid only if passive diffusion is the principal determinant for the rates at which a contaminant enters and leaves an organism (e.g. Connor, 1984). Terrestrial animals need to conserve water, but lipophilic compounds, by definition, do not dissolve readily in water and so need relatively large amounts of body water for their excretion. This problem is often overcome by first metabolizing lipophilic compounds to water-soluble products, so that metabolic rate can be an important determinant of excretion rate (Moriarty and Walker, 1987) but may not be proportional to the dose.

(3) No allowance is made for environmental effects on body burden.

Kenaga (1980) concluded that, in the preliminary screening of organic compounds, those that dissolved more than 25 parts $\times$ 10^{-6} in water were unlikely to be hazardous as persistent compounds in living organisms. Walker (1987) has suggested that when metabolic rate is the key determinant of body burden, in vitro tests with liver microsomes may yield useful predictions of bioaccumulation.

The rate at which equilibrium is attained may be as important as the equilibrium value itself (Leo et al., 1971) but, of more immediate relevance, it can be difficult to measure the partition coefficient of lipophilic pollutants that dissolve very little in water: the classical method is to measure the concentrations of the compound of interest when its distribution between the two solvents has reached equilibrium (de Bruijn et al., 1989). Alternative approaches include:

(1) Partition coefficients between water and n-octanol can be estimated from coefficients for water and other immiscible solvents, and the effect on the

partition coefficient of a range of substituents can also be estimated by the simple addition of appropriate constants (Hansch and Leo, 1979; Albert, 1985).

(2) Water solubility and chromatography can both be used to estimate the partition coefficient, but the error can be considerable (Esser and Moser, 1982: Brooke et al., 1986; Opperhuizen et al., 1987).

Connell (1994) discusses the accuracy of both practical and theoretical estimates of partition coefficients.

Partition coefficients are only suitable for screening tests. Ideally for any compound we would know the exposure/dose relationships (see Chapter 6) for each species of interest. The practical compromise is usually, when deemed necessary, to measure the bioconcentration factor after chronic exposure when a steady state exists. The OECD is developing standard tests for a range of species (Römbke and Moltmann, 1996).

This idea, that the degree of bioaccumulation depends primarily on a compound's physical chemistry, was challenged by views that developed from field studies on persistent organochlorine insecticides. Hunt and Bischoff (1960) found that deaths of western grebes (*Aechmophorus occidentalis*) in Clear Lake, California, had resulted from spraying the lake with DDD. They suggested, tentatively, that DDD reached the grebes by accumulation and concentration along the food chain. This involves two distinct ideas, that concentrations increase in successive trophic levels, and that the mass of pollutant passing along the food chain is conserved, so that concentrations increase. This second idea can be dismissed. Compartmental models imply that pollutants are excreted and or metabolized, and it is only a small proportion of a persistent pollutant that is retained within the body (e.g. Table 7.2). Clearly, once a steady state has been established, the proportion of pollutant that is retained during chronic exposure will steadily decrease, and before then the rate of retention will be decreasing. Hunt and Bischoff concluded that carnivorous fish accumulated more DDD than did plankton-eating fish of similar size and age, and that this difference was because of their positions in the food chain. However, these conclusions on biomagnification are of doubtful validity (Moriarty, 1972). The data showed that the concentration of DDD in muscle increased with age, and the only strictly comparable figures suggest that any difference between carnivorous and plankton-eating species is correlated with differences of fat content (Table 7.3). However, the idea that persistent pollutants concentrate and accumulate along food chains has become widely accepted, and it will repay detailed consideration.

The first important point is that comparisons should be based on compatible data. Data for organisms from lower trophic levels are often based on analysis of the whole animal, whereas specific tissues are commonly analysed from larger animals. This difference stems, at least in part, from the need for a suitable weight of sample for chemical analysis, but although

Table 7.2 The retention and concentration of dieldrin ingested by four song thrushes (*Turdus ericetorum*) as a result of being fed contaminated earthworms for 6 weeks[a]

Bird	Concentration of dieldrin in diet (parts × 10⁻⁶)	Total amount of dieldrin eaten with the earthworms (μg)	Total amount of dieldrin in body after 6 weeks (μg)	% of dieldrin from diet present in bird	Concentration of dieldrin in birds (parts × 10⁻⁶)	Concentration of dieldrin in birds as % of that in diet
A	0.15	203.6	1.54	0.8	0.02	13.3
B	0.32	436.8	7.11	1.6	0.09	28.1
C	3.06	3837.0	96.0	2.5	1.36	44.4
D	5.69	7297.0	318.0	4.4	4.03	70.8

From Jefferies and Davis (1968).
[a] All concentrations are based on wet weights.

Table 7.3 Analyses of muscle from composite samples of two species of fish, 4–6 months old, taken from Clear Lake, California

Species	Type of food	DDD in muscle (parts × 10^{-6})	Fat in muscle (%)
Sacramento blackfish (*Orthodon microlepidotus*)	Plankton eater	7–9	1.5
Largemouth bass (*Micropterus salmoides*)	Carnivore	22–25	6.0

Data from Hunt and Bischoff (1960).

comparisons between such data are sometimes made, they can be misleading. The choice of comparable tissues is not always straightforward. Reinert and Bergman (1974) found that different regions of the axial muscle from coho salmon contained different concentrations of DDT residues, although most of this variation could be attributed to different lipid contents in the different regions (Table 7.4). However, when concentrations were expressed in terms of lipid weight rather than fresh weight, there were still marked differences in concentrations between different tissues and organs, and there were also pronounced seasonal differences. Use of total body burden avoids some of these difficulties, and there is also some evidence to suggest that it is a better measure of chronic exposure than is the burden of an individual tissue (Moriarty *et al.*, 1984).

Secondly, it is important to distinguish between aquatic and terrestrial food chains. Food must be the principal source of persistent organochlorine insecticides, and many other pollutants, for most trophic levels in terrestrial systems, but aquatic species may also acquire residues direct from the abiotic environment. We can therefore deduce that, unless there is another, compensating, difference between aquatic and terrestrial food chains, concentrations will tend to increase less, or decrease more, along terrestrial food chains than along aquatic ones. Mechanisms for excretion may be relevant, given the generally greater need for terrestrial species to conserve water (see p. 180).

We will first consider aquatic food chains, both for the relative importance of food and water (or other parts of the habitat) as sources of pollutant, and to see whether concentrations increase along food chains. Reinert showed in 1967 (in a Ph.D. thesis, results of which were not published until 1972) for a simple food chain of an alga (*Scenedesmus obliquus*), a crustacean (*Daphnia magna*) and a fish (*Poecilia reticulata*), that when exposed separately to dieldrin in the water, concentrations still increased in successive members of the food chain, even though the crustacean and fish were given uncontaminated food (Table 7.5). Additional tests showed contaminated food to be unimportant as a source of dieldrin. Similar results were obtained for another simple laboratory food chain, when midge larvae and another fish species, the reticulate sculpin (*Cottus perplexus*), were exposed to 0.5 parts × 10^{-9} dieldrin in the

Table 7.4 Concentrations of DDT residues in axial muscle of coho salmon (*Oncorhynchus kisutch*) from Lake Michigan[a]

Time of sampling	Whole muscle	Loin region	Dorsal region	Tissue Medial region	Ventral region	Brain	Eggs
			Total DDT residues (parts × 10^{-6} fresh weight)				
21–8–68	14.9	5.6	61.1	41.1	66.0	1.2	7.4
18–10–68	16.3	4.7	106.1	59.8	106.4	2.4	10.2
31–1–69	18.4	10.1	59.6	41.3	68.5	5.0	7.9
			Total DDT residues (parts × 10^{-6} lipid weight)				
21–8–68	91.9	103.7	98.4	92.4	98.5	13.8	67.9
18–10–68	230.0	204.3	270.7	266.9	256.4	41.4	118.6
31–1–69	800.0	594.1	726.8	536.4	805.9	64.9	84.0

Original data from Reinert and Bergman (1974), with additional calculations from Phillips (1978).
[a] Concentrations are expressed in terms of both tissue fresh weight and lipid weight, for different parts of the axial muscle and for the brain and eggs, and for three different times of the year.

Table 7.5   Mean concentration factors for dieldrin residues in three species when exposed to dieldrin in water[a]

Species	Concentration factor
Scenedesmus obliquus (alga)	1 282
Daphnia magna (crustacean)	13 954
Poecilia reticulata (guppy)	49 307

Data from Reinert (1972).
[a] Concentration factors are based on dry weights of tissues, and are the steady-state concentrations of dieldrin in the whole organism, divided by the concentration of dieldrin in the water.

water. Fish reached steady-state concentrations within 2 weeks, attained similar concentrations whether they fed on contaminated or uncontaminated larvae, and after 3 weeks' exposure not more than 16% of the total body burden in the fish could have come from their food, even if all of the ingested dieldrin were retained (Chadwick and Brocksen, 1969).

A similar experiment and calculation with PCBs in a marine food chain showed that lipid content is not necessarily the dominant factor. Three species—an alga (*Dunaliella* sp.), a rotifer (*Brachionus plicatilis*) and newly hatched larvae of the northern anchovy (*Engraulis mordax*)—attained steady-state concentrations of PCBs within 5 days, when exposed to PCBs in both the water and, except for the alga, in their food. The anchovy larvae reached distinctly higher concentrations of PCBs, when based both on dry weights and lipid weights (Table 7.6), and would have needed 10 days to accumulate all of their PCBs from their food (rotifers). Another experiment made food even less likely as a significant source of PCBs. Two batches of anchovy eggs with known PCB concentrations were hatched and the larvae analysed either 2 or 3 days later, by which time they had absorbed their yolk sacs and would normally start to feed. The concentrations of PCBs in the eggs and the seawater differed in the two tests, but both tests gave bioconcentration factors similar to the bioaccumulation factor found for larvae which had been fed for 25 days (Table 7.7). Scura and Theilacker (1977) subscribed to the partitioning theory, but hesitated to conclude that PCBs are ten times more soluble in lipids from anchovies than from rotifers. The term lipid does cover a somewhat diverse range of compounds, characterized by their solubility in solvents such as chloroform. Little work of value appears to have been done on the significance of any differences in the relative solubility of pollutants in different lipids, either for the amounts of pollutants in different organisms or for the distribution of pollutants within organisms.

Jarvinen *et al.* (1977) made a more detailed study of the relative contributions from food and water when they exposed fathead minnows (*Pimephales promelas*) to combinations of DDT at a nominal 0.5 or 2.0 µg/l in water, and in food (five species of clam) which had reached a steady-state concentration when exposed to an average of 1.81 µg DDT/l of water (Fig. 7.3). The clams

Table 7.6 Average concentrations of PCBs, expressed as percentages of dry tissue and of lipid content of algal cells (*Dunaliella* sp.), rotifers (*Brachionus plicatilis*) and anchovy larvae (*Engraulis mordax*)

Organism	Lipid content (%)	Steady-state concentrations of PCBs (parts $\times$ 10^{-6}), based on:		Estimated concentration of PCBs in ambient seawater (for anchovy larvae) or (for other two species) in rearing medium (parts $\times$ 10^{-12} w/v)	Concentration factor ($\times$ 10^6) (based on concentration in lipid)
		Dry weights	Lipid weights		
Dunaliella sp.	6.4	0.25	3.91	8.2	0.48
Brachionus plicatilis	15.0	0.42	2.80	8.2	0.34
Engraulis mordax	7.5	2.06	27.46	2.0	13.70

From Scura and Theilacker (1977).

Table 7.7   Concentration factors for PCBs in anchovy larvae (*Engraulis mordax*), expressed as a proportion of the lipid content, compared with concentrations in the ambient seawater

	PCB concentrations			
	In seawater (parts $\times 10^{-12}$)	In eggs (dry weight)	In larvae (lipid weight) (parts $\times 10^{-6}$)	Concentration factor ($\times 10^6$)
Unfed 2-day old larvae	4.5	0.33	62.7	13.9
Unfed 3-day old larvae	2.5	0.36	37.3	14.9
Average for samples of larvae analysed every 5 days for 25 days after hatch	2.0	0.35	27.5	13.7

From Scura and Theilacker (1977).

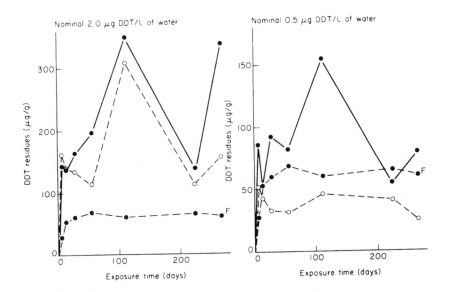

Fig. 7.3   Concentrations of *p,p'*-DDT residues (DDT, DDE plus DDD) in whole bodies of fathead minnows (*Pimephales promelas*) during exposure to DDT in food and or water. (●——●) Fish exposed to DDT in both water (nominally 2.0 or 0.5 μg/l, actually about 25% lower) and food (clams that had reached a steady-state concentration from exposure to an average of 1.81 μg DDT/l water); (○– –○) fish exposed to DDT in water only; (●– –●F) concentrations in fish exposed to DDT in food only (same results for both graphs). Residues in controls never exceeded 2.0 μg/g fish. (From Jarvinen *et al.*, 1977.)

were exposed to [^{14}C]DDT, so that the source of DDT residues (DDT, DDD and DDE combined) within the fish could be identified, and it was found that:

(1) Net intake from water occurred within 7 days, whereas net intake from food took 14–28 days.

(2) There were pronounced fluctuations in DDT residues, associated with the spawning that occurred between 112 and 224 days after the start of exposure.

(3) Where fish obtained DDT from both their food and the water, the authors calculated that about 60% of their total residues came from food when there was 0.5 μg DDT/l of water, and about 30% when there was 2.0 μg DDT/l of water. There are some difficulties in these calculations. Not only is there considerable variability in the data, but although concentrations refer to whole-body analyses, there are no data on growth rates, and all treatments killed fish, at different rates, so that results may be biased. Nevertheless, it is reasonable to conclude that a major part of the total DDT residues came direct from the water.

It is also likely that growth rate will affect the relative importance of food and water as sources of pollutant. Norstrom et al. (1976) started from the assumption that intakes of pollutants from food and water are both functions of metabolic rate, itself related to body size, and they found reasonable agreement between their predicted values and the observed concentrations of PCBs and methyl mercury in a population of yellow perch (Perca flavescens) from the Ottawa River.

The available experimental evidence does suggest that intake from food is unlikely to be the major source of residues for persistent organic pollutants in aquatic species, although this conclusion contains the critically important assumption that both prey and predator species are equally exposed to the pollutant in their environments. This may not always be true. There was a minimum of sediments and suspended particulate matter in these experiments, but it has been shown that most of the DDT applied to a marsh rapidly becomes associated with the sediments (Meeks, 1968), and often organic compounds are mostly adsorbed to particle surfaces (Crosby, 1975). Expressed in ecological terms, the important details of exposure depend on the niche, and contradictions between laboratory and field results may arise from the use, in laboratory tests, of unrealistically high concentrations in water (Harding et al., 1981). More information is needed on the actual exposures.

The alternative to the experimental approach is to consider the field evidence for increase in concentration along aquatic food chains. Field observations give conflicting accounts (Phillips, 1980). Often concentrations do increase along food chains, but not always. Concentrations of organochlorine compounds expressed in terms of fresh weight of organism increased with trophic level in samples of food chains taken from the Wadden and

North Seas, but there was little or no evidence of dependence on trophic level when concentrations were based on lipid content of organisms (ten Berge and Hillebrand, 1974). Organisms from the Atlantic Ocean showed no signs of increase with trophic level for concentrations in either fresh or lipid weights, and phytoplankton had the highest mean concentrations of PCBs (Harvey et al., 1974). The relevance of these observations can be disputed, because specimens were collected at different times from different sites, but Fowler and Elder (1978) avoided these two difficulties by sampling in the Mediterranean, 5 km off the French coast, for 2 h at depths between 0 and 100 m. Except for the lack of detectable DDE in microplankton, there was no indication of increase with trophic level (Table 7.8).

Even with careful selection of samples, it is still difficult to draw unqualified conclusions from field data. Not only is there the need for comparable tissues or organs from different species, but the assignation of species to particular trophic levels is often dubious, and whether each predator consumes prey whose residues are representative of the population is also debatable, particularly if exposures are sufficient to affect some individuals (e.g. Cooke, 1971). Moreover, there is the implicit assumption that residue levels have reached a steady state, and that immigration and emigration have little effect on the concentrations found in samples. Rosenberg (1975a, b) circumvented some of these difficulties when he applied dieldrin to a 1 ha slough. He concluded,

Table 7.8 Concentrations of organochlorine compounds in pelagic organisms collected within 2 h in November 1974, at depths of 0–100 m, from the Mediterranean, 5 km off the coast at Villefranche-sur-Mer, France

Organisms	Trophic level	Concentrations (μg/kg dry weight)		Concentration factor for PCBs[a]
		p,p'-DDE	PCBs	
Microplankton (principally copepods (small crustacea), chaetognaths (arrow worms), phytoplankton and detritus)	1	<0.5	4500	170 000
Meganyctiphanes norvegica (omnivorous macroplanktonic euphausiid—crustacean)	2	26	620	50 000
Sergestes arcticus ⎱ carnivorous	3	15	470	47 000
Pasiphaea sivado ⎰ shrimps	3	5	210	20 000
Myctophus glaciale (fish feeding principally on pelagic crustacea)	3–4	1	50	6 000
Surface water		not detected	0.0025	

From Fowler and Elder (1978).
[a] Concentration in wet weight of organism divided by concentration in water.

from both his own results and the literature that, especially for invertebrates, organochlorine insecticides do not increase particularly with trophic level: habitat and mode of life are likely to be more important determinants.

Many factors can influence pollutant availability (Moore, 1981; Luoma, 1983). Bryan (1979) concluded that although food is often the most important source of pollutants for marine species, it does not always follow that predators at higher trophic levels contain the highest pollutant concentrations, and Prosi (1979) reached similar conclusions for heavy metals in marine species. Most studies with metal radionuclides indicate a decrease in concentration along food chains, and the initial high concentration in phytoplankton, detrital material or sediments is probably an adsorption phenomenon whose mechanism has no biological significance (Preston *et al.*, 1972). In some instances of increased concentration along food chains, length of life may be significant too (Bryan, 1979; Prosi, 1979). For both organic compounds and metals, the evidence suggests that pollutants sometimes do, and sometimes do not, increase in concentration along food chains (Biddinger and Gloss, 1984).

In summary, the results suggest that biomagnification is too simple an idea for aquatic habitats: the relative importance of intake from food and direct from the abiotic environment is a key factor, and the approach implied by compartmental models seems more appropriate. Attention is then focused on the rates at which pollutants enter and leave organisms, instead of their position in the food chain. In essence this is a question of physiology and biochemistry, although ecological aspects influence the details greatly. Livingstone *et al.* (1992) give a recent account for marine ecosystems. At the other extreme, the use of *n*-octanol : water partition coefficients for organic compounds, though useful for preliminary assessment, does ignore the important differences that exist between individuals of different species.

The question, whether concentration factors—the ratio of pollutant concentration in predator to that in its prey—are greater than unity is more easily considered in terrestrial species, where most if not all of an animal's pollutant will often come from the food. Analysis of field specimens for organochlorine insecticides suggested that residues depend very much on the nature of each species' particular diet, because these compounds, though persistent and widespread, are not distributed uniformly throughout the environment (Barker, 1958; Moore and Walker, 1964; Stickel, 1973). It is now often suggested that persistent pollutants attain highest concentrations in species at the ends of food chains, by analogy with the results of Hunt and Bischoff (1960) at Clear Lake (p. 181). However, if one attempts to compare like with like, interspecific comparisons from field specimens are not easy to make (Moriarty, 1985a):

(1) Comparisons should be for similar tissues or organs (see p. 181).
(2) With species from very different taxonomic categories, such as earthworms and birds, it is questionable whether any tissues or organs are

really comparable, and there is much to be said for the use of concentrations in whole organisms. This may not be feasible for very large animals, but one can argue that it is the only valid comparison. If five tissues and organs are analysed from both predator and prey, 25 concentration factors could be calculated, and although, say, concentrations in the brain of an earthworm and of a bird may be a more valid comparison than that between concentrations in an earthworm's brain and a bird's muscle, it is still debatable how valid the comparison is: there are considerable differences between the two brains in both structure and function. Really, the question should be put the other way: what good reasons are there for comparing single tissues and organs instead of whole bodies?

(3) Concentrations are expressed in a variety of ways. Commonly, they are based on fresh or dry weights, but lipid and protein weights are also used. Again, unless there are good reasons to the contrary, fresh weight of the whole animal seems to be the most appropriate measure. There are good reasons for preferring to measure the mass of pollutant rather than its concentration (some are discussed on pp. 162–163): concentration is a secondary unit, based on mass of animal and of pollutant, and its primary function is to compensate, albeit sometimes unsuccessfully, for the differences in size of different individuals.

(4) Concentration factors are difficult to interpret unless the predator has attained a steady-state concentration of pollutant, and unless items of food have a reasonably constant concentration of pollutant.

(5) It is not easy to know what a predator's actual exposure is. Diet varies, and predators do not necessarily take a random sample of their prey (e.g. Hopkin, 1989).

(6) It is difficult to ensure that the samples taken for analysis represent the population (Hunter et al., 1987).

(7) Correlations between concentration and position in the food chain could result from movement along the food chain or be a function of animal size.

It is difficult to circumvent some of these problems, but two types of evidence can be adduced to help formulate an opinion about concentration factors.

One can argue from compartmental models. The concentration factor in a predator with a steady-state concentration and which feeds solely on one prey species, whose individuals have a constant and uniform amount of pollutant, depends in particular on the rate of food intake as a proportion of body weight, the proportion of ingested pollutant that is assimilated into the body, and the rate of metabolism plus excretion (Moriarty, 1975b). Whether or not the concentration factor will exceed unity depends on the values of these parameters, but one general point is of interest. In general, the larger an animal, the smaller the proportion of its body weight that it consumes as food per day. It follows that if successive species in a food chain assimilate similar proportions of ingested pollutant, and also excrete and metabolize similar proportions of the assimilated pollutant, then concentration factors for successive transfers of pollutant along the food chain will decrease in value. One might expect, then, that pollutant concentrations would increase in concentration in successive species of a food chain, albeit at a decreasing rate, or that species in the middle of the chain have higher concentrations than those that come before or after, or that concentrations decrease at every transfer, at an increasing rate. For purposes of prediction, the important point from this analysis is that we need information on the rate of assimilation for pollutants, and on the factors that affect it.

One rider is necessary to these conclusions about concentration factors: it may be invalid to assume that the rate constants of metabolism and excretion are independent of trophic level. Specimens of the shag (*Phalacrocorax aristotelis*) from around the Farne Islands had 5–40 times more dieldrin in their livers than did samples from seven other vertebrate species (Robinson *et al.*, 1967). Walker (1975) found that the shag also metabolized dieldrin relatively slowly, and quotes a suggestion by Wit and Snel that herbivores will have more efficient systems for metabolizing foreign compounds than will carnivores because of the nature of their diet. We have also seen (p. 169) that there is some evidence that heavier species do have longer half-lives for organic compounds.

The alternative approach to concentration factors is to consider the results from chemical analyses of field samples. For all of their imperfections, one salient feature is the considerable variation between individuals that frequently occurs (e.g. Table 8.1), with a few individuals having relatively high concentrations. Newton and Bogan (1978) studied a population of sparrow hawks (*Accipiter nisus*) in Annandale, southern Scotland, and found that although average residue levels of DDE in eggs decreased by 10% per annum during the period 1971–1974, eggs from a minority of clutches throughout that period had levels sufficient to reduce the index of shell thickness by over 20%, which would reduce breeding success. Moreover, eggs from different individuals had significantly different levels of contamination, these differences between individuals were not consistent from one year to the next, nor were there consistent trends for eggs laid in the same territory by different birds in different years. This example suggests that sparrow hawks obtain occasional relatively high exposures. One could argue that such an exposure pattern occurs because the sources of DDT (and its metabolite DDE) were fairly discrete and localized, but this mosaic, intermittent, pattern of exposure is probably fairly general (see Chapter 1). In fact, it is usually impracticable to calculate meaningful concentration factors.

I would conclude, tentatively, that for prediction the question of increase in concentration along the food chain is of secondary importance, and is far from an invariable rule. It is more important to attempt to predict a pollutant's pathways: which species, in what circumstances, are likely to be exposed to a potential pollutant. The peregrine falcon (*Falco peregrinus*) disappeared from southern England in the late 1950s, in all probability from dieldrin ingested with pigeons (Jefferies and Prestt, 1966) (see p. 256). This can be attributed to concentration along the food chain of dressed seed→ granivorous bird→predatory bird. However, many pigeons also received lethal doses. The more important point about food chains in general is that species at the ends of them are often K-species, and are relatively slow to recover when their populations have been reduced.

Effects on organisms

Ideally, one would predict both the types of effect that a potential pollutant would produce, and the exposures needed to produce them. Study of effects implies knowledge of the mode of action, and more is known about mode of action, and the consequent effects on individuals, for organophosphorus and carbamate insecticides than for any other group of potential pollutants. Even with these compounds, as for all of the pesticides currently in use, essential-ly all of the new molecules with biological activity have been discovered either by random synthesis of compounds with subsequent screening for biological effects, or as variations on a molecule with known activity (Corbett et al., 1984). Despite a number of attempts, no pesticide has yet been devel-oped by designing a molecule that will act on a specific site (Pillmoor and Foster, 1994). One may deduce, conversely, that in the foreseeable future the degree and type of biological activity of pollutants in general is unlikely to be predictable from general principles alone. Experiment and observation will be essential for well-informed opinions, let alone certainties.

Short of studying fate and effects in the proposed areas of use, the ideal is to test a compound in the field or, less realistically, in simulated ecosystems, although such data can be difficult to interpret (see pp. 203–210). Often there are only chronic or acute toxicity data, and sometimes for new compounds just molecular properties. Consequently there is considerable interest in the correlation of biological activity with molecular properties. We discussed (pp. 177–181) the use of partition coefficients as a measure of molecular proper-ties for compounds without a specific mode of action, and it has been sug-gested that, for compounds with the same specific mode of action, their toxicities (measured as the log LC_{50}) form a negative linear regression on the logarithm of the partition coefficient (see Schultz et al., 1986). Groups of com-pounds with other specific modes of action form similar linear regressions, but with different constants in the equation. The same reservations apply to these relationships as for those between bioconcentration and partition coef-ficient, but one has also in addition to consider the measure of toxicity. The LC_{50} and similar measures are crude indices of biological effect (see Chapters 2 and 6). Moreover, unless the regressions of percentage kill on concentration have similar slopes for all of the compounds being considered, then use of, say, the LC_{10} could yield a different relationship between toxicity and parti-tion coefficient.

Measures of molecular properties have become more sophisticated. Collander (1954) showed for single cells of the alga *Nitella* that their permeability to 70 non-electrolytes not only increased with the partition coefficient but also decreased as molecular weight increased. Hansch and his colleagues (Hansch *et al.*, 1963; Hansch and Fujita, 1964) then argued that, for a series of chemically related compounds, their relative degree of biological activity would depend on speed of passage to the site of action, indicated by the partition coefficient, and by the ease of attachment of molecules to the receptor. They developed the equation

$$\log (1/C) = -a\pi^2 + b\pi + c\sigma + d$$

which is rather similar to that in Fig. 7.1 except that:

(1) $\pi = \log (P_X/P_H)$, where P_X is the partition coefficient with the substituent X, and P_H is the partition coefficient for the related compound where hydrogen replaces the substituent X.
(2) σ is a measure of electron density, initiated by Hammett and now becoming rather complicated (Hansch, 1978). Again, as for the partition coefficients, the effect of electron density in a particular compound is assessed relative to that in the parent compound where hydrogen replaces the substituent of interest.

Regressions of this type between chemical structure and degree of biological activity have come to be called QSARs (quantitative structure–activity relationships): they are developed empirically (e.g. Schultz *et al.*, 1988) and are usually calculated for homologous series of chemicals (Könemann, 1986). Other aspects of molecular structure are now included in some studies, and a range of parameters of molecular structure has been proposed (Hansch, 1978; Hermens, 1989; Samiullah, 1990a; Donkin, 1994). One example, not strictly structural, should suffice to illustrate both the approach and its limitations.

Biochemical and bacterial tests are commonly used for the preliminary screening of sewage and industrial effluents for their toxicity (Kilroy and Gray, 1995), at least in part because of their speed, ease of use, reproducibility and low cost (Liu and Dutka, 1984). The marine *Photobacterium phosphoreum* is probably the most widely used of all the bacteria (Mayfield, 1993). It can emit light by the activity of the enzyme luciferase, and is used in the "Microtox" test to estimate the EC_{50} for test substances: that concentration of the test substance that reduces light emission by 50% (Bulich, 1982, 1984). This test has good reproducibility, but results can depend critically on the test conditions and they do not always agree with those of other screening tests (Luoma and Ho, 1993). However, the main question is, what relevance do EC_{50} values have to possible ecological effects?

Indorato *et al.* (1984) tackled this question by comparing the EC_{50} values of 19 organic compounds with their toxicity to the fathead minnow (*Pimephales promelas*). They assumed for the 96-h LC_{50} value that

$$LC_{50} = a(EC_{50})^b$$

where a and b are constants. By using a logarithmic transformation they could plot the linear regression of LC_{50} on EC_{50} (Fig. 7.4a), and could then calculate for any given degree of probability that EC_{50} value which would indicate an LC_{50} value greater than any specified level.

This procedure involves no more than standard statistical calculations (Sokal and Rohlf, 1995), but does therefore contain various assumptions which must be met if the significance tests are to be valid. Regression assumes that the value of the independent variable is known without error and that values of the dependent variable conform to a normal distribution about the regression line. The data appear to conform reasonably well with the latter assumption, with one compound out of nineteen falling outside the 95% confidence limits ($P \sim 0.028$).

A "trigger-point" of 10 mg/l was chosen for the LC_{50} value. This regression can then be used with new test compounds to predict, from their EC_{50} values, whether there is a probability of more than 5% that the LC_{50} value is less than 10 mg/l. Only then would further tests be needed. If one accepts the assumptions of a linear regression, then one may conclude that, at the 95% probability level (one-tailed test), a compound with an EC_{50} value of at least 1 g/l has an LC_{50} value above the "trigger-point" of 10 mg/l and requires no further testing. This EC_{50} value is much higher than the best estimate from the regression that an EC_{50} value of 8.1 mg/l indicates an LC_{50} value of 10 mg/l.

More important is the assumption that there are no "outliers", compounds that do not conform to the same, normal, distribution. This is a recurring problem with ecotoxicological data (see pp. 227–228). To put the question more specifically, does proprionitrile belong to the same population as the other 18 compounds? If it does, one would conclude from the EC_{50} value of 5.2 g/l that no further testing was needed, although the estimated LC_{50} value is at about the "trigger point".

There are no entirely satisfactory techniques for detecting "outliers" (Sokal and Rohlf, 1995), but if the regression is re-calculated without the values for proprionitrile (Fig. 7.4b) such an extreme position has a *nominal* probability of about 0.0045 (1 in 222) of occurring by chance, which suggests proprionitrile may belong to a different population of compounds, with a different relationship between their EC_{50} and LC_{50} values.

Screening tests with QSARs may avoid the need for more difficult or expensive tests on some compounds, but unless the further tests are made one does not know whether the compound being screened can properly be assessed in the same way as the other compounds. For example, the isomer γ-hexachlorocyclohexane would be classified as an inert class 1 compound

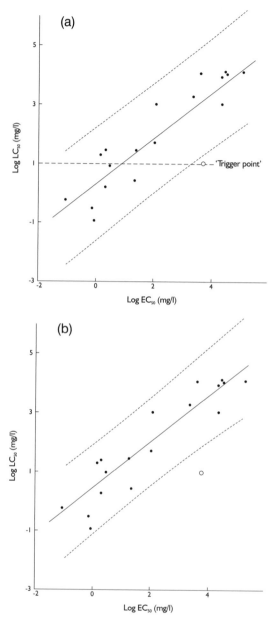

Fig. 7.4   (a) Linear regression for 19 compounds of the 96-h LC$_{50}$ value for the fat-head minnow (*Pimephales promelas*) on the EC$_{50}$ value from the Microtox test (see text). ○ indicates proprionitrile, − − indicates the 95% confidence limits for predicted LC$_{50}$ values. (The 95% confidence limits shown in the original paper are for values estimated from the regression, not for individual values predicted from the regression.) (b) A similar regression, excluding the data for proprionitrile. (Data from Indorato *et al.*, 1984.)

from its chemical structure, but it is the insecticide lindane (sometimes misnamed benzene hexachloride or BHC) (OECD, 1995a). Does a new compound conform to the QSAR, or is it an outlier (Turner et al., 1987)? Nirmalakhandan and Speece (1988) had similar difficulties when classifying 53 petrochemicals into toxic and non-toxic compounds by cluster analysis. Not all compounds conformed to their predicted group. Moreover, QSARs take no account of possible effects when more than one compound is present (Yang, 1994a).

Apart from these reservations about the logical validity of QSARs one must also ask what is the ecological significance of the measured or estimated LC_{50} value? The seeming precision is illusory because no account is taken of sublethal effects, of the possibility of more than one mode of action, nor of the influence of environmental conditions in either the test or in field conditions.

Alabaster et al. (1972) discuss one of the most thorough applications of this type of measurement. Many laboratory studies have been made of the 48-h LC_{50} (that concentration which is lethal to 50% of a test population) to rainbow trout (Salmo gairdneri) for many of the common pollutants found in British rivers, and equations have been derived that relate changes in these LC_{50} values to environmental variables such as temperature, pH and dissolved oxygen. These LC_{50} values were applied to water samples taken at intervals in 1968 and 1969 from 73 sites on rivers and streams drained by the River Trent. These water samples were analysed for ammonia, phenols, cyanide, and for the heavy metals copper, nickel, zinc and cadmium. Environmental factors were also measured, and the concentration of each contaminant was expressed as a toxic unit (p. 155), a fraction of its 48-h LC_{50} value, after due allowance had been made for the observed environmental variables. The total toxicity of each water sample, to rainbow trout, was then expressed as the sum of the separate fractions of the 48-h LC_{50} values.

This procedure is of doubtful validity. It was assumed that the observed fluctuations in the concentrations of individual pollutants (Fig. 7.5) could be summarized adequately by using mean concentrations and that interactions between pollutants in their effects were of relatively minor importance, but tests with mixtures of pesticides and of toxicants in industrial sewage found toxicities were sometimes several-fold greater than additive (Alabaster et al., 1988), and unpublished long-term studies suggest that the toxicity of mixtures may be "markedly more than additive" (Alabaster et al., 1994). There were also more fundamental assumptions: that acute toxicity is a reliable indicator of chronic toxicity and of sublethal effects, that food and other relevant species are not more susceptible, and that the important form of exposure is being tested.

Despite these reservations, it was found that, in general, water samples from sites with fish had lower calculated toxicities than samples from sites that lacked fish, and sites with the game species trout and grayling (Thymallus thymallus) had lower calculated toxicities than those with coarse fish. In 1968, sites whose median toxicities were less than about 0.1 of the

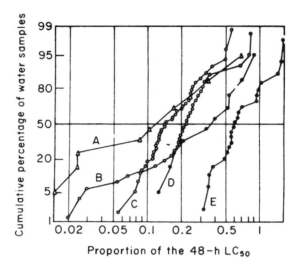

Fig. 7.5   The cumulative frequency distribution for estimates of toxicity to rainbow trout (*Salmo gairdneri*) of water samples taken at intervals during 1968 from five sites in the drainage basin of the River Trent. ($\triangle - \triangle$) One site (A) where game fish were present; ($\bigcirc - \bigcirc$) sites B and C where coarse fish only were present; ($\bullet - \bullet$) sites D and E that lacked fish. (From Alabaster *et al.*, 1972.)

estimated 48-h LC_{50} supported game fish, and values above about 0.2 indicated total lack of fish. The corresponding values in 1969 were about 0.03 and 0.12.

These results are not too surprising. Absence of fish suggests an intolerable habitat, and water sufficiently toxic to kill 50% of fish within two days needed to be diluted less than five-fold to simulate the river water from which fish were absent. However, the mere presence of fish, without more quantitative detail, gives little evidence for lack of ecological effects. Individual fish may migrate into and out of polluted stretches of river, and population size and structure could still be affected.

There have been few similar studies, but Alabaster *et al.* (1994) adduced results from a study by van Loon and Beamish (1977) to support this approach. A series of Canadian lakes, contaminated to different degrees by aerial deposition of heavy metals from a local smelter, were examined for both quantity of fish and amounts of metal in the water. It is not easy to interpret the data (Table 7.9) but the results do suggest that mere presence of fish is a poor indicator of pollution and that it is difficult to evaluate the impact of variable exposures when fish move between different waters. It was also difficult to determine how much of the metal was actually in solution: the standard procedure was followed of measuring that metal which passed through a filter of 0.45 μm aperture. The use of LC_{50} values to predict ecological effects on freshwater fish remains of equivocal value.

We have already discussed exposure, dose and the need for as much detail as possible when studying pollutant effects on organisms (pp. 157–163). In

Table 7.9 Summary of data on fish, and metal concentrations, in three Canadian lakes contaminated by heavy metals deposited from a smelter in Flin Flon, Manitoba. Water samples taken in spring, summer and autumn, 1973–1974

Lake	Mean metal concentration in water (μg/l)		Fish
	Zinc	Copper	
Cliff Lake	85	10	Abundant—little fishing
Hammell Lake	300 (range 130–360; range in 1973 130–160)	11*	Not rare—population not estimated—adverse effects apparent
Ross Lake	8000	450	Fish could not survive, but some species present. Springtime immigration of one species

Data from van Loon and Beamish (1977).
* Alabaster *et al.* (1994) quote a value of 15 μg/l, which in the original table is the value for iron, in the adjacent column.

most experimental work exposure occurs during either a single short period of time, or at a uniform rate, when the magnitude of both amounts of pollutant within the organism and effects are likely to be related to the dose, for acute exposures, or dose rate, for chronic exposures. The results of LD_{50} and LC_{50} experiments demonstrate this quite clearly. Predictions are aimed at field situations, where exposures are unlikely to be so standardized, but will fluctuate, which may pose two related difficulties: little is known for fluctuating exposures of the relationships between the exposure and either the amount of pollutant at the site of action, or the biological effects, but these relationships could be more complicated.

Classical theory supposes that the degree of effect is proportional to the proportion of receptors that are occupied by a drug (p. 157). Although this gives an adequate qualitative explanation for drug action, quantitative details of some drug effects are better explained by the supposition that the important feature is not the number of receptor groups that is occupied, but the rate at which receptors are occupied by drug molecules (Paton, 1961). This could explain, for example, the finding that an animal dosed with twice the LD_{50} dose of a cholinesterase-inhibiting organophosphorus compound will die within a few minutes, but the degree of inhibition in the brain is less than that caused by repeated small doses that produce no observable adverse effect (Barnes, 1975). Albert (1985) considers this rate-dependent mode of action to be rare, but some pollutants may be different.

Destruction of the ozone layer increases exposure to ultraviolet radiation (p. 109), whose biological effects depend on the wavelength (p. 127). UV-B (wavelengths from 320 to about 285 nm) affects photosynthesis, and Cullen and Lesser (1991) exposed cultures of the marine diatom *Thalassiosira pseudo-nana* to visible radiation plus one of six different intensities of UV-B radiation. Exposures were for 0.5, 1.5 and 4 h, in both nutrient-replete and nitrate-limited culture solutions. For equal cumulative doses of UV-B (J/m²) high irra-diance (dose-rate) (W/m²) inhibited photosynthesis more than did lower more prolonged irradiance (Fig. 7.6). The environment also affected the response: nitrate-limited cultures were much more sensitive to UV-B. This dependence on dose-rate rather than total dose probably results from the activity of repair mechanisms, which then intervene in the exposure–dose relationship.

Other intervening mechanisms may also complicate the exposure–dose relationship. For example, trout (*Salmo gairdneri*) acquire copper at a faster rate during exposure to intermittent sources than from continuous exposure (Seim *et al.*, 1984). For an example with effects, plants may be more susceptible to

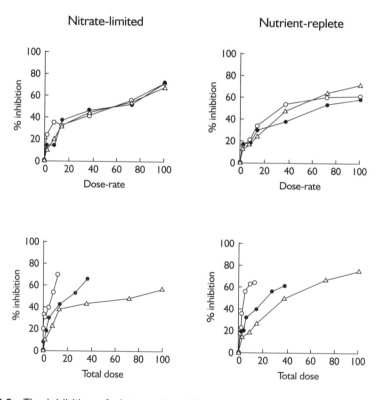

Fig. 7.6   The inhibition of photosynthesis in cultures of *Thalassiosira pseudonana* exposed to visible light and also to one of six irradiance levels of UV-B. Dosages of UV-B expressed as both total dose (as % of that in the treatment with the maximum dose) and dose-rate (as % of that in the treatments with the maximum dose-rate), for cultures in both nutrient-replete and nitrate-limited culture solutions. ○—○, ●—● and △—△ indicate exposure to UV-B for 0.5, 1.5 and 4 h respectively. (From Cullen and Lesser, 1991.)

intermittent than to continuous high levels of ozone, because plants produce more ethylene when exposed intermittently than with continuous exposure (Mehlhorn and Wellburn, 1987). and there is evidence to suggest that it is the compounds formed by chemical reaction between the ozone and ethylene that damage plants (Elstner, 1987).

Perhaps the nearest approximations that we have to responses to fluctuating exposures in the field are for the amounts of pollutants needed to kill, or to affect eggshells, in the field. For example, it is normally considered that 10–20 parts $\times$ 10^{-6} or more of dieldrin in a bird's liver indicate dieldrin as the cause of death (Moore, 1965; Robinson, 1969). Similarly, the degree of eggshell thinning in birds increases with residues of p,p'-DDE within the egg (Cooke, 1973). However, both of these biological responses to exposure occur within a relatively short period of time, and can for present purposes be more properly regarded as the consequences of an acute exposure. We do not know what patterns of fluctuating exposure will, or will not, produce these responses. Moreover, for biological responses that may be manifest for longer periods of time, such as reduced growth rate, the determination of critical levels within the body may be less straightforward, too.

The earliest attempts to predict doses and effects on a field population were for the effect of radionuclides on human beings, where the toxicity resides in radiation from the atom, not in the entire molecule. In the late 1950s the United Nations Scientific Committee on the Effects of Atomic Radiation (UNSCEAR) developed a different measure of dose, the "dose commitment", which takes account of how little of a dose may be retained within the body. In essence, this was a calculation to give the total amount of radionuclide that would occur within the body during a specified period of time after a proposed release of radionuclide (Butler, 1978). This dose commitment is expressed not in units of mass but in units of mass (or concentration) $\times$ time, and is calculated from sources and rates of the proposed release, rates of transport in the abiotic environment, which can be seen as a series of compartments (e.g. Fig. 7.10), and of rates of intake and loss by man. It is the integral of mass of radionuclide estimated to be present within the body, plotted against the duration of exposure. Similar calculations have been employed for other potential pollutants (Foulkes, 1989), and for other species (Miller, 1978), although the transfer rates between compartments can be difficult to estimate if the pollutant's distribution has not attained a steady state (Bennett, 1981). Fumigation against insect pests gives some credence to this approach. The dosage is usually equated with the product of ambient concentration and time, and in practice the dose needed for effective fumigation is taken to be constant. In fact, this simple type of relationship only holds over a limited range of concentrations (Busvine, 1971), although it is possible to allow for threshold levels (Lefohn and Benedict, 1982). Krupa and Kickert (1987) review the mathematical models used to measure plant exposure and response to air pollutants.

It would be unwise to assume for most pollutants that effects are proportional to total dose, or that neither rate nor duration of exposure is relevant alone, apart from its effect on total dose. We lack information on the occurrence of chronic sublethal effects in the field, and do not know how to measure the dose so that it indicates the likelihood of effects. Butler (1980) concluded that dose should be defined as the amount or concentration, at the site of action, at all times from the start of exposure, so that the peak, average and integral over time of concentration can all be estimated, but added the rider that we do not yet know what time to specify for a meaningful calculation of dose.

This is likely to be more than just an academic quibble: Table 6.1 showed that plants of *Lolium perenne* grew better after almost 4 months of intermittent exposure to polluted air than they did with either continuous or no exposure to polluted air. The explanation that appears to involve least upset to conventional views is that a relatively slight exposure to sulphur dioxide stimulated growth, at least in those experimental conditions, whereas greater exposures decreased it. There is some evidence for other biologically active chemicals to suggest that there is an intermediate phase between no effect and adverse effect when the opposite response, in the instance of *L. perenne* the response of increased growth, is obtained (see pp. 133–134).

Hormetic responses are of particular relevance for ionizing radiation, which can cause cancers and mutations in man. Initially safe exposure levels were set, by analogy with engineers' safety factors, at an appropriate value below the presumed threshold exposure (Sagan, 1994). This changed with the development of the nuclear industry after the Second World War to a linear hypothesis, when the risk of harm is directly proportional to the total dose. It is probably the least unsatisfactory way to extrapolate from observed results of high doses to estimate the effects of low doses, although some have disputed that the data support this assumption (Weinberg, 1978), and there is some evidence to suggest that small exposures are not harmful and may even be beneficial (Sagan, 1989).

One final practical point needs to be mentioned: the need to distinguish between initiation and response time. Adult laying hens (*Gallus domesticus*) were dosed orally each day for up to 60 days with leptophos (an organophosphorus insecticide that is relatively persistent in the environment) at dose rates of 0, 0.5, 1.0, 1.5, 5.0, 10.0 or 20.0 mg/kg body weight/day (Abou-Donia and Preissig, 1976). Ataxia (paralysis of the legs, see p. 139) developed with all but the lowest dose rates. The higher the dose rate, the quicker ataxia developed (Fig. 7.7a), but the total dose ingested before the onset of ataxia increased with the dose rate too (Fig. 7.7b). The latter fact is not quite so surprising when it is remembered that there is a lapse of 1–2 weeks after single doses of organophosphorus insecticide before ataxia becomes apparent, and it is quite clearly possible that, in fact, with higher dose rates, less leptophos is ingested before ataxia is initiated than with lower dose rates.

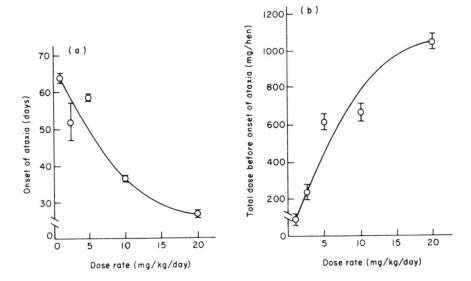

Fig. 7.7 Effect of dose rate of leptophos, administered orally each day to laying hens (*Gallus domesticus*), on (a) time for symptoms of ataxia to appear and (b) total dose ingested before symptoms of ataxia appear. (From Abou-Donia and Preissig, 1976.)

Effects on ecosystems

Our main concern in this book is with the sixth of Korte's questions (p. 175), of effects, but the answers to the first five questions do help to predict the likely exposures that organisms will receive. Both physical models of ecosystems (microcosms and mesocosms) and mathematical models have been developed over many years to describe and predict the movement of pollutants in the environment, which is seen as a series of linked compartments (Samiullah, 1990a). In the most general terms concern is with the mechanisms and rates of transfer across the interfaces between compartments, and of transport within compartments. Care is needed when applying predictions from such models. They may be either specific to a particular situation (e.g. Fig. 7.8) or evaluative, when no attempt is made to simulate the real world, but the aim is to measure the compound's behaviour in known conditions. Thibodeaux (1996) argues that the factors controlling chemical movement in the real world are so complex that models are highly simplified versions. In addition, the effective exposure of individual plants and animals may depend critically on such minutiae as the size of particles that are retained within the lung from inhaled air (Butler, 1978).

Baughman and Lassiter (1978) suggested that first-order kinetics are usually adequate to estimate the rates of the more significant environmental factors such as hydrolysis, photolysis and volatilization. One should then be

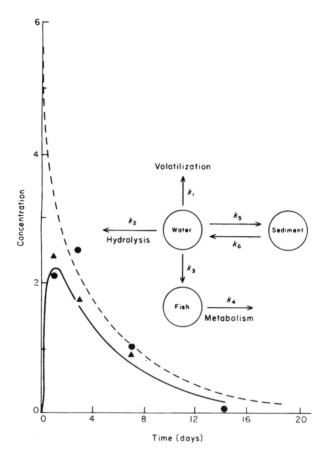

Fig. 7.8   The concentrations of chlorpyrifos in fish (expressed as parts × 10⁻⁶) and water (expressed as parts × 10⁻⁹) during the first 3 weeks after application. (●) Observed concentrations in fish; (▲) observed concentrations in water; (—) estimated concentrations in fish; (– –) estimated concentrations in water. These estimates are derived from the model shown. (From Neely and Blau, 1977.)

able to predict, by compartmental models similar to those discussed on pp. 158–162, the exposure of individual organisms that results from a given rate of entry by a chemical into an ecosystem (Branson, 1978). Neely and Blau (1977) used this method to estimate the concentrations of chlorpyrifos (an organophosphorus insecticide) after a single application to earthen ponds of 404 m² surface area and of 0.25–2.0 m depth. Three compartments only were considered—water, fish, and soil plus plants—but observed concentrations of chlorpyrifos in fish and water agreed quite well with the predicted concentrations (Fig. 7.8). The model, though crude, was effective, although it should be noted that this compound was not very persistent, and that the ponds were of recent construction. It would have been more difficult to simulate a more persistent compound in a more complex ecosystem (Scheunert and Klein, 1985). We discuss this topic further on pp. 208–210.

This approach requires the same proviso that we made for toxicity tests: the degree of effect, be it a chemical process or a biological response, depends on many details of the environment. Roberts and Marshall (1980) stress that persistence of a chemical depends not only on the properties of that chemical, but also on the characteristics of the particular ecosystem. For example, the rate of hydrolysis in water will be influenced by pH. So predictions for one ecosystem will not necessarily be valid for others.

Considerable efforts have been made by the OECD to develop standard agreed tests that are acceptable to national regulatory authorities, and descriptive models are not recommended, because they lack generality (Hushon et al., 1983). The OECD recommends that exposure be estimated by an evaluative model that, in its simplest form, assumes a closed environment of five compartments—air, water, sediment, soil and aquatic biota—of known fixed volumes, with concentrations of pollutant in the different compartments at equilibrium (Fig. 7.10 illustrates a similar model).

A key concept is fugacity (f), the tendency of a chemical to leave a compartment, which is expressed in units of pressure (MacKay, 1979). For any one compartment, at the low concentrations relevant to potential pollutants,

$$f = \frac{c}{z}$$

where c is the concentration and z is the fugacity capacity constant. When two compartments (1 and 2) are in equilibrium, then

$$f_1 = f_2$$

when

$$\frac{c_1}{z_1} = \frac{c_2}{z_2}$$

or

$$\frac{c_1}{c_2} = \frac{z_1}{z_2} = K_{12}$$

where K_{12} is the partition coefficient. Fugacities are preferred to partition coefficients because they can be estimated from related thermodynamic data and, for systems with many compartments, fewer parameters need to be estimated: one fugacity per compartment instead of one coefficient for each pair of compartments. In practice the relevant physicochemical data are sometimes lacking or of poor quality (Calamari and Vighi, 1992).

This simple model is not dynamic. It gives no indication of the time needed to reach equilibrium, and only relative concentrations between compartments are relevant. Actual concentrations depend not only on total mass of pollutant present but on compartmental volumes. MacKay and

Patterson (1981) developed more complex forms of this model to allow for continuous input of and non-equilibrium distribution of pollutant, loss by chemical reactions and varying rates of processes. Many models of this type have now been developed (OECD, 1989b; Samiullah, 1990a; MacKay, 1994).

∗

The transition from prediction of exposure to pollutants to prediction of effects by pollutants introduces an additional complexity—the number of species. Even the simplest ecosystems contain a considerable number of species, which makes it difficult to predict effects of pollutants on communities, even if the effects on some individual species are known. Toxicological data are inadequate, because even multigeneration chronic exposure studies only test a species' ability to survive under given conditions, and also give no direct measure of the pollutant's impact on competitive ability (Crow and Taub, 1979).

However, legislative demands caused Kooijman (1987) to use acute toxicities. The LC_{50} values for a range of species were fitted to a log-logistic distribution curve and the concentration calculated at which, for a specified probability level, a specified, small, percentage of species would be harmed. This hazardous concentration was taken to be that above which a community would be harmed. Kooijman did express some misgivings, because one is extrapolating far beyond the limits of one's knowledge, but defended this estimate as the best available response. Wagner and Løkke (1991) replaced the log-logistic by the log-normal distribution curve to simplify the calculation. They also replaced LC_{50} values with NOEC (no observed effect concentration) values, which had been suggested by van Straalen and Denneman (1989). The latter change appears to be logically unsound, because NOEC values have no confidence limits. The whole approach appears to be driven by the need for reproducible results, and assumes that relevant answers are possible with that type and amount of information.

The alternative approach has been to use physical models. We have already discussed the complexities of interactions between populations of different species (Chapter 3), and we must discuss briefly one widespread view, which can be summarized as "diversity begets stability". Stability has a range of meanings (see Connell and Sousa, 1983; Emlen, 1984; Pimm, 1984), but for our purposes can denote no more than a lack of change, although it often has additional dynamic meanings such as the ability to resist perturbations and the speed of return after disturbance by a perturbation (Orians, 1975). This view was developed clearly by Elton (1958), who deduced from what we know of natural communities that populations tend to be more stable in complex, or diverse, communities than in simple ones. Tropical rain forest is often quoted as the prime example of both complexity and stability. The evidence though is only correlative, and May (1974) argued persuasively that diversity can only develop in stable environments, and also showed that mathematical models of ecosystems become less stable as their complexity increases. Conversely, some simple ecosystems appear to be very stable.

Ecosystems are complex, so most studies have introduced at least some degree of simplification, and Römbke and Moltmann (1996) discuss the range of approaches that has been developed. Metcalf *et al.* (1971) described a model ecosystem, or microcosm, for their studies on pesticides. Plants of *Sorghum halepense* grow, in an aquarium, on a bank of sand that slopes down into 7 litres of standard water, which contains a second plant species, the alga *Oedogonium cardiacum*. Eight biotic components are considered in the system, and it is supposed that radiolabelled pesticides applied to the *Sorghum* plants are distributed along the following pathways:

Sorghum halepense
eaten ↓ by
Estigmene acrea
(caterpillar)
  ↓

excreta → Diatoms → Plankton → *Culex pipiens* → *Gambusia affinis*
  ↓   (mosquito larva)   (fish)
Oedogonium cardiacum
eaten ↓ by
Physa sp.
(snail)

It is a very ephemeral system. Pesticide and caterpillars are placed on the plants 20 days after the microcosm is first set up, and the caterpillars eat most of the treated plant surfaces within the next 3–4 days. Mosquito larvae are introduced 6 days after the pesticide, and the fish are introduced after another 4 days. Three days later the experiment ends. Water and sand are sampled at intervals, but organisms are sampled for analysis only at the end of the experiment, 33 days after the start. In later studies organisms were exposed to pesticides for up to 33 days (Lu *et al.*, 1975). Data are obtained on the occurrence and distribution of the insecticide and its metabolites in the water and organisms. Conclusions are then drawn about the environmental fate of the pesticide, measured in particular by its "ecological magnification" and "biodegradability index". Lu and Metcalf (1975) described a variation— a model aquatic ecosystem—for studies with relatively volatile compounds, in which experiments last for 2 or 3 days.

One virtue of this system is that it examines metabolites produced by a range of organisms. On the other hand, it is difficult to attribute much significance to the quantitative data (Moriarty, 1977). It is impossible to determine the exposure that individual organisms have had. All one knows is the total amount of pesticide put into the system, amounts in the water and sand at intervals, and amounts in organisms at one time.

However, the fundamental criticism is that this collection of species within an aquarium is not an ecosystem: none of the species reproduce themselves, let alone maintain a population. Nor does the system bear much relevance to field situations. For example, pure sand is used instead of soil, because sand gave much more reproducible results, and soil had "uncertain effects" and interfered with the extraction and determination of metabolites (Metcalf et al., 1971). It is also rather limited biologically. Although this is a relatively complex microcosm, it is nevertheless limited in the range of niches that is represented. There are, for example, no soil-inhabiting, or benthic, organisms. Metcalf (1977) has concluded that these model ecosystems are of limited value. They do appear to have the worst of both worlds: they are too complicated to give results that are easily interpreted, but too artificial to be of immediate relevance to field situations.

Because of these deficiencies, more complex less unrealistic mesocosms have been developed, which are larger than microcosms, outdoors, and with less control. These have taken two principal forms, either artificially-constructed streams designed to simulate real streams (e.g. Rodgers et al., 1996) or enclosures in lakes and marine ecosystems (Clark and Cripe, 1993). Terrestrial mesocosms have been little used for organic compounds, and often differ little from experimental field plots (Gillett, 1989).

Various groups of marine biologists have developed controlled aquatic ecosystems, columns of seawater enclosed within plastic bags. The details vary considerably (Boyd, 1981), but in essence the bags hang from the surface of the sea, and may be either open or closed at the bottom. The largest CEPEX* enclosures contain 1700 m^3 of seawater, sufficient to enclose populations of at least three trophic levels—plants, herbivores and small carnivores—which can sustain themselves for long experimental periods of several weeks. These systems have the additional advantages that, initially, different species occur in their natural proportions, and that interactions can occur between species of the same trophic level, although some organisms, particularly the larger fish and mammals, will be excluded. Moreover, these controlled ecosystems can be replicated, although as the size of the individual unit increases, replication becomes more costly and difficult.

Davies and Gamble (1979) concluded, from experiments with the addition of mercury at 1.5 and 10 µg/l to controlled ecosystems, both at Loch Ewe, Scotland, and in Saanich Inlet, Vancouver Island, that there were few marked changes in rate processes of population structure after exposure to 1 µg mercury/l seawater, apart from a transient reduction in the rate of photosynthesis. They therefore suggested that, because surface waters around the United Kingdom are usually reported to contain only 0.001–0.022 µg/l, these experimental results suggest that there is at present little risk of damage from mercury to populations of phytoplankton and zooplankton in British seas.

---

* CEPEX is the acronym for "Controlled Ecosystem Population Experiment", originally called "Controlled Ecosystem Pollution Experiment" (Parsons, 1978).

This deduction may well be correct, but some reservations apply to the evidence. The controlled ecosystems do not simulate the sea in all respects (Fig. 7.9), and the walls of the container do impose some artefacts. Eddy diffusivity is reduced by at least an order of magnitude compared with that of the open sea, which may affect the sinking rate of phytoplankton; lateral exchange by advection and diffusion is minimal; organisms grow on the walls of the ecosystem, blocking light and competing with planktonic organisms. Finally, tests with a simple marine food chain of phytoplankton, a benthic bivalve (*Tellina tenuis*) and juvenile flatfish (*Pleuronectes platessa*) in large tanks showed that 0.1, 1.0 and 10 μg mercury/l seawater all reduced photosynthesis, bivalve condition and fish growth (McIntyre, 1977). Both systems differ in some respects from the open sea, so it becomes difficult, without further investigation, to decide how applicable either result is.

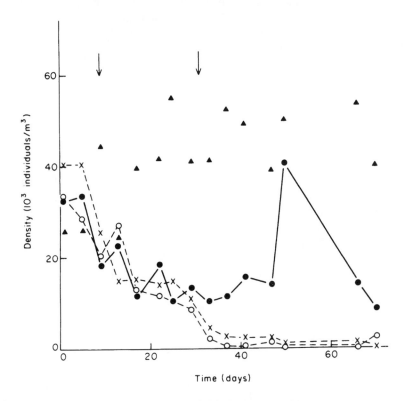

Fig. 7.9   The density (10³ individuals/m³) of all zooplankton in controlled ecosystems and in the ambient sea at Loch Ewe, Scotland, autumn 1976. (▲) Data for the sea; (●) controlled ecosystem with no mercury added; (○) and (×) controlled ecosystems in which mercury levels were raised first to 1, later to 10, μg mercury/l seawater, at the times indicated by ↓. All three controlled ecosystems had additional inorganic nutrients added, to replace those lost in detritus sinking to the bottom. The amounts of nutrient in ecosystems ● and ○ were three times that in ×. (From Davies and Gamble, 1979.)

Steele (1979) concluded that these large experimental ecosystems have taught us more about the interactions that occur within such systems than about the subtle long-term effects of pollutants: the addition of pollutants, and the effects of constraints imposed by these controlled ecosystems, impose qualitatively similar changes on the enclosed community. The aspect of these studies with most immediate relevance is of the transfer of pollutants, which may involve changes in chemical speciation, intake by organisms, and movement to the sediments. Boyd (1981) suggests that the chief value of these systems is not as simulations of nature but as experiments on selected components of the ecosystem. Similar conclusions have been reached for the use of enclosed freshwater systems (Sanders, 1985) and of artificial streams (Shriner and Gregory, 1984).

Gearing (1989) concludes that model ecosystems help us to understand the ecological effects of pollutants when combined with other types of study, but how useful are they for prediction? Many structural and functional characteristics can be measured within a mesocosm (Adams and Giddings, 1982), but critical tests are needed. Doubtless mesocosms can detect acute toxicity, but it is unlikely than an effect such as that from low levels of tributyl tin (pp. 141–144) would be detected. How then do regulators assess new potential pollutants?

Regulation of new chemicals

In many countries the state now has to give permission before significant quantities of any new synthetic chemicals can be marketed. Concurrently some countries are also assessing chemicals that are already in use. At present rather simple tests are used to assess possible ecological effects. The common aim of these national regulations is to protect man and his environment from adverse impacts, but the details of the regulatory procedures, and the results, can vary considerably between countries, for at least three good reasons: differences of culture (Jasanoff, 1986), differences of circumstances, and also because the uncertainties of prediction mean that there is no one best approach and that mistakes can occur (Korte et al., 1985).

It is always the manufacturer's or supplier's responsibility to provide the test data. Testing is expensive, and excessive testing may hinder trade, so there has been a considerable effort to harmonize—to minimize the differences between—the tests required by the different national authorities. Detailed guidance, with recommendations for appropriate tests, is given by the Organisation for Economic Co-operation and Development (OECD, 1998), and follows the rationale behind the six questions outlined on p. 175. It is now generally accepted that any such tests should, indeed often must, accord with Good Laboratory Practice (Stiles, 1993).

Tests are phased ("sequential hazard assessment"), with the results from the first set assessed before the need for any additional tests is decided, and this stepwise or "tiered" procedure continues until either a decision about permitted uses, if any, is made, or the manufacturer withdraws his application. In essence the risk assessment compares the expected environmental exposure to a chemical with the amount needed for a harmful effect: the "predicted environmental concentration" (PEC) is compared with a "predicted no effect concentration" (PNEC). Both of these predicted values are affected by many uncertainties (OECD, 1989a). This appraisal also involves considerable extrapolation from laboratory results to field conditions and there are fundamental doubts about the scientific basis of such extrapolations. These doubts and uncertainties are usually allowed for by a safety (assessment) factor, commonly within the range 10–1000.

This assessment continues until it is clear that the PEC : PNEC ratio is or is not adequately high. Sometimes tests stop when a specific characteristic such as toxicity or persistence exceeds an acceptable level. The final decision on acceptability is likely to include expert interpretation of the data, which may also include field trials, particularly for pesticides, and—for chemicals already in use—field data on environmental concentration.

The amount of chemical to be released into the abiotic environment is estimated from the amount to be marketed, its uses and methods of disposal. The environment is then seen, as a first approximation, as three compartments (Fig. 7.10) (Scheunert and Klein, 1985), which differ in the speed with which chemicals mix within and transfer between them, and in the vulnerability of their biota (Table 7.10). Relative distribution between compartments is assessed principally from the physicochemical data, in particular the solubility in water and rate of hydrolysis, vapour pressure and boiling point. Loss

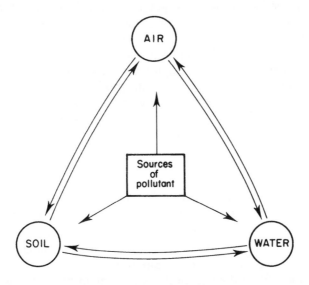

Fig. 7.10 Scheme to describe the movement of pollutants from one or more sources into environmental compartments.

Table 7.10 Comparison of environmental compartments. +, ++ and +++ indicate a slight, moderate or considerable degree respectively of the factor assessed

| Compartment | Transport of pollutant | | Uniformity of distri-bution | Degree of dilution | Residence time | Vulner-ability of organisms |
	Distance	Amount				
AIR	+++	++	+++	+++	+	+
WATER	++	+++	++	++	+++	+++
SOIL	+	+	+	+	+++	++

Adapted from OECD (1981).

from degradation by micro-organisms and abiotic factors is assessed from standard laboratory tests.

Mathematical models can then be used to predict the amounts and concentrations of the chemical in these compartments. Which model is used will depend on the quantity to be released and over how large an area it will disperse. Airborne persistent compounds could need a global model, but local models usually suffice for new compounds. Results are best used not to predict actual amounts and concentrations but to indicate a chemical's potential behaviour: the relative amounts in different environmental compartments, the dominant chemical reactions, principal transport routes and overall persistence (Sheehan *et al.*, 1985).

Different regulatory schemes have different objectives, so their details inevitably differ (OECD, 1995b), although they rely on the same basic approach. To see how it works in practice, we will consider the system for hazard assessment of new chemicals ordained by the "Sixth" and "Seventh" Amendments" (EEC Council, 1979, 1992) for use within the European Community (Vosser, 1986; Fielder and Martin, 1993). Haigh (1992, 1995) discusses this system in greater detail, Römbke and Moltmann (1996) discuss its implementation and how it differs from those enacted by other countries and Harwell (1989) describes the American system.

New substances (elements, compounds, additives and impurities) must be notified to the regulatory authorities before they are marketed, and the amount of information required depends on both the amount to be produced (Table 7.11) and the apparent potential hazard. Additional information can always be requested if appropriate. A limited announcement contains very sparse information, which includes data on acute mammalian toxicity. When the quantity to be marketed exceeds 1 tonne/annum, the base set gives information on:

(1) The identity of the substance, e.g. structural formula, spectral data, methods of detection and determination.
(2) Information about the chemical, i.e. uses, amounts and how to handle.
(3) Physicochemical properties, including the partition coefficient between *n*-octanol and water.

Table 7.11 Effect of amount of substance within the European market on the initial amount of information needed from the manufacturer or supplier for hazard assessment

Quantity marketed (tonnes)			Amount of information needed
Per annum	Total		
<1	–		Limited announcement
1 or more	<50	Full notification	Base set
10 or more	50 or more	Full notification	Base set *or* Level I
100 or more	500 or more	Full notification	Level I
1000 or more	5000 or more	Full notification	Level II

(4) Toxicology (for details see Fielder and Martin, 1993).
(5) Ecotoxicology.
   (a) The acute toxicity (LC_{50}) for one species of fish.
   (b) The acute toxicity (LC_{50}) for one species of *Daphnia*.
   (c) The rate of biotic and abiotic degradation processes, measured in standardized aerobic conditions. Anaerobic degradation may be more important for some chemicals, for which there are suitable tests, although they have not yet been agreed internationally. Results from such tests are accepted by national authorities.
   (d) Absorption and desorption.
(6) Possible methods for rendering the substance harmless.

Aquatic species are used for the ecotoxicological tests because release via sewage into rivers is a major pathway for many chemicals. Additional tests on other, terrestrial, species are required when species in other habitats are thought to be at risk. For the aquatic environment, the predicted environmental concentration is compared with the lower of the two LC_{50} values. If there is an adequate safety factor (often taken to be a 100-fold difference) then the proposed uses and amount of chemical are deemed to present an acceptable environmental risk. One has to accept that the choice of species for testing is necessarily arbitrary: no few species can be selected to represent the typical response of any taxonomic group. Species are therefore selected that differ physiologically and/or in their function in the ecosystem.

For level I tests, the ecotoxicological assessment can become more searching, both by extending the base-set tests and by requiring additional tests:

(a) A longer toxicity study, of at least 14 days, for one species of fish.
(b) A test of bioaccumulation, usually with a species of fish.
(c) A longer toxicity test, of at least 21 days, for one species of *Daphnia*.
(d) A toxicity test with one species of earthworm.
(e) A toxicity test with a "higher" plant. A species of *Lemna*, a small free-floating aquatic species, is commonly used. This meets the requirement, but cannot be regarded as a very representative genus.

(f) Further degradation studies.
(g) Further studies on absorption and desorption.

No tests are mandatory. The regulatory authority may waive, alter, or ask for additional, tests, and the notifier may also give reasons why certain tests are unnecessary for a particular compound. This flexibility becomes still greater for level II tests, where the nature of any additional tests will depend on the results of the earlier tests. The underlying aim of any further tests is to assess the likelihood and degree of biomagnification along food chains, the biological effects of prolonged exposure to the compound, and on the compound's fate in the environment if it is not easily degraded.

Many uncertainties exist in this process, not least because it involves extrapolations from acute to chronic exposures, from one species and one taxonomic group to another, from species to communities and to other environmental conditions. It is also difficult to determine how well these systems of hazard assessment work. Most of the data used in current procedures are confidential and so the system cannot easily be evaluated (Römbke and Moltmann, 1996). It is generally accepted that mistakes could occur, but we do not know how many past or present ecological effects have gone unnoticed. This uncertainty, plus the possibility that most new chemicals may not have serious ecological effects, could make the system seem more effective than it really is.

The question is also being asked, which of the many chemicals already in use need most urgently to be assessed for their potential effects on human and environmental well-being (EEC Council, 1976, 1993; OECD, 1986; Karcher, 1998). The emphasis is on rational, practical and cost-effective methods of selection, based on the same type of approach as that used for new chemicals, and Hedgecott (1994) describes the current approach.

The gist of this chapter is that our current ability to predict ecological effects is rudimentary. At best, experience with pesticides suggests that compounds that become widely distributed in the environment, that are lipophilic with less than 2 parts $\times$ 10^{-6} dissolving in water, that are persistent, and that are acutely toxic at less than 10 $\mu g/g$ body weight, are likely to cause problems.

At present, for practical purposes, it would be incautious to suppose that a sublethal exposure that affects individual organisms adversely is not close to that which will affect the population. There is no good reason to suppose that there is a constant relationship, for different pollutants or different species, between the dose needed to kill and that needed to impair an organism. Therefore, given the difficulties of studying an ecosystem, the most effective way to predict biological effects is likely to be by discerning the least exposure that produces a deleterious response in individual organisms (e.g. Moriarty, 1968), and then examining the extent to which different environmental conditions alter this minimum exposure. Figures 2.10 and 6.7 give simple examples of the significance of environmental conditions, and this theme is developed further on pp. 149–153. The likelihood of danger to other species, by concentration along food chains, is best assessed by measurements of the percentage assimilation and of persistence within organisms. Work of this sort would enable appropriate, tentative, maximum permissible exposures to be deduced, with safety factors appropriate to the perceived balance of potential benefit and potential loss.

8 Monitoring and Assessment

It is clear that, in the present state of knowledge, prediction alone is inadequate: we need to assess the distribution and effects of pollutants in the environment. The purpose of such an assessment may range from reassurance or demonstration that predictions are correct, through compensation for inadequate ability to predict, to safeguard against unappreciated hazards or failures of control. Three terms are commonly used, to distinguish three distinct but related activities. A survey, a single set of observations or measurements, will assess the situation at one particular time. Surveillance—repeated surveys—will indicate changes with time, whereas monitoring, in its restricted legal sense, indicates repeated observations or measurements that check whether that which is studied conforms to an already-stated standard. The distinction between these last two terms is quite clear in theory, but in practice the two activities are not always so distinct and tend to merge one with the other (Holdgate, 1979).

In the jargon of administrators, measurements of all three types of activity deal with either targets or factors, but the distinction between the two is subjective rather than an attribute of that which is measured. Targets may be physical systems, such as an estuary or a building, or biological systems, such as a population or a community. Factors, too, may be physical (or chemical) or biological. The critical distinction is that targets are of interest in their own right, whereas factors are of interest for their effects on targets. These distinctions and definitions, though useful, may appear rather unexciting. However, they do raise the question that should be central to every monitoring scheme: what is its objective?

Objectives do need to be determined at the very start of any programme, because the objectives dictate the details of any sampling programme that is to be adequate for reliable conclusions (Moore, 1975). Holden (1975) and Holdgate (1979) enumerate the types of information that can be gained by well designed and executed monitoring: the detailed applications are numerous, but in essence information of three types may be sought:

(1) Rates of release of pollutants into the environment.
(2) Degree and changes of environmental contamination, both biotic and abiotic.
(3) Biological effects.

Whatever the objectives may be, monitoring schemes should when possible be designed so that a specified degree of change can be detected with an appropriate degree of statistical significance (Segar and Stamman, 1986). Often the question arises, when a change does occur, whether this results from normal ecological variability or from one or more pollutants (e.g. Hawkins and Hartnoll, 1983). Control sites, uncontaminated sites comparable ecologically to those contaminated, are sometimes used for this purpose, but it is not always easy to select control sites that are really comparable to the contaminated sites (Eskin and Coull, 1984).

There is considerable disagreement about the relative usefulness of monitoring the amounts of pollutants in the environment and of monitoring for biological effects. Preston (1979) argues that, because of the difficulties of detecting biological effects in the field, it is more realistic to measure chemical residues, and to relate these to standards. The ideal monitoring scheme then depends on two preliminary activities. Critical pathway analysis should first determine the pathway or pathways and target or targets which, if safeguarded, ensure that there is adequate control for all targets. Then the critical dose–response level needs to be established, below which no harm, or an acceptable degree of harm, will occur (Holdgate, 1979). Once the critical dose–response level has been determined, it is possible to establish criteria, for the quantitative relationship between exposure and effect, from which can be derived:

(1) Primary protection standards—those amounts of pollutant in the target that must not be exceeded.
(2) Derived working limits, or environmental quality standards—levels in factors on the pathway to the target that must not be exceeded.
(3) Derived discharge limits—those levels of emissions from sources that will not cause environmental quality or primary protection standards to be exceeded.

In practice, critical pathway analysis frequently means to assess the contribution of food species to pollutant intake by man. Calculations are made from given (known or presumed) rates of discharge of the contaminant into the environment, of amounts that will occur in species eaten by man, and this exposure rate for man is compared with that which toxicological studies suggest is the maximum permissible exposure to avoid harm. Regular monitoring of amounts of contaminants in these food species then ensures that the permitted amounts of contaminant being released into the environment are not, in fact, endangering human health. This approach is suited to situations where damage to human health is the main concern. However, it is primarily an application of toxicology, and where damage to animal or plant populations is unacceptable, one enters the realm of ecotoxicology. There is, in theory, no reason why limits and standards should not be set for the protection of species other than man (Woodhead, 1980). However, the critical species, those species that are the first to be affected by a pollutant, are often unknown, and we have already discussed the problems of dose–response relationships (pp. 193–202).

Reliance on measurements of pollutant alone, without also assessing biological effects, does ignore some potential problems (Price, 1978):

(1) It is sometimes impracticable to measure all contaminants.
(2) Routine analytical techniques may be too insensitive.
(3) The biological significance of pollutants, at the levels found, may not be fully known.
(4) Combinations of pollutants may interact.
(5) Regular chemical measurements may miss occasional, significant, high values.

For these reasons effluents are sometimes monitored with plants or animals. For example fish behaviour can be recorded automatically and continuously, and any deviation beyond the normal, critical, limits of variation initiates a warning that something is amiss (e.g. Carlson and Drummond, 1978; Dickson et al., 1980).

Derived limits for emissions and environmental quality both imply that the environment has an "assimilative capacity" for pollutants, that degree of contamination which is not unacceptable in either degree of biological effect or extent of area affected. Pollutants that do not degrade, or do so very slowly, pose a different question. We have already considered the assumption of dispersion beyond the locality in which the contaminant is released (pp. 211–212). Two somewhat divergent views have now developed within the European Union (EU), and in other states, on the appropriate form of control for substances that are toxic, persistent and tend to accumulate in biota. Some countries, including the United Kingdom, favour environmental quality objectives (EQOs)—acceptable amounts in the environment—and some countries favour emission standards—acceptable amounts in discharges (Mance, 1987). Thus, for cadmium, the EQO for fresh water is that the total concentration of cadmium (dissolved plus particulate) must not exceed 5 μg/l, while the emission standard for cadmium used in pigments is that not more than 0.03% of the cadmium used may be released into the environment, and also there must not be more than 0.2 mg cadmium per litre of effluent (Gardiner and Mance, 1984). Toxicity is of course a function of both amount of pollutant present and duration of exposure: concentrations can therefore be stated in a variety of ways, such as levels that are never to be exceeded, averages, or percentiles—values that are not to be exceeded by a specified proportion, commonly 90 or 95%, of all samples taken.

It has been argued that emission standards, unlike EQOs, make economic competition between states within the EU "fair". This is perhaps a somewhat unusual concept in commerce and industry, and could have some interesting consequences if applied consistently to all aspects of economic activity. The more persistent a pollutant is, the greater the potential difference between EQOs and emission standards. Thus cadmium, once released into the environment, will be potentially available to biota for a very long time, when total amount released, rather than amounts near the point of release, becomes of increasing importance.

Emissions standards do have some advantages: compliance is relatively easy to monitor, and if uniform for all sources they are easy to administer. However, they may not always be appropriate for specific situations, and can therefore use resources inefficiently. EQOs, though, do need extensive reliable data on the condition of the environment into which pollutants are to be discharged and of current discharges and amounts of pollutant present. Abel (1996) discusses in detail the application of these two approaches for freshwater habitats.

Historically, for less specific environmental issues, most effort so far has been devoted to measurements of amounts in the environment, with relatively little attention paid to the detection or measurement of effects (e.g. Holdgate, 1979; Cairns, 1980). Concern about pollutants derives from the effects that they cause, not from their mere presence in the environment, and Cairns argues that it was neglect of effects that led to the American view that emissions of pollutants should be controlled by the "best available technology" (BAT). This is empirically attractive but can, according to circumstances, be either unnecessarily restrictive and expensive or still be damaging to ecosystems. The use of "best practicable means" (BPM) in the UK for the control of atmospheric pollution could also have the same result, although the trend now is towards the "best practicable environmental option" (BPEO) or the "best available technology not entailing excessive cost" (BAT-NEEC) (O'Riordan, 1989), that combination of methods of control and waste disposal that minimizes all forms of environmental damage from a source within the constraints of what is technically feasible and economically acceptable. Haigh (1995) describes the overall picture for recent UK practice, which is shaped partly by directives from the European Union and partly by specifically British legislation, and Abel (1996) discusses the practical problems that can arise. McLoughlin and Bellinger (1993) describe the ways in which different countries have controlled pollution.

An adequate monitoring system would assess both the effects and the amounts and distribution of pollutants. Munn (1981) discusses in detail the objectives, design and interpretation of data from monitoring networks for atmospheric pollution, and Livett (1988) develops these themes for the atmospheric deposition of heavy metals. We will not consider measurement of pollutants in abiotic samples any further, except to reiterate the point made in Chapter 1, that the way in which measurements are usually taken, with little monitoring of discharges, makes it impossible sensibly to relate observed levels in the environment to the rates at which pollutants are released (Preston, 1979). Spellerberg (1991) discusses monitoring in the wider context as measurement of ecological change, both natural and man-made.

Organisms may be used in ecotoxicological monitoring schemes to assess either exposures to pollutants, or effects of pollutants. Such organisms are often called biomonitors or biomarkers, although there is considerable confusion about the precise definition of the latter term, which can mean anything from molecular changes in an organism's DNA to changes in the whole organism. Monitoring and Assessment Research Centre (1985), Burton (1986) and Samiullah (1990b) describe many such schemes, and both Phillips and Rainbow (1993) and Doust et al. (1994) discuss the difficulties of monitoring for aquatic pollutants.

Amounts of pollutants in organisms

Many schemes, from local to international, analyse organisms for pollutants. Often these schemes are part of a governmental system of regulation to ensure that discharges do not exceed an acceptable limit. The second major purpose is to assess the amount and distribution of a contaminant in the environment, by measuring the amount of contaminant within individuals of selected species. A third possible purpose is to estimate the exposure of species that feed on the monitored species, supposing of course that the species monitored is a major source of the pollutant. One can then avoid killing animals from a possibly endangered or otherwise protected species, but one does need to know something about the absorption of pollutant by the predator or herbivore. For example, carnivorous marine molluscs may not absorb much of the heavy metals in their prey species (Nott and Nicolaidou, 1990). The sample collected for analysis should also represent the individual plants or animals that are selected as food: often herbivores and predators do not select a random sample from their food population.

Direct analysis of samples such as sediments or air does give a more direct measure of environmental contamination than does analysis of organisms. Phillips (1980) reviewed the use of aquatic organisms for monitoring amounts of pollutants in the environment, and cited three advantages for analysis of biotic rather than abiotic samples:

(1) Some pollutants, notably heavy metals and organochlorine compounds, occur in much higher concentrations within aquatic organisms, sometimes by as much as a factor of 10^3–10^6, which can make chemical analysis much easier.

(2) Analysis of organisms measures the pollutant's availability, which is more important for biological effects than to measure the total amount of pollutant in the environment.

(3) Organisms integrate the amount of pollutant present during a period of time. Each combination of species and pollutant is unique, with its own degree of integration over time.

The same arguments can be advanced for the use of terrestrial species (Martin and Coughtrey, 1982), and this approach, of using living organisms for chemical analysis, is sometimes taken a stage further in studies on the effects of air pollutants on plants. Manning and Feder (1980) suggest that symptoms of damage to foliage be used as a measure of environmental concentration. They quote as an example the tobacco variety *Nicotiana tabacum* Bel-W3, which is particularly sensitive to ozone, and argue that the degree of damage to the plant can be used to measure amounts of ozone in the air. It is an open question how reliable this would be: the degree of damage is proportional to the exposure in carefully controlled environmental conditions (Menser *et al.*, 1963), but similar work with another tobacco variety showed that many factors affect the plant's response to ozone (MacDowall, 1965), although most of this variability could be accounted for by incorporating the coefficient of evaporation into the equation that relates degree of exposure to damage (MacDowall *et al.*, 1964). Manning and Feder (1980) go on to argue that plants are cheaper and easier to use, especially at remote sites, than are instruments for chemical analysis. This may be so, although convincing evidence for this assertion is desirable. They stress the need for careful breeding and selection of plant material, and careful control and measurement of environmental conditions at the monitoring sites, if meaningful results are to be obtained. Moreover, many other factors can produce symptoms in plants similar to those caused by pollutants, and diagnosis is a skilled task (Taylor *et al.*, 1986). At a simpler level, plants are sometimes used just to detect when and if exposures are sufficient to produce symptoms. For example, the tobacco variety Bel-W3 has been used to detect episodes of high ozone levels in the British Isles (Ashmore *et al.*, 1980).

These advantages of biological monitors can often be more apparent than real. It is indisputable that organisms can sometimes be easier to analyse than abiotic samples, but other important needs should also be met:

(1) The sample of plants or animals taken for analysis should represent the population in a consistent manner. This may be difficult to achieve. We have already discussed seasonal fluctuations in amounts of pollutants in animals (pp. 163, 188), and it may also be difficult to obtain a random sample of animals. Animals already dead are not necessarily representative (e.g. Table 8.1). If exposures are high enough to kill some individuals, then there is a cut-off point—a concentration that will not be exceeded because the organism will be dead. For lesser exposures that

have sublethal effects, those individuals most affected may be under- or over-represented in the sample. For example, carabids (ground-dwelling beetles) are commonly sampled by pitfall traps: the probability of capture increases with the individual's degree of activity, so biased samples will be obtained if the higher exposures alter activity. Accumulator species, those species which acquire high levels of pollutant, are sometimes recommended for monitoring without awareness of their potential limitations.

(2) The amount within the organism should be a function of that in the environment. This relationship differs between both species and pollutants. Baker and Walker (1990) described three patterns of metal intake by plants (Fig. 8.1), with no simple connection between pattern of uptake and degree of tolerance to the pollutant. Field data for copper and zinc levels in the marine worm *Nereis diversicolor* illustrate two of these patterns of uptake (Fig. 8.2). Levels of zinc were much more regulated than those of copper, and were relatively independent of ambient concentrations. Some species do regulate their content of essential heavy metals, when, for aquatic habitats, sediment analysis is indubitably to be preferred for a measure of environmental contamination. Other factors, too, can sometimes be important. Luoma and Bryan (1978) collected specimens of the bivalve *Scrobicularia plana* from 17 estuaries in south and west England, kept them for a week in clean seawater to allow undigested sediments to be excreted (see Bryan *et al.*, 1985), and then analysed the bodies—minus shells—for their total lead content. Samples of the fine particles (<100 μm diameter) in the surface sediment layer were also taken, digested with 1 N hydrochloric acid, which dissolves much but not all of the lead, and this dissolved lead was measured. The results gave a positive correlation with the amounts of lead in the bivalves ($r=0.69$) (Fig. 8.3a), with less variation than for analyses for the total lead content of sediments after digestion with nitric acid ($r=0.61$). Nevertheless, there was still a considerable degree of variation, of which a large part was due to differences between bivalves from different estuaries. A much closer correlation between the lead contents of bivalves

Table 8.1 Effect of sampling method on the range of mercury concentrations found in the livers of pheasants (*Phasianus colchicus*)

Type of sample	Number of birds	Percentage of birds within various ranges of mercury concentrations (mg/kg)						
		0–1.0	1.1–2.0	2.1–5.0	5.1–10.0	10.1–20.0	20.1–40.0	>40.0
Found dead	102	49.0	18.6	15.7	10.8	2.0	1.0	2.9
Shot	78	59.0	16.7	15.4	6.4	2.6	–	–

Data from Borg *et al.* (1969).

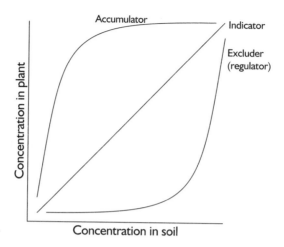

Fig. 8.1   Three basic strategies for the uptake of metals by plants. (Adapted from Baker and Walker, 1990.)

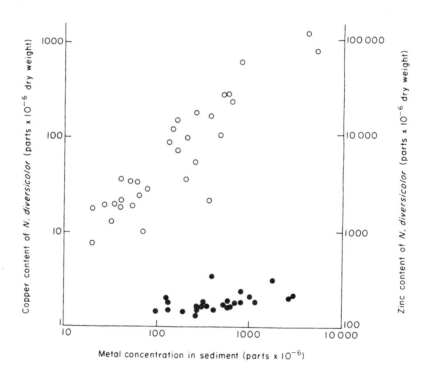

Fig. 8.2   Concentrations of zinc (●) and copper (○) in the tissues of the polychaete worm *Nereis diversicolor* (expressed as dry weights of tissue) taken from sites in more than 20 estuaries in Devon and Cornwall, and in the sediments at those sites. (From Bryan, 1976.)

and sediments occurred when the lead content of the sediments was expressed as the ratio of the concentration of lead to that of iron, multiplied by 10^3 (Fig. 8.3b). The correlation between the lead content of *S. plana* and the lead/iron ratio in sediments was less good if the metals were extracted with nitric acid. Hydrochloric acid extracts less than half of the total iron, and Luoma and Bryan suggested that the method using extraction with hydrochloric acid estimated that part of the total iron content in sediments that controls the biological availability of lead to *S. plana*. Be that as it may, the lead/iron ratio gave good predictions of the lead concentrations in specimens of *S. plana* taken from five other estuaries in England and north-west France.

Jones and Hopkin (1991) found better correlations for cadmium levels between species of terrestrial isopods and snails than between these species and amounts in the soil. It is again likely that the proportion of available cadmium in soil samples differs from site to site, but age and body weight may also be relevant—certainly levels of copper in isopods increase with both (Hopkin, 1989).

This work illustrates two points of relevance for monitoring. The amount of pollutant within an organism does not necessarily give a very clear indication of the ambient concentrations unless a great deal is known about the particular pollutant, species and habitat (e.g. Newman and McIntosh, 1982). Moreover, the amounts in one species do not necessarily give much indication of the amounts in another species: Luoma and Bryan found virtually no correlation between concentrations of lead in *S. plana* and in samples of the seaweed *Fucus vesiculosis* taken from the

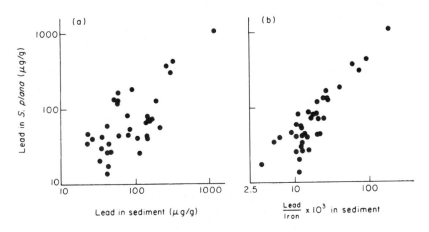

Fig. 8.3 Concentrations of total lead (μg/g dry weight) in the soft tissues of 37 samples of the bivalve *Scrobicularia plana* collected from 17 estuaries in south and west England, plotted against (a) the lead content of sediment particles and (b) the ratio of the concentration of lead to that of iron, multiplied by 10^3, in sediment particles. (From Luoma and Bryan, 1978.)

same sites ($r=0.04$). They suggested that *Fucus* might indicate the amount of available lead in the water. Again, studies on amounts of several heavy metals in *F. vesiculosis* suggest that relationships between exposure and metal content are complex (Bryan and Hummerstone, 1973); the part of the plant that is analysed, season, height in the intertidal zone, light, and competition from other metals can all affect this relationship.

In brief, a pollutant's distribution in the environment is best studied by analysis of the samples of direct interest. Analysis of biota as an indirect measure introduces additional complexities, and difficulties of interpretation. However, the chemical analysis of biota is sometimes easier because pollutant concentrations are higher, and in some circumstances this may become the overriding consideration, e.g. "mussel watch" (Goldberg *et al.*, 1978; Widdows and Donkin, 1992).

(3) The argument that analysis of organisms indicates a pollutant's availability can be used both ways. We have already seen (Fig. 8.3) that analysis of pollutant in the environment may give only a partial idea of availability, but conversely, different species do not necessarily follow the same pattern. This is perhaps an extreme example: the presumption is that lead in *S. plana* is derived primarily from that in the sediments, whereas lead in *F. vesiculosis* is derived from that in the water. However, even species with more similar exposures are likely to differ in some relevant aspects of their niches. Bryan (1980) concluded that there is no universal indicator of environmental contamination, and that one should therefore analyse several species, with different types of exposure, e.g. a seaweed, a filter feeder and a deposit feeder. Moreover, it may be difficult to interpret a change of pollutant level.

Eggs of the shag (*Phalacrocorax aristotelis*) taken from colonies on the Isle of May, off the south-east coast of Scotland, and on the Farne Islands, off the north-east coast of England, were analysed for p,p'-DDE and dieldrin each year from 1964 to 1971 (Coulson *et al.*, 1972). The general trends, indicated by the regression lines, were consistent with the known uses of dieldrin and of DDT, the precursor of DDE, at that time. However, in 1968, concentrations of both compounds in eggs from both colonies were much lower (Fig. 8.4). This was not thought to indicate decreased amounts of environmental contamination, but was tentatively explained as the result of a temporary change in the ecosystem in that year, when many sea-birds, fish and molluscs in the area died by poisoning from neurotoxins, produced by a dinoflagellate bloom ("red tide"). Neurotoxins may have interacted with pesticide residues in surviving birds to reduce levels in their eggs.

(4) The final argument for analysis of organisms in preference to the abiotic environment is that organisms integrate the amount of pollutant that is present. In the strict sense, that all of the pollutant that enters the organism is retained, this is untrue. Organisms give some sort of an average

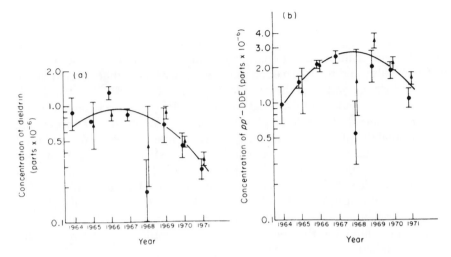

Fig. 8.4 Quadratic regressions for the concentration of (a) dieldrin and (b) p,p'-DDE in samples of eggs taken from two colonies of shags (*Phalacrocorax aristotelis*) during the years 1964–1971. The results for 1968 are excluded from the regression. (●) Eggs from the Farne Islands; (▲) eggs from the Isle of May; confidence limits (of unstated probability values) are shown by the vertical lines. (From Coulson *et al.*, 1972.)

measure for fluctuating exposures, although the influence of the most recent exposures will contribute more to the amounts of pollutant present than will earlier exposures. Some abiotic systems, such as some air samplers, give truly integrated samples. It is debatable though whether integrated samples are always to be preferred. In many situations exposures fluctuate, and the fluctuations are of interest both for understanding the distribution and kinetics of a pollutant in the environment, and for understanding biological effects. There were intermittent failures of oyster larvae to develop and settle satisfactorily in the Conway estuary in North Wales (Thornton, 1975). One contributory factor appeared to be zinc, released from water draining from old mine workings into the rivers. Water samples taken at the hatchery had marked periodic fluctuations in concentrations of soluble zinc, with values ranging at the most extreme from 0.5 to 470 µg/l. These fluctuations corresponded to the cycle of neap and spring tides, and would in all probability be important in determining the biological effects of zinc contamination. Occasional, or truly integrated, samples would not have given a very realistic picture of the actual pattern of exposure.

Two other difficulties also need to be noted: the problem of "outliers", and the effect of age and body weight on body burden of pollutant.

It is a common experience, even with well-designed sampling programmes, to find that a small proportion of the biological samples contain aberrant amounts of pollutant when compared with the range of values found in other seemingly similar samples (e.g. Wright *et al.*, 1985). Such samples are often excluded from further consideration. Values were excluded from one major international programme if the difference between the extreme value and the mean of all the samples was more than three times the standard deviation for all the sample values (OECD, 1980). Sokal and Rohlf (1995) describe more sophisticated variations of this approach, and also point out that if such tests are to be objective the population should approximate to a normal distribution. Some samples may contain no detectable pollutant, when the limit of detection needs to be stated. Many analytical techniques give slightly variable responses for blank samples, when the limit of detection can be calculated, for any given degree of probability, as a multiple of the variability of responses to blank samples (Welz, 1985). In the OECD (1980) programme samples below the limit of detection were assigned values of one-half the limit, but the appropriate way to handle such data will depend on the detailed circumstances.

There is a great deal of field data to suggest that an organism's whole-body concentration of pollutant is influenced directly by weight (size or age) (Phillips, 1980). Relationships can be complex: the concentration of total mercury in herring (*Clupea harengus*) increases with age, but, during just the first two years of life, within groups of similar age, larger fish have lower concentrations (Braune, 1987). It is therefore common practice in monitoring programmes to specify a standard weight for individual specimens of the biota. This approach can run into difficulties of interpretation if the regression coefficients of body burden on body weight differ from different samples. When the regression coefficients differ, the rank order of samples for degree of contamination could be altered by the choice of standard weight (Moriarty, 1985b). The best approach is to accept that these regression coefficients may differ, and to test the possibility by analysis of covariance (Moriarty *et al.*, 1984). If the regression coefficients are not significantly dissimilar, then a common coefficient is estimated for all of the data, and differences between samples can be estimated efficiently without need to select a standard weight. Comparisons between samples are of course not easy to make if the different samples have different regression coefficients or, as sometimes happens, have significantly different body weights (Moriarty, 1985b). Thus, the values for body burden of DDE in Fig. 8.5 were taken to indicate that the degree of environmental contamination by DDE increased in 1975, but the data may equally well indicate simply that food suitable for young starlings was scarce in 1975.

The overall implication of these arguments is that analysis of organisms for pollutant content is to be preferred only when that species is of interest in its own right, although a species may sometimes sensibly be used to monitor a pollutant's distribution in the environment if chemical analysis is thereby simplified. Conversely, the example of oyster larvae in the Conway estuary also

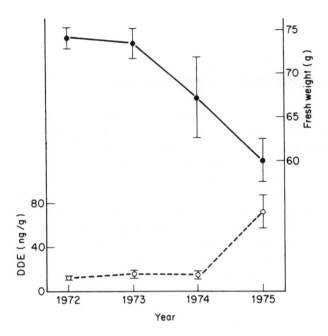

Fig. 8.5 Fresh weight (●—●) of 14-day-old nestling starlings (*Sturnus vulgaris*), and the *p,p'*-DDE content (○– –○) of their pectoral muscle, from samples of ten individuals taken for four successive years at Kvismaren, Sweden. The horizontal bars indicate the standard errors of the mean values. (Data from OECD, 1980.) (From Moriarty, 1985b.)

illustrates that, for biological effects, chemical analysis of abiotic samples alone is inadequate. The occurrence and relative abundance of different species does indicate, in biological terms, the nature of the abiotic environment.

A relatively recent development, in both the Federal German Republic and the USA, was to establish an environmental specimen bank, a store of biotic and abiotic samples from the environment, for indefinite storage at –190°C (Holden, 1979). These samples are to be stored in anticipation of future, as yet unforeseen, needs (Luepke, 1979). It is envisaged that this material will permit measurements of pollutants:

(1) As yet unknown or thought to be unimportant.
(2) In sites or specimens that are at present of no concern, but which may become so in the future.
(3) To give a historical record.
(4) With new improved analytical methods.

If these objectives can be fulfilled by this scheme, then it could be very useful. Doubts stem from the difficulties of designing an effective sampling programme when important details of the objectives are unknown.

Nevertheless, several countries have now established such banks, and Emon (1997) reviews the current situation.

Effects of pollutants on organisms

To monitor for effects has two aspects: the ability to detect changes, and the ability to ascribe causes to those observed changes. Many approaches are used. We have just discussed amounts of pollutant within organisms as a measure of amounts in the environment; they can also be used to assess the likelihood of effects (see pp. 193–202). It is generally appreciated that many other variables can affect the relationship between the amount of pollutant within an organism and the effect (e.g. Heinz *et al.*, 1979), but the most direct relationship will presumably exist between effect and the amount at the site of action. Unfortunately, even if the site of action is known, it may be difficult or impracticable to measure the amount of pollutant that is present. Amounts in other parts of the body can only give an index of the amount at the site of action if sufficient is known about the relationships between amounts in different parts of the body, or compartments.

Both the magnitude and the duration of previous exposure to a pollutant determine an organism's body burden. The compartmental model (pp. 158–162) implies, and experimental data illustrate (Table 8.2) that, in the

Table 8.2 Concentrations of dieldrin in three tissues of female rats fed in the laboratory on a diet with 50 parts × 10^{-6} dieldrin[a]

| Days fed | Concentration in liver | | Concentration in fat | |
| | Concentration in blood | | Concentration in blood | |
	Observed concentrations	Concentrations from equations	Observed concentrations	Concentrations from equations
1	21.5	19.3	137	39
2	24.6	22.6	122	240
4	21.7	25.3	207	362
9	25.6	29.3	407	505
16	36.6	32.2	667	620
31	53.6	33.9	1182	710
45	33.5	34.0	769	726
60	28.4	34.0	566	729
95	24.7	34.0	678	730
183	30.4	34.0	689	730

Original data from Deichmann *et al.* (1968); calculations that are based on compartmental models are derived from Moriarty (1975b).
[a] Concentrations are expressed as ratios of that in the blood, and have been calculated from both the experimental data and from the equations, derived from one-compartment models, that gave the best fit to the experimental data.

simplest situation of constant exposure, the relative amounts of pollutant in different tissues are near-constant only when each tissue has attained a near-steady state. Wild animals in terrestrial habitats are unlikely to have anything like a constant exposure, so that amounts in one tissue will be in varying proportions to those in any other tissue. In contrast to this deduction, Capen and Leiker (1979) concluded that samples of blood serum taken from 42 adult white-faced ibis (*Plegadis chihi*), collected at different times during the breeding season and analysed for DDE, predicted the concentrations of DDE found in subcutaneous fat and breast muscle (Fig. 8.6). Similar conclusions

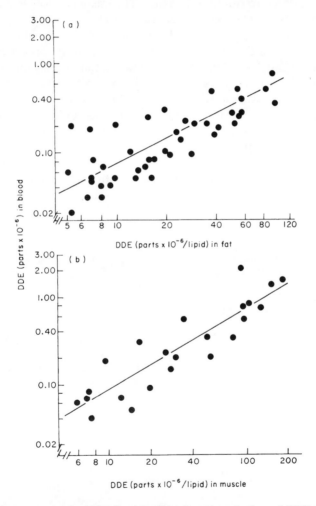

Fig. 8.6   Linear regressions for the logarithmic concentration of DDE in the blood serum of white-faced ibis (*Plegadis chihi*) on the logarithmic concentration of DDE in (a) subcutaneous fat and (b) breast muscle. Concentrations are expressed in parts $\times$ 10^{-6}: those for serum are based on fresh weight and those for fat and muscle on their lipid content. (From Capen and Leiker, 1979.)

were reached from experiments in which p,p'-DDE, PCB, dieldrin or methyl mercury was fed to mallard ducks (*Anas platyrhynchos*) (Friend *et al.*, 1979; Heinz, 1980), and from analysis of total mercury in field specimens of tawny owls (*Strix aluco*) and barn owls (*Tyto alba*) (Stanley and Elliott, 1976). However, to argue from a significant correlation or regression of this sort is a little misleading. Figure 8.7 shows the experimental data from which Table 8.2 was derived, with the concentrations of dieldrin in both liver and fat plotted against the concentration in blood. The correlations are obviously significant, although Table 8.2 shows that the relative concentrations in different tissues change as concentrations increase. The mean ratio of DDE in blood to that in fat or muscle in Fig. 8.6 changes by much less than a factor of two over the range of concentrations found, but the residual variation about the regression line, for individual pairs of concentrations, is appreciably greater, and it is the magnitude of this residual variation that is relevant. A large part of this variability is likely to result from different birds having had different past histories of exposure to DDE, which will have consequences of the type indicated in Table 8.2. Variability in past exposures could explain the lack of correlation found between the PCB concentrations in some tissues of the common dolphin (*Delphinus delphus*) (O'Shea *et al.*, 1980).

To put it another way, as the duration of a constant exposure increases, so do the concentrations in all tissues increase, until a steady state is attained. It follows that concentrations in different tissues are correlated. At the same time, the relative concentrations in different tissues are changing, until a

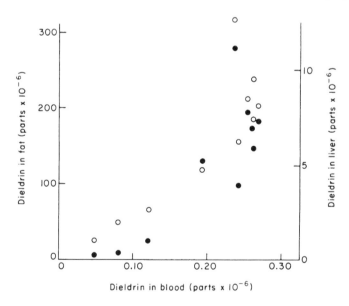

Fig. 8.7 Correlations between the concentration of dieldrin in the blood of female rats fed on a diet containing 50 parts × 10⁻⁶ dieldrin and the concentration of dieldrin in the liver (●) and the fat (○). (Data from Deichmann *et al.*, 1968.)

steady state is attained. Therefore, birds with different levels of exposure, which also have similar concentrations in one tissue, are likely to have different relative concentrations in other tissues. The degree of these differences would be indicated by the residual variation about the regression line.

In summary, two deductions may be drawn. In particular, the concentrations of pollutant in one tissue cannot, except in conditions of sustained constant exposure, give a very precise measure of amounts in other parts of the body. In general, more consideration might be given to estimating the total body burden of pollutant; if concentrations are measured in specific tissues or organs, the reasons for that choice need to be clearly decided at the outset.

To know the amount of pollutant in an organism has some value, but many variables can affect the dose–response relationship. Moreover, pollutants cannot always be readily detected within organisms, and effects are sometimes indirect, mediated through other species or the abiotic environment. Other indices or measures of effect are also needed.

Such measures are most useful if they reflect exposure to a specific pollutant or group of pollutants. For example, inhibition of AChE in the brain can be a useful biomarker for the risk of death from organophosphorus insecticides (pp. 134–141). However, the degree of inhibition of other esterases does not indicate degree of effect from these insecticides very well, because these other esterases are not part of the toxic biochemical lesion (Sanchez *et al.*, 1997).

At a higher level of organization, imposex (pp. 141–144) indicates the presence of TBT. Imposex develops during the individual mollusc's life, so in the field the degree of imposex is a function of the time, duration and degree of exposure. Two quantitative measures of effect have been developed for the dog-whelk (*Nucella lapillus*) and other molluscs:

(1) Size of the female penis (Fig. 8.8a). For concentrations of TBT in seawater of less than about 0.2 ng Sn/l one can record the incidence of females with a penis. At higher concentrations virtually all females have a penis, and the mean penis length of females within a population is compared with that of the males. This relative measure allows for differences of size between populations. Only over a limited range of concentrations does the response alter with exposure. Above 3 ng Sn/l the response is unaffected by changes of concentration.

(2) Development of the female vas deferens (Fig. 8.8b). Again the response only increases with exposure over a limited range of concentrations.

These two measures of imposex have been used extensively with several species both to survey the occurrence of TBT in coastal waters and to monitor long-term changes in the degree of pollution by TBT (Gibbs and Bryan, 1994).

Not all biomarkers are so specific to particular pollutants. A considerable amount of research has been published on measures of health (see p. 150)

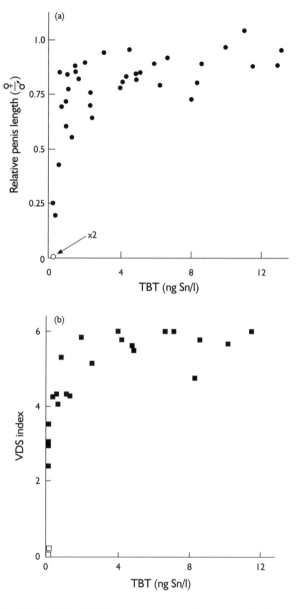

Fig. 8.8 Effects of tributyl tin (TBT) in ambient seawater on populations of dog whelks, *Nucella lapillus*, taken from sites around south-west England 1984–1986. Open symbols indicate values for two uncontaminated sites in the Isle of Mull, west Scotland. (a) The mean relative lengths of male and female penes in different populations. Usually the cube of these values is given, expressed as a percentage, to indicate the relative penis volumes, but the variance then increases markedly with the index. (b) The degree to which the females have developed a vas deferens. Six easily identifiable developmental stages are scored successively 1–6 and a mean value or index calculated for the vas deferens sequence (VDS). (Adapted from Gibbs and Bryan, 1994.)

that may detect degrees of, or effects from, sublethal exposure to pollutants, but the wider the range of possible causes for a change in the biomarker the greater is the care needed to determine whether pollutants are involved. Thus mono-oxygenase activity (p. 169) is induced by a wide range of pollutants, but also by a wide range of naturally occurring compounds, and other environmental variables also affect the degree of activity. Induced activity may indicate effects by one or more pollutants, but additional information is needed.

In essence, our concern is with the size, structure and distribution of populations of individual species, and ideally we would monitor the impact, if any, of a pollutant on all species within a community.

Most attempts to monitor whole communities have been made with aquatic habitats. Interest in the impact of pollutants on these communities goes back at least to the beginning of this century, when the principal effects came from organic pollution—sewage in particular (Hellawell, 1978). This is a complex situation: sewage introduces a whole range of organic and inorganic compounds, suspended solids may have simple physical effects, and the water can become deoxygenated (Hynes, 1960). The normal biological processes decrease the magnitude of these factors as one moves downstream from a discrete source, and Kolkwitz and Marsson described three main zones in this process of self-purification:

(1) The polysaprobic zone, where the water contains little if any dissolved oxygen, and chemicals are usually in a reduced state. Bacteria are abundant.
(2) The mesosaprobic zone, where complex organic compounds are still present, but the water contains some dissolved oxygen and oxidation reactions occur.
(3) The oligosaprobic zone, where oxidation is complete.

These ideas were first published in 1902, and Kolkwitz and Marsson (1908, 1909) later enumerated extensive lists of plant and animal species characteristic of these zones, when the mesosaprobic zone was further subdivided into α and β (strongly and weakly mesosaprobic) zones. Sládeček (1973) detailed the subsequent elaboration of this system, in which Wuhrmann eventually described 15 zones. For any one river, with a single source of sewage, one could continue this process of subdivision indefinitely. Indeed, the purpose of Kolkwitz and Marsson's scheme was to measure the degree of water purity ("Reinheitsgrades") by the species present. They were using species as indicators of the degree of organic pollution.

Several attempts have been made to quantify the data from investigations of this type (see Bick, 1963). This is desirable scientifically, to facilitate assessment of data, in particular comparisons between sites and within sites at different times. The use of numerical, or biotic, indices is sometimes justified also by the need to convey the relevant information to non-specialists, indeed to non-biologists, although to some extent this laudable attempt appears doomed to failure: with any system of units, unless it is understood by the user, misunderstandings will arise (Buffington and Little, 1980). The simplest approach has been to number the saprobic zones as 1–4 instead of oligosaprobic-polysaprobic, but several more sophisticated methods have been proposed. For example, Pantle and Buck developed an index of saprobity (S):

$$S = \frac{\Sigma sh}{\Sigma h}$$

where each species is given a score (s) of 1–4 according to which zone it characterizes, and h is the measure of relative abundance (1, 3 or 5, for rare, frequent or abundant). Washington (1984) gives a comprehensive review of saprobic indices, and they are still widely used in continental Europe.

All such schemes, qualitative or quantitative, are only as good as their assumptions are valid. Indicator species are not equally sensitive to all pollutants, and they do not represent community structure. The original scheme of Kolkwitz and Marsson developed from observations on slow, evenly flowing rivers. In practice, these schemes presuppose detailed knowledge both of the river's ecology and of the pollutant effects, but many rivers and many instances of pollution do differ considerably in detail. Hynes (1960) concluded, from several European studies, that although chemical and biological assessments of pollution have always agreed in distinguishing between clean and severely polluted rivers, the detailed results have not fitted so well. The same discrepancies still exist (see Tables 9.2 and 9.3, and pp. 269–274).

The nub of the difficulty rests in the concept of pollution or, conversely, of water purity. Chemical analyses can yield information on the amounts, distribution and movement of contaminants and biological samples can yield information on the distribution and numbers of species, but the link between the two is neither constant nor simple.

Several attempts have been made to overcome this incongruity by relating the species present in a wide range of unpolluted sites to a wide range of physical and chemical features of those sites. It might then be possible, given the environmental characteristics of any site, to predict the species that should be present if there were no pollution. Wright et al. (1984) sampled macro-invertebrates by standardized sampling methods at 228 sites on 41 British rivers. Although free from serious pollution, it is unlikely that any of these sites were completely unaffected by man's activities. Most organisms were identified to species, although those species difficult to identify were combined into larger groupings. Multivariate analysis of the species lists from individual sites yielded 16 groups or communities, whose occurrence was then correlated with 28 environmental variables. The predictive ability of this analysis was then tested by comparing the predicted and observed results for 40 other sites that had been sampled at the same time. For half of the 40 sites, predicted and observed groups were identical and for another quarter the correct allocation was to the second most probable group. Conversely, of course, for one-quarter of the sites environmental variables did not predict particularly well which group of macroinvertebrates was present. This relative degree of success must be qualified: only a limited range of species was considered, and no account was taken of population size.

This approach is now being simplified. The Biological Monitoring Working Party (BMWP) identified macroinvertebrates not to species but to family, which simplifies identification enormously. Each family has a score within the range 1–10: the lower the score assigned to a family the more tolerant it was deemed to be of organic pollution and the consequent depletion of oxygen. The total BMWP score obtained for any site can then be divided by the number of families recorded to give the average score per taxon detected (ASPT). Predicted and observed values were then compared for the same 40 sites (Fig. 8.9). There is a significant correlation, better for averages than for scores, but any predicted value has a considerable degree of uncertainty. That is to be expected: no distinction is made between species within families, number of species per family is ignored and great care is needed to ensure a standardized sampling technique because the samples are unlikely to represent all of the families that are present (see also Underwood, 1994). Advantages are that the system does take some account of environmental differences apart from pollution, and data are easy to obtain. Proponents hope this approach may be applicable to a very wide range of catchments (Abel, 1996). A computer program (RIVPACS, the River Invertebrate Prediction and Classification Scheme) has now been developed to calculate observed and predicted BMWP scores, ASPT, and similar measures. One application is discussed on pp. 271–274.

✳

Attempts to devise biotic indices of pollution, based on the abundance, presence or absence of different species, suggest the more general question, quite

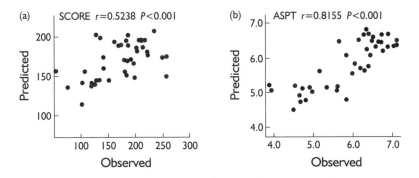

Fig. 8.9 Correlations between predicted and observed values for macroinvertebrates at 40 sites of "good" or "fairly good" quality in British rivers. Values based on three seasons' combined data. (a) The Biological Monitoring Working Party (BMWP) score. (b) The average score per taxon (ASPT). (From Armitage *et al.*, 1983.)

apart from any effects that pollutants may have, of whether there are any general laws that govern the composition and structure of communities that contain many species (Pielou, 1975). If so, one might hope to describe communities by equations, and perhaps use these to detect effects by pollutants. Preston (1948) made one of the earlier studies, in which he concluded, from two sets of bird counts and three collections of moths caught in light-traps, that the data for numbers of different species seen or caught conformed to a log-normal distribution (Fig. 8.10). Preston used musical terminology, in which the number of individuals determined the "octave" to which the species belonged. Each octave, frequency class or interval covered a range of frequencies up to double the maximum of the previous octave. The essential features of such data are:

(1) Invariably, the community that is studied excludes many other species that occur in the study area. It is therefore, for ecotoxicological studies, a very incomplete census of species present.
(2) The larger the sample, the area sampled, and/or the longer the time for which sampling continues, the more species that will be found, though at an ever-declining rate. Rarely is it possible to make a complete census of all species and individuals present.
(3) In general terms, one or two species are relatively very abundant, and Preston argued that, if the sample is large enough, there will also be few very rare species, although these will tend to be discovered only when

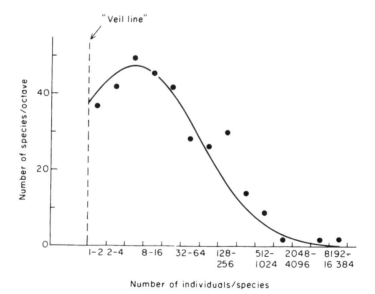

Fig. 8.10 The frequency distribution of numbers of individual moths of each species caught at a light-trap at Orono, Maine, 1931–1934. The total number of moths caught was 56 131. (Original data from Dirks; taken from Preston, 1948.)

relatively large samples are taken. Usually, these rare species will be to the left of the "veil line", not represented in the sample, which will contain few specimens of a relatively large number of species. Most species occur towards the middle of the whole frequency distribution curve.

(4) No attention is paid to the characteristics of the individual species, although there is good evidence to suggest that if one species is removed from a community, it may be replaced by another species (Heatwole and Levins, 1972).

Preston argued that the frequency distribution is log-normal, so that the number of species (y) that occurs in the Rth octave either side of the mode is

$$y = y_0 e^{-(aR)^2}$$

where y_0 is the number of species in the modal octave and a is a constant. Thus the species frequency distribution can be described by two parameters (y_0 and a), and Preston (1962) later condensed these to a single parameter. This equation does appear to describe a general rule when the total number of species is large enough, but is as much a reflection of the statistics of large numbers, with many independent factors that interact in a multiplicative manner, as of any essentially biological features (May, 1975), although Stenseth (1979) has developed a biological theory to explain this distribution.

Patrick (1973) applied this model to diatoms in freshwater streams, and found that for a total count of 5000–8000 specimens a typical curve of this form was obtained, with the mode of more than 20 species in the second to third interval, and a total range of 10–12 intervals (Fig. 8.11). Organic pollution first caused some species to become more abundant, so that the curve covered more intervals, usually 13–15, and above a certain degree of pollution the height of the mode decreased. Toxic pollutants simply reduced the height of the mode, with little effect on the number of intervals.

Gray and Mirza (1979) and Gray (1981a) then developed this approach with data from coastal benthic communities. The log-normal distribution was transformed to a straight line by plotting probits of the cumulative percentage of all species found against frequency classes, when increasing degrees of organic pollution cause the line first to exhibit two distinct sections of different slope, and then for all of the data to conform to a single new line of lesser slope (Fig. 8.12). This approach may be effective in detecting some types of change in community structure, but to explain such changes may be difficult if no other information is available. Natural events such as storms or episodes of low oxygen concentrations may cause similar responses. Conversely, studies on similar Norwegian communities exposed to pollution from heavy metals showed no such effect on the frequency distribution of species (Rygg, 1986).

Much sophisticated mathematical theory has been devoted to frequency distributions, all models concentrate on the number of species present and

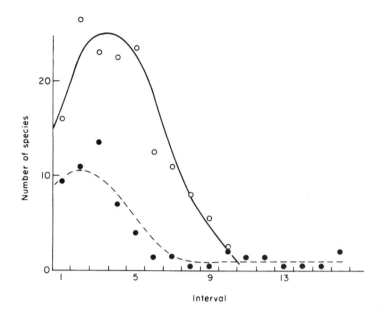

Fig. 8.11   The frequency distribution of numbers of individual species of diatoms in two freshwater streams. (O—O) A sample from an unpolluted stream, Ridley Creek, Chester County, Pennsylvania; (●– –●) a sample from a stream with organic pollutants, Back River, Maryland. The first interval contains species with 1–2 individuals. The maximum number of individuals in each successive interval is double that of its predecessor, so that the maximum number of individuals in the xth interval is 2^x. (From Patrick, 1973.)

their relative abundance, and there is much disagreement about the use and value of the various models that have been proposed (e.g. Taylor *et al.*, 1976). Southwood (1978) emphasizes that much thought is needed if one is to avoid using an inappropriate model or index. The objectives should be clearly stated, and it is necessary to define carefully the boundaries of the community, the period of time within which the community is to be measured, and which species are to be considered.

A complete census would give the total number of species, or species richness, sometimes called diversity, and the numbers of each species, again called the diversity, or evenness, for that community. One must note that diversity has recently acquired some additional meanings. Rosen coined the term biodiversity in 1985, synonymous originally with biological diversity, but now used as a generic term for three distinct levels: genetic differences within species, the traditional meaning of differences in number of and/or between species within a community, and differences between communities in species present (Harper and Hawksworth, 1994).

Maximum diversity occurs when all species have the same number of individuals. Several non-parametric indices have been developed, which combine these two attributes in a single number. These indices have the virtue

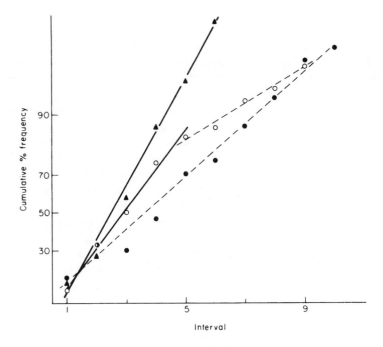

Fig. 8.12 Linear regressions for the frequency distributions of the benthos in Loch Eil, Scotland. All species with only one individual were allotted to the first interval, and the maximum number of individuals in the xth interval is $(2^x - 1)$. The cumulative percentage frequency of all species found is expressed on a probit scale. (▲) Results for 1966, before a pulp mill started; (○) results for 1968, a transitory phase with a pronounced deviation from a log-normal distribution; (●) results for 1970, the polluted phase. (Original data from Pearson, 1975; taken from Gray and Mirza, 1979.)

that they make no assumptions about the theoretical distribution to which the data conform, although it then becomes rather difficult to say what biologically meaningful attributes they do measure. The most widely used of these indices, the Shannon–Wiener index (H') (sometimes miscalled the Shannon–Weaver index (Krebs, 1994)), is defined as

$$H' = -\sum_{i=1}^{S_T} p_i \log_e p_i$$

where p_i is the proportion of the community belonging to the ith species, and S_T is the total number of species. This index will distinguish between different types of frequency distribution when S_T is large (some hundreds of species), but given the variability that is often associated with estimates of H', it is an insensitive measure of the frequency distribution when S_T is small (May, 1975). To a lesser extent, the same comment applies to other diversity indices. Pielou (1975) stressed that standard errors should be estimated. In practice, for pollution studies, this is rarely done, although there are likely to be at least two important components of error. Not only will there be variation between samples taken in the same place at the same time, but in at least some situations there is likely to be a pronounced seasonal effect. In practice, the sample is equated with the whole community, a dubious procedure.

These non-parametric indices can be criticized on several grounds (Hurlbert, 1971). A central difficulty is the appropriate relative weights to be given to species richness and to the relative abundance, or evenness, of different species. There is no general agreement about this: hence the plethora of indices. Moreover, these measures take no account of the relative importance or ecological function of different species, but assume that all species are of equal significance. Finally, these indices have to be restricted to roughly comparable groups of species. Otherwise, differences between different groups may cancel each other out, with one group's diversity increasing as another group's decreases: Morris (1971) found that cessation of grazing on grassland can eliminate annual and low-growing plants, but increases the diversity of some insect groups. Moreover, the smallest, most numerous species, such as bacteria, would determine the relative abundance of different species.

However good or bad the theoretical bases of species-abundance distributions and of diversity indices may be, the important question in ecotoxicology is whether these measures will detect changes in community structure caused by pollutants. Clearly, these measures will detect gross changes, but it seems likely that the effects of pollutants on a relatively limited number of species may be rather difficult to detect by such measures, especially if the species are relatively uncommon. One might expect different indices to give similar trends when the species-abundance distribution changes, but this is not always so (Hellawell, 1978, 1986; Chadwick and Canton, 1984). Gray (1981b) has concluded that, for benthic communities, changes in diversity are only likely to become statistically significant when half the species are lost, by which stage the changes are obvious anyway.

To summarize, biotic and diversity indices have many defects:

(1) Diversity indices are unaffected if one species replaces another, and are unlikely to be affected if only a few species, especially uncommon species, disappear.
(2) Biotic indices can only be devised if a great deal is already known about the ecosystem and the pollutant.
(3) Both biotic and diversity indices need accurate identification to individual species. Related species can react very differently to environmental changes (Resh and Unzicker, 1975), and identification to only generic or family level can invalidate the use of diversity indices (Wu, 1982).
(4) Only a limited range of species is included in any index of community composition.

(5) Communities are not supra-organisms (see Chapter 3), so it is difficult to attach any biological meaning to an index of diversity.

(6) Any changes in an index will need to be explained in terms of the dynamics of individual populations before the effect of a pollutant can be established.

We have already discussed the classification of terrestrial communities by degrees of similarity (p.58), and indices of similarity are now being developed for the assessment of water pollution (Abel, 1996). They need further evaluation to test whether they are more sensitive to changes in community structure than are diversity indices (Washington, 1984).

The possibilities for monitoring range between relatively superficial observations on a wide range of species that are frequently assumed to represent the whole community, and relatively intensive studies on one or a few species.

Monitoring of named species

We discussed in Chapter 3 the structural and functional analyses of ecosystems—study of the number and sizes of populations, or of the flow of energy and cycling of nutrients. Sheehan (1984) regards an ecosystem as essentially an energy-processing unit, and Cairns and Pratt (1986) point out that effects on the usually overlooked micro-organisms can cause important changes in energy flow. Römbke and Moltmann (1996) then argue that both functional and structural measures of community well-being are needed, because it is often difficult to predict effects on ecosystems from changes in populations. It is true that the links between community structure and function are complex, but one must first answer two questions. First, of values—what do we consider important, and second of practicalities—how can we best detect changes? One has to study the system of interest at the appropriate level of complexity, and I would suggest that study of structure is usually more appropriate than study of function (see p. 66).

Monitoring of named species is the more incisive approach. Although most species within the community are then likely to be ignored, clear objectives can be stated from which practicable programmes of monitoring can be devised, and any changes are relatively easy to interpret.

The lack of information about most of the species within a community is less serious an omission than might at first appear, for two reasons. First, changes in one population will sometimes have consequences for other populations. Secondly, in practice, there is relatively little concern about more than a very limited range of species. For example, the American Clean Water Act sets the goal of restoring and maintaining the biological integrity of the nation's waters (United States Environmental Protection Agency, 1990). This is a nebulous concept, whose only practical interpretation could be to ensure that aquatic biota are unaffected by man's activities. That is an impracticable goal: eutrophication, sewage effluent, control of river flow and many other factors have altered the aquatic environment significantly. In practice, this goal was interpreted as the protection of fish, shellfish and wildlife, when one is again concerned with individual species. The only alternative would be to measure aspects of a community's activities, such as the standing crop and flux rates for energy and nutrients, but such measurements are difficult both to take and to interpret in terms of individual species. Mann and Clark (1978) argued that individual organisms, with their great variability, compensate for each other in indices of community performance such as diversity and productivity, so that changes in the values of indices of the whole ecosystem can be explained in meaningful terms. It is doubtful whether this is true for any but the most obvious impacts.

One point does need to be emphasized. Detection of change, by itself, does not necessarily tell one anything about the biological effects that pollutants may be exerting (Glover, 1979). Populations and communities do change from time to time, and it is usually difficult to discover the causes of these changes. It is because of these difficulties that the need for base-line data is frequently urged. If an index or measure of a population or community is taken for a long period of time, before the risk of impacts from pollutants, it should then be less difficult to interpret changes that occur when pollutants may be having an effect. Figures 2.11 and 9.1 illustrate this point for two single populations, although in both of these examples the degree of random fluctuation from year to year is rather small. However, data of this sort cannot, by themselves, constitute complete proof or disproof that a species is being affected by pollution. In essence, at best, one acquires correlative data where changes in the chosen measure of biological performance correspond to changes in the measure of contamination. Absolute proof that the observed biological changes are caused by the observed changes in contamination is rarely obtainable.

Many, sometimes conflicting, criteria can affect the choice of species for a monitoring scheme. Moore (1966b), Butler et al. (1971) and Bryan et al. (1980) list desirable characteristics in species selected for monitoring pollutant levels. Individual organisms should be:

(1) Widely distributed, and abundant where they occur. This minimizes the risk that the populations will be affected if samples are taken.
(2) Easy to identify and collect.
(3) Of a convenient size. The definition of convenient is not straightforward. Small organisms can be grouped together to give adequate material for chemical analysis; large organisms necessitate, or permit, removal of specific organs or tissues for analysis. Large animals are usually easier to observe and identify, and tend also to have stable populations, which makes it easier to detect significant changes in population size.

(4) Sedentary at most stages of their life cycle. It can be argued that more mobile species will integrate the degree of pollution over a wider area, but detailed understanding of results is likely to be difficult.

(5) Easy to age.

(6) Contain levels of pollutant that are neither so low as to make chemical analysis difficult nor so high as to either affect or kill individuals, when samples may be biased.

Except for the last qualification to point (3), these criteria refer solely to species selected for measurement of pollutants. When monitoring for effects is considered, it is also highly desirable that the species chosen should:

(1) Be suitable for laboratory studies, because most problems of pollution need a combination of field observation and laboratory experiment before they can be resolved.

(2) Belong to species of aesthetic, economic, educational, scientific or sporting interest, because these are the species for which we have most biological information. They are also the species that we most wish to protect from pollutants.

(3) Where possible, in particular where we have sufficient knowledge, be amongst the first species to be affected by the pollutant. Whether or not this means species at the ends of food chains will depend on many factors (see, in particular, pp. 181–193).

There are, in general terms, three ways in which effects on a species population can be detected: changes in the structure and performance of individual organisms (see Chapter 6), changes in size, structure and distribution of populations (see Chapter 2), and changes in the gene pool (see Chapter 4). We will consider, briefly, examples and possibilities of each approach.

The British population of the peregrine falcon (*Falco peregrinus*) illustrates not only effects by pollutants on both individuals and populations, but also the need for monitoring. Shell thinning (see pp. 144–149) in birds was first discovered in response to complaints by the owners of racing pigeons in Great Britain that the population of peregrine falcons had increased, and that they were killing more racing pigeons than previously. In fact, a survey showed the very opposite, that the numbers of peregrine falcons had declined during the previous 10–15 years, and it was the subsequent investigations to discover why the number of breeding pairs of falcon had halved that led to the discovery of thin eggshells (Ratcliffe, 1967) (see pp. 255–260 for a more detailed account). If we did not readily become aware of such an obvious change as the virtual elimination from some parts of England of a species of great interest to naturalists, in a country renowned for its interest in natural history, there seems very little likelihood that any but the most devastating impacts by pollution will be spotted unless we are consciously on the alert, and have suitable monitoring schemes.

The degree of shell thinning could be monitored (e.g. Fig 9.1) to assess the degree of impact on a population by DDT. Provided that there is a reasonable relationship between degree of exposure and degree of thinning (e.g. Fig. 6.7), then thinning is a useful index. However, there are two complications:

(1) DDT (or DDE) is not the sole possible cause of thin eggshells (Cooke, 1973). Other factors, such as starvation, can have the same effect.
(2) The effect on populations can differ for different species. Some North American species of predatory bird slowly declined in numbers because of reduced breeding success, associated at least in part with thin eggshells, whereas European species that had thin-shelled eggs tended to maintain their population size until they suffered acute mortality from cyclodiene insecticides (Newton, 1979).

In the most general terms, the effects of pollutants on individual organisms are rarely uniquely different from the effects of all other factors, and the consequences of such changes can differ for different species or populations.

In a wider context, Bayne et al. (1979) point out that physiological responses such as growth integrate the effects of stressors on biochemical and cytological targets, and that such physiological properties are therefore unlikely to indicate the effects of specific stressors. This is the potential weakness of the use of effects on single organisms. Whatever measure is used to assess the organism's viability, a real damaging effect on the population may not be indicated, because no single measure can encompass all the biological activities that might influence the population. A secondary problem is that measures such as growth and lysosomal fragility do not identify particular stressors. Nevertheless, where modes of action or even pollutants are unknown, then this approach may prove to be very useful.

The simplest measure of population size is to record it as present or absent. Records on the occurrence of lichens in Great Britain extend back for nearly two centuries, and it is obvious that many species have become rarer or extinct during this period. Already, in 1859, Grindon invoked air pollution as the cause of the declining lichen flora in South Lancashire, and Nylander suggested in 1866 that lichens could indicate air quality (see Hawksworth and Rose, 1976). Since then, many studies have been made on the distribution of lichens, and Hawksworth and Rose (1970) devised a scale, based on the presence and abundance of lichens on the bark of trees, by which sites could be allocated to one of 11 zones (Table 8.3). These zones were related to mean winter concentrations of sulphur dioxide within the range of "pure" to > 170 μg sulphur dioxide/m³, and the scale was compiled from field observations of lichen species in areas with known degrees of contamination by sulphur dioxide. There is limited experimental evidence to show that these levels of sulphur dioxide do affect these species, but predictions of sulphur dioxide levels, from the lichen flora at specific sites, have been confirmed in some instances

Table 8.3   Zones, based on presence and abundance of lichens on bark, which indicate mean winter concentrations of sulphur dioxide in England and Wales

Zone	Moderately acid bark	Basic or nutrient-enriched bark	Mean winter SO$_2$ (μg/m^3)
0	Epiphytes absent	Epiphytes absent	?
1	*Pleurococcus viridis* s.1. present but confined to the base	*Pleurococcus viridis* s.1. extends up the trunk	>170
2	*Pleurococcus viridis* s.1. extends up the trunk; *Lecanora conizaeoides* present but confined to the bases	*Lecanora conizaeoides* abundant: *L. expallens* occurs occasionally on the bases	≈ 150
3	*Lecanora conizaeoides* extends up the trunk; *Lepraria incana* becomes frequent on the bases	*Lecanora expallens and Buellia punctata* abundant; *B. canescens* appears	≈ 125
4	*Hypogymnia physodes* and/or *Parmelia saxatilis,* or *P. sulcata* appear on the bases but do not extend up the trunks. *Lecidea scalaris, Lecanora expallens* and *Chaenotheca ferruginea* often present	*Buellia canescens* common; *Physcia adscendens* and *Xanthoria parietina* appear on the bases; *Physicia tribacia* appear in S	≈ 70
5	*Hypogymnia physodes* or *P. saxatalis* extends up the trunk to 2.5 m or more; *P. glabratula, P. subrudecta, Parmeliopsis ambigua* and *Lecanora chlarotera* appear; *Calicium viride, Lepraria candelaris* and *Pertusaria amara* may occur; *Ramalina farinacea* and *Evernia prunastri* if present largely confined to the bases; *Platismatia glauca* may be present on horizontal branches	*Physconia grisea, P. farrea, Buellia alboatra, Physcia orbicularis, P. tenella, Ramalina farinacea, Haematomma ochroleucum* var. *porphyrium, Schismatomma decolorans, Xanthoria candelaria, Opegrapha varia* and *O. vulgata* appear; *Buellia canescens* and *X. parietina* common; *Parmelia acetabulum* appear in E	≈ 60
6	*P. caperata* present at least on the base; rich in species of *Pertusaria* (e.g. *P. albescens, P. hymenea*) and *Parmelia* (e.g. *P. revoluta* (except in NE), *P. tiliacea, P. exasperatula* (in N)); *Graphis elegans* appearing; *Pseudevernia furfuracea* and *Alectoria fuscescens* present in upland areas	*Pertusaria albescens, Physconia pulverulenta, Physciopsis adglutinata, Arthopyrenia gemmata, Caloplaca luteoalba, Xanthoria polycarpa* and *Lecania cyrtella* appear; *Physconia grisea, Physcia orbicularis Opegrapha varia* and *O. vulgata* become abundant	≈ 50
7	*Parmelia caperata, P. revoluta* (except in NE), *P. tiliacea, P. exasperatula* (in N) extend up the trunk; *Usnea subfloridana, Pertusaria hemisphaerica, Rinodina roboris* (in S) and *Arthonia impolita* (in E) appear	*Physcia aipolia, Anaptychia ciliaris, Bacidia rubella, Ramalina fastigiata, Candelaria concolor* and *Arthopyrenia biformis* appear	≈ 40
8	*Usnea ceratina, Parmelia perlata* or *P. reticulata* (S and W) appear; *Rinodina roboris* extends up the trunk (in S); *Normandina pulchella* and *U. rubiginea* (in S) usually present	*Physcia aipolia* abundant; *Anaptychia ciliaris* occurs in fruit; *Parmelia perlata, P. reticulata* (in S and W), *Gyalecta flotowii, Ramalina obtusata, R. pollinaria* and *Desmazieria evernioides* appear	≈ 35
9	*Lobaria pulmonaria, L. amplissima, Pachyphiale cornea, Dimerella lutea,* or *Usnea florida* present; if these absent crustose flora well developed with often more than 25 species on larger well-lit trees	*Ramalina calicaris, R. fraxinea, R. subfarinacea, Physcia leptalea, Caloplaca aurantiaca* and *C. cerina* appear	< 30
10	*L. amplissima, L. scrobiculata, Sticta limbata, Pannaria* spp., *Usnea articulata, U. filipendulla* or *Teloschistes flavicans* present to locally abundant	As 9	"Pure"

From Hawksworth and Rose (1976); adapted from Hawksworth and Rose (1970).

by subsequent chemical analysis for sulphur dioxide. The presence and abundance of individual lichens is influenced by many factors, so sites have to be selected with care. The trees should be in open situations, within a specified size range, away from main roads, not in areas of known pollution other than sulphur dioxide, and not liable to spray drift from pesticides. Ideally, trees are all of one species, but in practice different species can usefully be separated into groups with similar bark characteristics. For example, trees with rough acid bark (particularly oak and ash, *Quercus* and *Fraxinus*) form a group. Smoke is commonly associated with sulphur dioxide, but smoke levels did not correlate very well with the type of lichen flora (Hawksworth and Rose, 1976). Other air pollutants, notably fluorides, also affect lichens, but appear to be relatively local in their impact, so that from a survey of the lichen flora it was possible to construct a contour map for England and Wales that indicated mean winter concentrations of sulphur dioxide (Fig. 8.13).

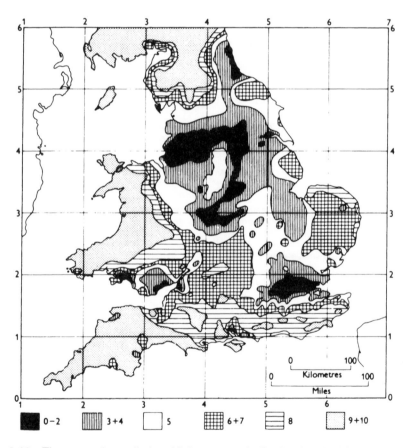

Fig. 8.13 The approximate limits of lichen zones in England and Wales during the years around 1970. The lower the number of the zone, the higher the degree of air pollution, principally from sulphur dioxide. (From Hawksworth and Rose, 1976.)

Lichens are relatively sensitive to sulphur dioxide, and can be taken as an example of the use of indicator species, but they do illustrate the qualifications made earlier (pp. 55–66). The lichen flora of England and Wales is well known, with a good understanding of the important environmental factors. One cannot, with safety, transfer this scheme to other areas without question: it has been found that similar exposures to sulphur dioxide may have different degrees of effect in eastern and western Europe. We also lack information on the dose–response relationship, which could conceivably be altered, for example, by relatively short exposures to relatively high levels in humid conditions (Treshow and Anderson, 1989). Recolonization when sulphur dioxide levels start to decrease again can be relatively rapid. It began within a few years for an urban area the size of London when mean winter levels of sulphur dioxide had fallen to 130 $\mu g/m^3$ (Rose and Hawksworth, 1981). The subsequent decline by the years 1985–1987 to levels of 29–55 $\mu g/m^3$ enabled lichens typical of zones 6 and 7 to reappear, although some species typical of the less sensitive zones 4–5 were still absent. This "zone skipping", recolonization without the return of species progressively lost when sulphur dioxide levels were rising, is attributed to the speed at which sulphur dioxide levels declined. There was insufficient time for conditions to be established that are suitable for some species typical of zones 4–5 (Hawksworth and McManus, 1989). In brief, unless a species is chosen because it is of interest in its own right, a lot of information is needed before a species can be of much value for monitoring biological effects.

We have already seen that the relationship between amounts of pollutant and their biological impacts is neither constant nor simple (p. 236). Similarly, pollutants act on individual organisms, but the focus of interest is on the population, and many variables can affect the consequences for the population of pollutant-affected individuals. However, there is one approach that may possibly integrate both the individual and the population aspects: effects on the gene pool. If the appropriate genetic variation exists within a population, and if that population is affected by a pollutant, then the nature of the gene pool will change. Conversely, if pollutants kill or otherwise affect some individuals, but these deaths or other effects have no effect on the population, then there will be no effect on the gene pool. This approach could allow one to differentiate between two populations of the same species, both affected to the same extent by a pollutant, but otherwise different in the extent to which other variables affect them. We discussed several examples of effects on the gene pool—melanism, tolerance of plants to heavy metals, and insecticide resistance—in Chapter 4. This approach does avoid the doubts that usually arise from the fact that many factors other than pollutants can change population size, but has so far been almost completely ignored as a suitable way in which to monitor biological effects.

Three important points need to be remembered:

(1) If the appropriate genetic variation does not exist within a population, then the gene pool will be unaffected by a pollutant even if the population is affected. The evidence from metal tolerance in plants suggests that a minority of species contains the appropriate genetic variability, although insect pests have perhaps a higher likelihood of appropriate variability, linked possibly to their mobility and pest status.

(2) Species differ in both their susceptibility, the degree of exposure needed to affect them, and, within communities, in their vulnerability, the likelihood of receiving a given degree of exposure. If the gene pool of one species is affected, it does not necessarily indicate that the numbers of all other species in that habitat that lack the appropriate genetic variability will have been reduced, but it probably does imply that some other species, those whose susceptibility and vulnerability render them at least as much at risk, will have been reduced in numbers or eliminated.

(3) An increase in the proportion of a particular genotype as the degree of exposure to a pollutant increases does not, by itself, prove that the pollutant has caused the change in the gene pool. There are several examples of correlations between the frequency of a particular genotype and environmental variables (Hedrick *et al.*, 1976), but additional evidence is needed to establish that the observed variables control the genotype frequency. Sometimes this caution is ignored, and correlations between gradients of "pollution" and changes of allozyme frequencies are taken to suggest cause and effect (e.g. Battaglia *et al.*, 1980). This may lead to false conclusions.

Many populations of the peach-potato aphid (*Myzus persicae*) are resistant to organophosphorus insecticides, and that resistance is associated with increased amounts of an enzyme, a carboxylesterase designated esterase 2. This correlation does not prove that the allozyme esterase 2 causes the resistance, but Beranek and Oppenoorth (1977) showed that this enzyme is in the same electrophoretic fraction as an enzyme, from a resistant strain, that hydrolysed an organophosphorus insecticide *in vitro*. A susceptible strain had low esterase 2 activity, and lacked the hydrolytic activity. Moreover, the two enzymes were affected by the same inhibitors, and it was therefore concluded that the two enzymes were identical. In this instance one could reasonably ascribe the occurrence of the allozyme esterase 2 to selection pressure by organophosphorus insecticides.

Melanism in *Biston betularia* is genetically controlled, and has been shown to be associated with air pollution, particularly smoke and sulphur dioxide (see pp. 80–86). The incidence of melanism might then be an effective way of monitoring at least one of the biological consequences of reducing the degree of air pollution. The relationship appears in fact not to be straightforward (Fig. 8.14): a number of smokeless zones were established on the Wirral peninsula in 1962 and 1963, and although there was an immediate slight decline in the incidence of melanism, the major decline came about 15 years

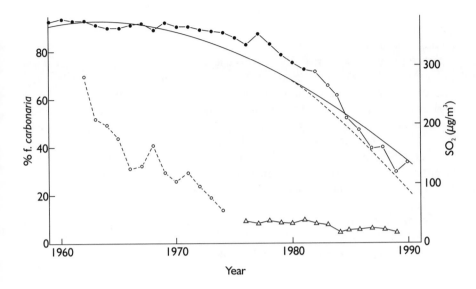

Fig. 8.14 The frequency of the melanic form (f. *carbonaria*) of the moth *Biston betularia*, and the mean atmospheric concentration of sulphur dioxide during the winter, on the Wirral peninsula, Cheshire, England. Between 122 and 994 moths were trapped each year, at Caldy (●—●) from 1959–1981 and at West Kirby (○—○), 0.5 km away, from 1982–1990. The smooth curve indicates the predicted incidence of f. *carbonaria* when it is assumed that the level of sulphur dioxide was constant from 1976 onwards, and the dotted line indicates the predicted incidence if sulphur dioxide declined at a steady rate from 1976 onwards (see p. 86). Levels of sulphur dioxide were measured initially at seven stations on the Wirral (○– –○), and later at one station only (West Kirby) (△– –△). (Data from Clarke *et al.*, 1985, 1990; Mani and Majerus, 1993.)

later. This delay may indicate the time needed for lichens to re-establish themselves (Brakefield, 1987).

Blanck and Wängberg (1988) applied a variation of this theme to whole communities. They argued that sufficient exposure to a pollutant would remove the more susceptible species. The surviving species would be either naturally tolerant or would have developed genetic resistance, and would be joined by new less susceptible species. Experimental exposure of such a pre-exposed community to the pollutant would not affect community functions such as the rate of photosynthesis, whereas it would in a community not previously exposed to the pollutant. Experiments with arsenate supported this argument. Cultures of periphyton (algae attached to firm substrate) were pre-exposed to a range of arsenate concentrations. Individual cultures were then transferred to a range of test solutions to estimate that concentration of arsenate that would inhibit photosynthesis by 20%—the IC_{20}. The pre-exposed cultures were less affected (Fig. 8.15). This "pollution-induced community tolerance" (PICT) was achieved by loss of some species and a lower total biomass.

Subsequent work with TBT, which is used to control algal growth on ships' hulls (pp. 141–142), found that tolerance to TBT of photosynthesis by periphyton increased when samples in coastal waters were taken from sites with higher ambient concentrations (Blanck and Dahl, 1996) and that this tolerance decreased again when ambient concentrations of TBT decreased (Blanck and Dahl, in press). Although increased community tolerance indicates effects by a pollutant, co-tolerance induced by exposure to another pollutant, or exposure to a compound that modifies the availability or effect of a pollutant, might also increase tolerance in some situations and complicate the detailed interpretation.

In general, one might expect that, if the use of genetic adaptation is developed for monitoring and survey, changes in resistance would be the most common index of genetic change. It is likely to be a difficult and uncertain

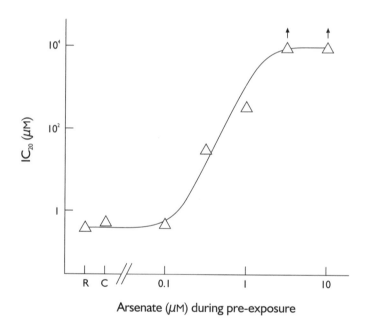

Arsenate (μM) during pre-exposure

Fig. 8.15 The effect of exposure to arsenate on photosynthesis by communities of periphyton from Gullmar Fjord, Sweden. Individual cultures were pre-exposed for 2 weeks to one of a range of arsenate concentrations, and the IC_{20}, that concentration needed to inhibit photosynthesis by 20% during 15 min exposure, was then estimated. Some cultures were taken direct from the fjord (R, for reference) and some of the laboratory cultures were pre-exposed for 2 weeks to uncontaminated water from the fjord, which had < 0.01 M arsenate (C, for control). ↑ indicates that the test concentrations inhibited photosynthesis by less than 20%. (From Blanck and Wängberg, 1988.)

process to demonstrate that other types of genotypic difference are caused by different degrees of exposure to a pollutant, whereas the adaptive value of increased resistance in the presence of a pollutant is self-evident. Moreover, once it has been established for a species that its variations in resistance are genetically controlled, it is necessary only to measure the degree of resistance, which is a relatively well-understood and straightforward process.

Earlier chapters have emphasized the need for a quantitative approach: for precise measures of exposure and dose, and of the degree of confidence that can be attached to predictions of ecological effect. Little attention has been paid in monitoring schemes so far to their sensitivity or accuracy, apart from chemical analyses. This chapter started by emphasizing the importance of clear-cut objectives for schemes of monitoring. In practice, objectives are often not clear-cut (Moore, 1975), and it becomes difficult or impossible to state either the actual or the desired levels of accuracy and sensitivity. This topic will need more attention in future.

9 Case Studies

Much of the stimulus for research in ecotoxicology comes from past, present and potential future practical problems. I have selected four well-known examples of pollution to illustrate how, in practice, problems have been detected, how and why the problems have occurred, and what has or can be done to alleviate the problems. These examples should also help to support and amplify many of the points and conclusions from earlier chapters.

The peregrine falcon

During the 1950s and 1960s there was fierce controversy in Britain about the effects that pesticides may have had on wildlife. A major concern was the acute toxicity of organochlorine insecticide to birds, including the peregrine falcon (*Falco peregrinus*) (see Moriarty, 1975a, and Newton, 1979, for general accounts). Concern for the peregrine was stimulated by a rather odd chain of events.

By 1960 there were complaints from some of those who fly racing pigeons (*Columba livia*) that the number of peregrine falcons was increasing, and that they were killing more racing pigeons than hitherto. There was little or no direct evidence for this assertion, but it could not be positively dismissed as untrue. The peregrine's principal prey species, south of the Scottish Highlands, is the domestic pigeon, although it can and does prey on a wide range of other birds too.

However, a survey showed to the contrary that there had been a dramatic decline in the number of peregrines, a decline which started in southern England in about 1955, and then spread northwards to parts of the Scottish Highlands. The breeding population of British peregrines had been reasonably stable until about 1955, probably for some centuries, but by 1962 92% of known pre-war territories in southern England were deserted, whereas from examination of the pre-war data one would have expected only about 15% to be unoccupied (Ratcliffe, 1963). Moreover, for the whole of Great Britain, there were successful nestings in only 26% of the occupied territories. What was causing this rapid decline in numbers and lessened breeding success?

One has first to consider the possibility of natural causes, which can be summarized as changes in climate, food, other species (including man and diseases) and places in which to live. There was no good evidence for significant change in any of these factors, and pollutants seemed to provide by far the most likely explanation. Two compounds were implicated: p,p'-DDE, a persistent metabolite of p,p'-DDT, and dieldrin. Regional differences in residues of both did correlate with the degree of effect observed (Table 9.1): only birds in the East and Central Highlands of Scotland appeared to be unaffected. The reduced breeding success, and reduced population size, appeared then to have been caused by insecticides, but it was not a simple matter either to suggest the causes of these reductions, nor to produce an argument to convince sceptics (Ratcliffe, 1993).

It is now generally agreed that the population decline was caused by dieldrin, which came into widespread use after 1955, especially as a seed dressing for cereal seed. There were many incidents of birds, including pigeons, found dead and dying in the spring from acute toxicity after eating dressed seed. It was not feasible to do laboratory tests on peregrines to demonstrate directly the toxic dose of dieldrin for this species—it is not an easy species to keep in captivity, and was moreover by then an endangered species. However, the evidence suggested that consumption of two or three heavily-contaminated pigeons could be sufficient to kill a peregrine (Jefferies and Prestt, 1966).

The reduced breeding success, which started some years earlier, is perhaps of greater scientific interest. Nesting peregrines habitually lay one clutch of eggs per year, with almost invariably either three or four eggs. The mean clutch size is 3.5 eggs, with an average of 2.5 fledged young. The other egg was usually either infertile or the embryo or chick died in the nest, and only 4% of clutches had any broken eggs. Ratcliffe discovered that peregrines had started to lay eggs with thinner shells in 1947–48 (Fig. 9.1), such shells are

Table 9.1   Residues of two insecticides in eggs of the peregrine falcon (*Falco peregrinus*)

Region	Number of eggs analysed	Concentration (μg/g fresh weight) (means and ranges)	
		p,p'-DDE	HEOD (the active principle of dieldrin)
East and Central Highlands of Scotland	16	3.25 trace – 8.6	0.29 trace – 1.6
Rest of Britain	42	13.67 0.2 – 33.0	0.57 0 – 2.6

Data from Ratcliffe (1970).

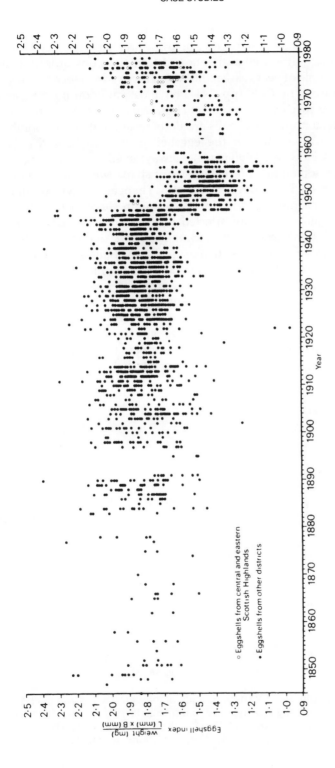

Fig. 9.1 Changes from 1845 to 1979 in Ratcliffe's eggshell index (shell wt, mg)/ (length, mm × breadth, mm) of British peregrine falcons (*Falco peregrinus*). This is the most extensive published set of data for eggshell thickness in birds, and indicates the abrupt thinning that started in 1947. (From Ratcliffe, 1993.)

mechanically weaker (Fig. 9.2), and the incidence of broken eggs within clutches rose from 4% to 39%. We do not know precisely when eggs started to break more often. It is well authenticated from 1951 onwards, but could have started a couple of years earlier. Although there is evidence that thyroid function is disturbed by DDE (pp. 144–149), most study on mechanisms of shell thinning has focused on events much nearer to the actual formation of the shell, where abnormal eggshells are perhaps associated with inhibition of the enzyme Ca^{2+}-ATPase in the eggshell gland (Cooke, 1973, 1979b; Lundholm, 1987). However, DDE may not always have a direct effect on shell thickness: it may sometimes affect shell size or shape, with concomitant secondary effects on shell thickness (see Fig. 9.3). Then, factors which influence egg size may be of some relevance for studies on modes of action (Moriarty *et al.*, 1986). Not all species are affected in the same way. DDT thickens eggshells in the bengalese finch (*Lonchura striata*) (Jefferies, 1969), and eggshells of gallinaceous species such as quail, pheasant and the domestic hen, species that are often used for laboratory experiments, are relatively unaffected (Cooke, 1973).

Several features of this classic episode are still of general relevance:

(1) The decline in the peregrine population only became generally known so soon after the event because of the complaints, albeit misplaced, by owners of racing pigeons. The thinning of eggshells was only discovered

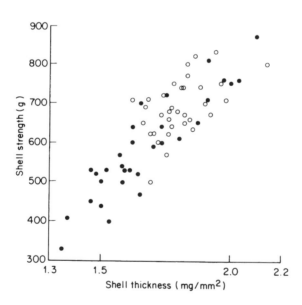

Fig. 9.2 Shell strength and thickness in eggs of the peregrine falcon (*Falco peregrinus*), for two samples of eggs: those laid from 1850 to 1942, before the advent of DDT (○), and those laid from 1970 to 1974 (●). Strength was assessed by the weight needed to pierce the shell under standardized conditions. (From Cooke, 1979a.)

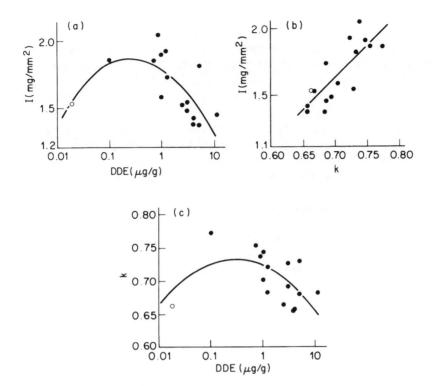

Fig. 9.3 Regressions of Ratcliffe's eggshell index (I) on (a) DDE content (based on fresh weight) and (b) shape (k) (maximum breadth/maximum length), and (c) regression of shape on DDE content, for heron eggs (*Ardea cinerea*) taken from a colony at Troy, Lincolnshire, in 1973. The open circle in each graph indicates one particular eggshell which perhaps illustrates that low doses of DDE have the opposite effect to high doses (see p. 201). The significance of DDE and of shell shape for shell thickness varied appreciably between samples taken in different years and from different colonies. (From Moriarty *et al.*, 1986.)

15 or so years after it had started, and only then because of very detailed investigations. Even fairly dramatic events in a species of considerable conservation interest are not readily observed.

(2) Ratcliffe's survey of the peregrine's status was greatly facilitated by the existence of data on the population before synthetic insecticides were developed. Base-line data are invaluable for monitoring a population.

(3) It was not easy to establish the causes of the peregrine's decline. Field data consisted of correlations, it was not feasible to use this species for laboratory experiments because it was endangered and is very difficult to maintain in captivity, and many of the early laboratory experiments used standard laboratory species such as the chicken, pheasant and quail, which appear to be immune to shell-thinning by DDE.

(4) Shell-thinning was caused by a metabolite of DDT. For prediction of poss-
     ible effects, one needs always to consider not only the compound of
     interest, but also possible metabolites and breakdown products.
(5) The effect of DDE was completely unexpected. Indeed, the exact mode
     of action is still uncertain (Cooke, 1973; Moriarty et al., 1986).
(6) It was a surprise that dieldrin should affect peregrines and other species
     of bird. It had been thought that to apply dieldrin as a seed-dressing was
     highly efficient, putting the insecticide exactly where it was needed to
     protect the crop. With the wisdom of hindsight this environmental path-
     way (seed → seed-eating bird → predator) is not too surprising, but our
     ability to predict the detailed environmental distribution of pollutants is
     still limited.
(7) It is a nice question what the effect of dieldrin on the population would
     have been if breeding success had not already been impaired by DDE.
     This was perhaps an instance where it was the combined effect of two
     pollutants that affected the population.
(8) The debate in the 1960s about the possible effects of DDT on peregrines
     and other species of birds was often heated. That emotions ran high on
     both sides of the debate matters little, but there was, and still is, an
     underlying problem. Decisions about the possible limiting or banning of
     the use of DDT did need to be made, without absolute certainty about
     the real situation. One had to weigh the balance of probable advantage
     and disadvantage of two or more different decisions. The lack of con-
     sensus about the value of wildlife made it much more difficult to agree
     what ought to be done. The more recent debate in both North America
     and Europe about possible reductions of acid deposition illustrates the
     same problems, with the added difficulty in that example of the need for
     international cooperation and agreement.

Control of the spruce budworm (*Choristoneura fumiferana*)

Pesticides are, by definition, biologically active compounds. They are also,
with few if any exceptions, non-specific, although they do differ widely in
the range of species affected. Their use therefore always carries the possibility of ecological effects quite apart from any consequences of reducing the
pest population. This risk of "side-effects" is sometimes greatly increased
because pesticides are less effective than might be expected: rapid kill of a
large proportion of a pest population does not necessarily guarantee the end
of the pest problem. On the contrary, total kill of the pest population is usually impossible, so, unless environmental conditions change, it is unreason-
able to expect a single application of pesticide to eliminate a pest problem
permanently.

    An early example of pesticide use on a large scale was for the control of
spruce budworm, studied in detail in the Canadian province of New

Brunswick, where it is still a serious forest pest. This moth is indigenous to many of the forests in North America, and usually occurs in such low densities that it is barely detectable, and indeed could be called a rare species (Morris, 1963). In New Brunswick the life cycle usually takes one year, starting with eggs laid on pine needles during July and August. The preferred host species is the balsam fir (*Abies balsamea*), the dominant species in these forests (Way and Bevan, 1977), but the larvae will also feed on some species of spruce (*Picea*). The adult moths emerge the following July, live for about two weeks, and the females lay an average of about 200 eggs. From these eggs, in a stable population, on average only one female will survive to emerge as an adult, mate, and lay eggs for the next generation, and there are two principal phases of mortality (Fig. 9.4): firstly of starvation soon after hatch and secondly of predation the following spring.

Historical records show that occasionally, every few decades or so, the budworm population increases greatly, when many trees are severely defoliated. Five to eight successive years of severe defoliation are usually sufficient to kill a tree, and lesser degrees of attack can reduce the yield of timber and render the trees more susceptible to attack by other insect pests. Two features combine to favour the cyclical nature of these outbreaks. Firstly, they tend to start in extensive and continuous areas of mature balsam fir; commonly in dry sunny

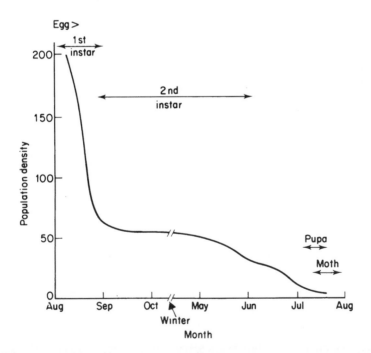

Fig. 9.4   Survivorship curve of the spruce budworm (*Choristoneura fumiferana*) from egg-laying to reproduction. Population density estimated as the number of individuals per 10 ft^2 of foliage. (From Miller, 1963.)

summers. Once a large population has developed moths disperse and then not only mature but also younger trees are attacked. Secondly, forest fires start readily in these consequent large areas of dead trees, when conditions become ideal for seeds to germinate and produce an even-aged forest (Mitchell and Roberts, 1984), which will therefore become vulnerable in due course to another outbreak of spruce budworm.

There had been an outbreak of budworm in New Brunswick during the years 1912–1919, and the next signs of an increase above the endemic population density were detected in 1947 (Morris, 1963). Two years later severe defoliation had occurred in several areas of northern New Brunswick. The forest timber had become an important part of the region's economy, so, starting in 1952, forests were sprayed from the air with DDT. Spraying, usually in June, was restricted to those areas of advanced damage where another attack by spruce budworm might be expected to kill the trees (6–40% of the total area suffering severe attacks), with a maximum of 2.1 million hectares sprayed in 1957 (Fig. 9.5). Spraying usually, compared with unsprayed areas, reduced the population of budworm by 80–90%. Spraying a few weeks later in the year gave a better kill of budworm but preserved little foliage (Webb et al., 1961). Populations began to decline in northern New Brunswick in 1956 and 1957, but patches of severe infestation persisted in central and southern New Brunswick, and spraying therefore continued.

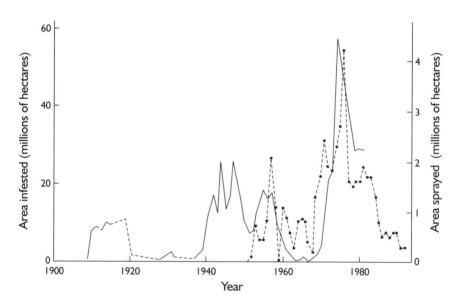

Fig. 9.5 The area of New Brunswick sprayed with insecticide to control infestation of coniferous forest by spruce budworm (*Choristoneura fumiferana*). (—), area moderately to severely infested; (●– –●), area sprayed with insecticide (see text for details of which insecticides were used). (Data from Freedman, 1995.)

The major reported adverse effects on wildlife were in fresh water (MacDonald and Webb, 1963), and the number of salmon declined dramatically in many rivers, including the Miramichi, one of the most famous Canadian salmon rivers (Kerswill, 1967). By 1963, concern about the environmental effects of DDT led to the use of alternative, less persistent organophosphorus insecticides (particularly fenitrothion and aminocarb) with less apparent effect on wildlife, and the use of DDT ended in 1968. These alternatives were less effective at controlling spruce budworm, and became part of the routine programme of forest management (Mitchell and Roberts, 1984).

In response to the problems posed by spruce budworm throughout the forests of North America, a six-year research programme was started in 1977 to develop better control methods that met the twin desiderata of economic viability and environmental acceptability (Sanders et al., 1985). Various alternative approaches were investigated in an attempt to reduce reliance on broad-spectrum pesticides. These included different silvicultural systems, the use of pheromones and biological control agents, and several Canadian provinces have now banned the use of chemical insecticides for pest control in forests, which have become a major market for the use of biological agents such as the bacterium *Bacillus thuringiensis*. This is usually less effective than insecticides but is preferred because of its relative specificity, limited mainly to Lepidoptera. Little was used in New Brunswick until 1984, but during the years 1987–1992 *B. thuringiensis* was applied on average to 28% of the total area sprayed (Freedman, 1995).

The debate on environmental effects was hindered by uncertainty of what were all the consequences. We have already discussed the difficulties of effective prediction and monitoring (Chapters 7 and 8), and the following example may serve to illustrate the point. Initial monitoring programmes suggested that fenitrothion had little impact on birds. Subsequent work showed that the numbers of some bird species that inhabit the top of the forest canopy, where aerial sprays make their first impact, do decrease significantly after spraying (Mitchell and Roberts, 1984). The change from DDT to organophosphorus insecticides did reduce the hazard to fish, but birds tend to be extremely susceptible to acute poisoning by organophosphorus insecticides (Wallace, 1992). Freedman (1995) discusses the impact on birds in some detail.

Is it good sense, ecologically and economically, to continue annual spraying indefinitely? Spruce budworm exemplifies one characteristic type of insect pest on long-lived crops: occasional, relatively local eruptions of population density which then spread over large areas (Berryman, 1986). This results from the interplay of two factors, the trees' resistance to attack by pests, and the pest population's ability to increase in size rapidly.

The resistance of individual trees to pests and diseases tends to increase with growth-rate. Individual trees, like most other organisms, have a sigmoidal growth curve, and commercial forests are commonly planted over large areas at one time. Plantations tend therefore to become susceptible to damaging attack by spruce budworm as they mature. In a young forest, newly planted or regenerating from an earlier attack by spruce budworm, the trees are small and have discrete gaps between their canopies, so that many larvae die during the initial wandering stage after hatch. As the forest matures and the canopy closes the percentage mortality of the early larval stage decreases, but the population still remains stable because of increased predation on the larvae by insectivorous birds. However, mature balsam fir trees produce many flowers, which are the preferred feeding site for bud-worm larvae and on which larvae develop quicker. In dry warm weather lar-val mortality then sometimes decreases to such an extent that predation is inadequate to control population size, which so increases that predation is unable to reduce numbers to the normal level again (Way and Bevan, 1977). Once an outbreak has started, if nothing is done to control it, the outbreak usually continues until the host trees have been killed.

In New Brunswick, spraying with DDT and, subsequently, other suitable insecticides reduced the pest population sufficiently to prevent trees being killed, but had to be repeated annually to keep trees alive. Ideally spraying would reduce the population of budworm to the level at which natural con-trols would prevent its resurgence, but this requires adequate treatment of the entire outbreak area, which in practice has not been feasible. Instead, an unsta-ble equilibrium was established, dependent on annual application of insecti-cide, which both prevents trees from being killed and, at the same time, thus maintains an adequate food supply for subsequent generations of spruce bud-worm. Outbreaks can therefore last indefinitely (Mitchell and Roberts, 1984).

One is bound then to consider the alternatives (Berryman, 1986). Ideally, forest management would maintain conditions such that pest outbreaks never occurred. In practice, this means felling trees before they reach the crit-ical over-mature stage, growing trees in small areas of different ages, or of growing mixtures of trees, with susceptible species forming a small propor-tion of the total. The forest industry in New Brunswick has found the per-sistent semi-outbreak condition of spruce budworm to be acceptable: during 25 years or so as much timber has been lost by budworm attack as would have been lost in a single uncontrolled epidemic, but the loss has been spread over many years, and annual harvests could always be taken (Mitchell and Roberts, 1984).

However, many other special interests are also involved. In essence there is a conflict of interest between those who earn their livelihood from forestry and those who suffer, or fear, an adverse effect on their interest, be it farm-ing, fishing or conservation (Mitchell and Roberts, 1984).

The River Thames: pollution of an ecosystem

Streams and rivers form an important part of the hydrological cycle, return-ing much of the rain that falls onto land back to the seas and oceans. They are also an important resource for mankind, with many, sometimes conflicting, functions, being for example both a source of drinking water and a repository for sewage.

We have already considered the effects that sewage can have on rivers (p. 235). Many of the major biological effects of organic human wastes in waterways result from the depletion of dissolved oxygen as bacteria metabolize the organic molecules of protein, fat, carbohydrate and nucleic acid to, mostly, water, carbon dioxide and nitrates. The magnitude of the effects depends on the balance between the rates of oxygen depletion by bacteria (measured in the laboratory as the biochemical oxygen demand (BOD)) and of replenishment. As a rule of thumb, coarse fish (species other than salmon, trout and grayling) in British rivers need a minimum of about 25% of the saturation value of oxygen to maintain sizeable populations, and young salmon (the most sensitive species) require a minimum of about 35% before they can traverse an estuary on their migration back to fresh water. If the oxygen content drops to about 5% of the saturation value bacteria start to obtain oxygen from dissolved nitrates and nitrites, and the innocuous gas nitrogen is released during this anoxic phase. Finally, if the BOD is high enough, conditions become anaerobic. Bacteria then obtain oxygen from sulphates, the resultant sulphides are toxic to aquatic life, and hydrogen sulphide may be released. There are also other impacts from sewage effluent: breakdown products, especially nitrates and phosphates, can be important plant nutrients, and the suspended solid matter can also influence the community structure. Industrial organic wastes may also produce toxic compounds as they are broken down. Hynes (1960) discusses in detail the physical, chemical and biological gradients that form below freshwater sewage outfalls.

The River Thames provides one of the best-documented examples of how a river can be affected by human activities, and of how effects can be lessened. The tidal ebb and flow in the Thames estuary stops at Teddington weir, 31.5 km above London Bridge and 245 km downstream from the source: we will consider first this tidal part of the Thames, around which London has developed since Roman times (Wood, 1982). Many influences have changed the nature of the estuary since then. Natural causes include the relative rise of mean sea-level in south-east England of about 0.36 m per century. Other changes during the last 500 years, such as artificial embankment, can be seen as functions of an enlarging human population (Fig. 9.6). The available evidence, principally quantities and species of fish caught, suggests that until 1800 the Thames itself was still relatively unaffected, or "healthy", with salmon and trout migrating along the river. At that time rubbish and human wastes from cesspools were collected and deposited on land outside the city, and a system of sewers had been developed to convey rainwater to the Thames and its tributaries. Latrines were allowed to discharge into watercourses, and the illegal dumping of rubbish into these watercourses was also commonplace. Consequently most of the Thames' tributaries in London had become so polluted that they had been covered over and formed part of the sewerage system.

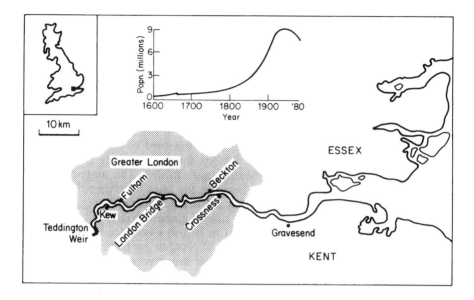

Fig. 9.6 Map of the tidal Thames, with the boundaries of Greater London and a graph of the size of London's human population.

After 1800, for the next 50 years, there was a serious deterioration in the biota of the Thames. The human population nearly doubled in that time, but there were two additional factors. Firstly, it became mandatory in 1815 to connect cesspools to the sewers, which had then to carry a much greater load of human wastes into the Thames. With hindsight the effects were predictable. By the 1850s the stench from the Thames was notorious and, of greater consequence, there were four serious epidemics of cholera, a water-borne bacterial disease, between 1831 and 1866. London's drinking water, rarely purified, came from local shallow wells and direct from the Thames, and was therefore easily contaminated. Secondly, with the development of the Industrial Revolution, the Thames received various more directly toxic (and sometimes more persistent) pollutants, such as phenols, tar oils, naphthalenes, pitch and ammonia from gas and coke manufacture. Symbolic of the decline, no more salmon (*Salmo salar*) were recorded from the estuary that century after 1833.

In response to this deterioration, London developed a piped water supply and it was forbidden to abstract drinking water from the tidal Thames. The sewerage system was coordinated by joining all existing sewers to new trunk sewers, so that by 1864 all those north of the Thames discharged at Beckton, and by 1865 all those south of the Thames discharged at Crossness, about 12 and 15 km respectively downstream of London Bridge. Effluent was discharged during the ebb tide only. Unfortunately the new system was not intended to cope with heavy flows of storm water, which could carry much sewage over weirs into the original sewers that still led to the river at many points within London. It was assumed that the storm water would adequately dilute the sewage.

Although these developments reduced London's problems, satisfaction was short-lived. Within a few years there were complaints of pollution and mudbank formation near the outfalls, and fish disappeared from a considerable distance above the outfalls to 25 km downstream. In consequence two current practices were initiated: removal of solids from sewage, and regular chemical monitoring. Firstly, the incoming sewage was treated at the sewage works before discharge, to precipitate and retain the solids (sludge), which was until recently mostly dumped further out in the Thames estuary or spread onto land. Current practice is to remove solids by settlement and filtration (Mason, 1996). Secondly, in 1893 Dibdin, chemist to the London County Council, introduced regular chemical monitoring of the whole of the tidal river. The first years' results demonstrated the effectiveness of the improved system of sewage treatment (Fig. 9.7), confirmed by the reappearance of fish species.

However, the population of London continued to increase (Fig. 9.6), more sewage works had to be constructed, the oxygen demand from sewage effluent exceeded the river's capacity to dilute and oxidize, and conditions

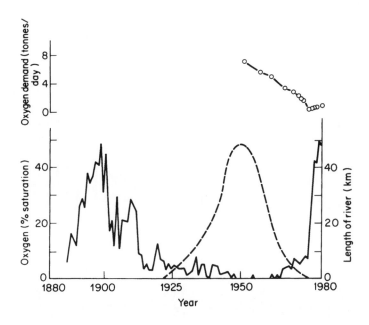

Fig. 9.7 Dissolved oxygen in the tidal Thames. (—) Samples taken at low water off Crossness during July–September, when conditions are usually at their most extreme; (– –) the length of river where the amount of dissolved oxygen is on average below 5% of the saturation value during July–September; (O—O) the estimated daily depletion of dissolved oxygen by the combined sewage effluent from Beckton and Crossness. (Data from Wood, 1982, and Andrews, 1984.)

steadily worsened until 1950. From 1920 to 1960 the eel (*Anguilla anguilla*) was apparently the only species of fish to be found throughout much of the tidal Thames (Wheeler, 1969). A comprehensive programme was then developed to again improve sewage effluents. Many small sewage works were closed down, and others were improved. The major sources of oxygen demand were the effluents from Beckton and Crossness; the sewage works at Crossness were rebuilt in 1964, and those at Beckton were extended in 1976. This led to concomitant improvements in the levels of dissolved oxygen (Fig. 9.7), the number of fish species present (Fig. 9.8) and the invertebrate fauna (Andrews *et al.*, 1982). The first recorded salmon for 141 years was caught in 1974.

The Thames was not alone among British estuaries in being contaminated. The Royal Commission on Environmental Pollution (1972) assessed the problems, and attempted to reconcile two conflicting attitudes: that environmental contamination should be stopped to prevent harm to wildlife; conversely that not only is the damage to biota of uncertain magnitude, but that discharge of industrial effluents via sewage or directly to estuaries reduces costs and helps employment. The Commission concluded that, in theory, the degree of pollution abatement should be such that the marginal cost of control equals the marginal cost of the harm being done. In practice of course it is not easy to cost all factors, and they therefore recommended that:

(1) Persistent contaminants that accumulate in the river sediments or biota should be removed to the maximum practical extent before effluents are

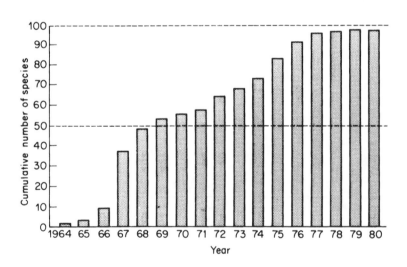

Fig. 9.8 The return of fish species to the tidal Thames (Fulham to Gravesend). Sampling effort increased in 1967, and was fairly uniform thereafter. (From Andrews, 1984; original data for 1964–1973 from Wheeler and Huddart (For details see Andrews and Rickard, 1980).)

discharged. This is an important distinction from less persistent pollutants, which are released up to an acceptable level. Releases of mercury and non-biodegradable surfactants into the Thames have both been rigorously controlled (Andrews, 1984).

(2) For biodegradable wastes, the acceptable level of release is that which avoids loss of the benthic fauna needed to sustain sea fisheries, and permits passage of migratory fish at all states of the tide. This environmental quality objective (EQO), of not inhibiting the passage of migratory fish, fixed an upper limit for the permissible degree of pollution, measured by the level of dissolved oxygen.

For the Thames this EQO was taken, for operational purposes, to indicate that the average level of dissolved oxygen in the third quarter of the year should not drop below 30% of the saturation value. A mathematical model had already been developed to predict the effect on dissolved oxygen of point discharges of organic matter (Barrett et al., 1978), and chemical monitoring for dissolved oxygen (and other chemical entities) had started and developed since the end of the nineteenth century (Fig. 9.7).

The question then arises, are the chemical data adequate to indicate the degree of biological effect? Certainly there is no a priori reason to suppose that the degree of effect is linearly related to the amounts of pollutants. Routine biological monitoring of fish species caught on the intake screens of cooling water for power stations and by trawling started in 1967. The data show clearly that the tidal river's fauna improved greatly (Fig. 9.8), but are insufficient to examine closely the relationship between degree of effect and degree of exposure. This is not surprising. At least some species populations fluctuated appreciably from year to year (Andrews and Rickard, 1980), and studies on the Hudson River, USA, showed that many years' data are needed for variable populations before even large impacts can be detected confidently (McDowell, 1986).

The relationship between chemical and biological indices of river pollution has been studied in more detail for the non-tidal Thames, along with most other non-tidal rivers in England and Wales. Each stretch of river was classified by two separate sets of criteria, chemical and biological, into one of four classes (Table 9.2). The degree of correspondence between the two assessments was less than had been anticipated, particularly for rivers of intermediate quality. A similar discrepancy occurred when the survey was repeated three years later (Department of the Environment, 1975), and a new assessment was therefore developed for the next survey in 1980 (National Water Council, 1981). Five chemical classes were defined, corresponding approximately to the original four but with class 1 subdivided into classes 1A and 1B. These classes were based primarily on BOD, amounts of dissolved oxygen and on defined toxicities of pollutants to fish (see p. 197). The biological assessment developed from the premise that it is impracticable to devise a biological classification that is adequate for all English and

Table 9.2 The criteria used for, and the degree of correspondence between, the chemical and biological assessments of pollution in the 1970 survey of non-tidal rivers in England and Wales

Biological class			Total length (miles)	Chemical class				% agreement[b]
Invertebrates	Fisheries			1 Unpolluted or recovered from pollution[a]	2 Of doubtful quality[a]	3 Of poor quality[a]	4 Grossly polluted[a]	
A Widely diverse	Good		10 545	10 241	284	20	0	97
B Varied	Good, although salmon and grayling absent even if ecological factors suitable		3441	2107	1228	99	7	36
C Restricted	Moderate to poor		1616	452	554	427	183	26
D None, or a few resistant species	No fish		930	27	129	202	572	62
Total length (miles)			16 532	12 827	2195	748	762	
% agreement[b]				80	56	57	75	

From Department of the Environment (1971).
[a] Chemical assessment based principally on known occurrence of polluting discharges, BOD, dissolved oxygen, turbidity, absence of fish and frequency of complaints.
[b] Percentage of river, for each biological or chemical class, that had been allotted to the corresponding chemical or biological class.

Welsh rivers, because the influence on the biota of factors such as flow rate, water quality and nature of the river bed are insufficiently understood. Sites were therefore sampled for their macroinvertebrates, all families found were listed, a score given for each family present (ranging from 1 for Oligochaetae (worms) to 10 for Perlidae (stone-flies)), and the BMWP cumulative site score obtained by combining the scores for all families found in the sample (see pp. 236–237).

The results for sites on the non-tidal Thames and its tributaries, and, for comparison, non-tidal rivers in South Wales, illustrate the loose correlation between the chemical and biological assessments, and the difference in biological scores for sites of the same chemical class from different rivers (Fig. 9.9). Different unpolluted rivers will, naturally, have different communities of macroinvertebrates, so that to interpret the biological scores requires expert knowledge. The intent was to indicate anomalous situations and to provide a basis for future comparisons. It has to be accepted that base-line data—the quality of pristine rivers—are missing. Thus eutrophic rivers can be in class 1B, although oligotrophic waters have very different fauna and flora.

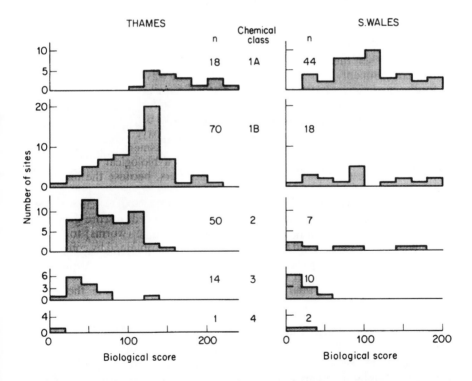

Fig. 9.9 Frequency histograms for biological scores at sites on the non-tidal River Thames and its tributaries, and on rivers in South Wales from Milford Haven to the Usk. Sites are divided into five classes by degree of chemical pollution. *n* is the number of sites examined in each class. (Data from National Water Council, 1981.)

Subsequent surveys in 1985 and 1990 heightened concern about the pollution of rivers in England and Wales. Overall there was a net decline in river quality, whereas quality had improved from 1958 to 1980. Much of this decline was in best-quality waters in the River Thames (Royal Commission on Environmental Pollution, 1992), and was ascribed to three principal factors: improved monitoring which gave more accurate results, dry weather in 1990 which reduced river flows and so diluted discharged pollutants less, and increased discharges.

The 1990 survey was conducted by the newly-formed National Rivers Authority (NRA), and used both the existing and also new methods for assessing chemical and biological classes, called grades in the new system (NRA, 1991, 1994). Chemical grades were based on three determinands only—dissolved oxygen, biochemical oxygen demand and ammonia—which indicate degrees of organic pollution from sewage, agriculture and industry. Biological grades were based on an environmental quality index (EQI), the ratio of observed to expected status, where

$$EQI_{BMWP} = \frac{observed\ BMWP\ score}{predicted\ BMWP\ score}$$

Although the BMWP score (described on p. 237) allows for environmental differences between sites, the results, like those in Table 9.2, lack close correspondence between the biological and chemical assessments (Table 9.3). Assessment by length of river in each grade can also give a misleading impression of the relative amounts of clean and polluted rivers. The former tend to be small fast-flowing upland rivers with small quantities of water flowing compared to the large downstream slow-flowing rivers.

The NRA concluded from the 1990 survey that the monitoring system needed to be more consistent and objective (NRA, 1991). Sometimes "subjective judgements had been used in place of actual chemical quality data" (NRA, 1994). The concept of "water quality" also needed to be reassessed. Traditionally the criteria of water quality had been determined by the intended uses, such as abstraction for drinking water or maintenance of fisheries. There is now more concern for the "sustainable" use of resources, defined by the Brundtland Commission (World Commission on Environment and Development, 1987) as "meeting the needs of the present without compromising the ability of future generations to meet their own needs". One indication of sustainable use is the ability to maintain and if possible to improve ecosystems. Hence the current demand for less restricted assessments that determine whether rivers contain appropriate communities.

Chemical and biological monitoring are complementary, and both are necessary. Problems can arise from daily, seasonal and episodic fluctuations, of which the last may arise by chance, accident or surreptitious design. Biological monitoring can indicate the cumulative effect of episodic events, unsuspected pollutants and of others present at low levels. Cost and

Table 9.3 The degree of correspondence between chemical and biological assessments of pollution in the 1990 survey of non-tidal rivers in England and Wales

Biological grade* (EQI_{BMWP}) grade	Total length (km)	Chemical grade†						% agreement‡
		Good A	B	Fair C	D	Poor E	Bad F	
≥1.0 a	8 061.2	2 390.6	3 370.1	1 519.6	576.6	194.1	10.2	30
≥0.8 b	6 280.4	1 379.8	2 370.8	1 569.6	585.1	341.6	33.5	38
≥0.6 c	5 263.6	703.1	1 639.1	1 480.4	837.1	551.4	52.5	28
≥0.4 d	4 008.3	261.2	715.9	1 382.4	873.5	671.9	103.4	22
≥0.2 e	3 044.5	137.1	351.8	723.5	790.3	851.8	190.0	28
<0.2 f	1 321.9	0.0	65.1	107.0	205.1	709.7	235.0	18
Total length (km)	27 979.9	4 871.8	8 512.8	6 782.5	3 867.7	3 320.5	624.6	
% agreement‡	49	49	28	22	23	26	37	

Data from NRA (1994).
* Based on the average of the BMWP scores obtained in the spring, summer and autumn at each site.
† Based on the percentage of samples that comes within prescribed ranges of concentration for dissolved oxygen, biochemical oxygen demand and ammonia.
‡ Percentage of river, for each biological or chemical grade, that had been allotted to the corresponding chemical or biological class.

practicability impose constraints on both, and the history of chemical monitoring in British rivers shows that samples have often been taken too infrequently, at unrepresentative times and places, and often not analysed to an acceptable standard (Royal Commission on Environmental Pollution, 1992). Desired degrees of precision and confidence should be set at the outset. The NRA (1994) therefore proposed use-related water quality objectives for local needs, and a national scheme for general quality assessment (GQA) to measure water quality irrespective of uses, to be repeated, like the previous scheme, every five years. This would measure chemical, biological, nutrient level and aesthetic characteristics when appropriate measures have been developed.

The Environment Agency (1996) implemented these recommendations in the 1995 river survey. Mean annual concentration of phosphate was taken as the measure of nutrient level, as a proposed measure of the risk from eutrophication. There was a slight improvement in the correlation between chemical and biological grades, explained perhaps by a net improvement in chemical grade for 28% of the total river length, although the details of the biological grades were also slightly different. For this survey the BMWP score was replaced by two measures, the number of taxa (usually families) and the average score per taxon (ASPT, p. 237). However, the discrepancy between chemical and biological grades for some stretches of river was still considerable.

In summary, several points deserve emphasis:

(1) The degree of pollution in the Thames estuary increased not only with the size of the human population, but also with the degree of its technology. Thus non-biodegradable surfactants were unacceptable in part because they reduce the rate at which atmospheric oxygen enters water and because they reduce the efficiency of the activated sludge process in sewage works. Technology can also alleviate pollution, at a cost.

(2) Almost invariably, and certainly for the Thames, we lack base-line data on the state of our environment, the conditions that prevailed before human activities had a major impact.

(3) The relationship between the degree of environmental contamination and the degree of biological effect is often not simple. One needs therefore as good an approximation as possible to base-line data, and adequate monitoring.

(4) Any monitoring system needs to have clearly defined objectives, with known precision and probability levels for numerical results, adequate quality control and inter-laboratory calibration exercises to ensure consistency of results. The Royal Commission on Environmental Pollution (1992) discusses this topic in some detail for British rivers.

(5) The first step before any rational system of pollution control is a value judgement (Portmann and Lloyd, 1986); what is the minimum degree of environmental conservation, or, conversely, the maximum acceptable

degree of damage? The Royal Commission settled for passage of salmon up the Thames, which cannot readily be justified by any solely economic criteria. In fact, by positive decision or by default we manage our whole environment, little if any of which can be regarded as "natural".

(6) One must question whether biological scores that ignore individual species are an adequate measure of biological effect (see p. 286).

Oil pollution: a world-wide problem

About 40% of the world's present energy supply comes from crude oil, the liquid form of petroleum (Brown and Skipsey, 1986), which is a mixture of many thousands of organic compounds, of which more than three-quarters are usually hydrocarbons (Whittle et al., 1982). The composition can differ not only from different sources, but also from one source at different times as extraction proceeds. Individual compounds differ greatly in their physical, chemical and biological properties, their identification is slow and costly, and only a small proportion is determined routinely. Crude oil is distributed very unevenly, with 55% of the world's proven resources in the Middle East, and half of those reserves are in Saudi Arabia alone. Considerable quantities of oil are therefore shipped around the world, with many consequent possibilities for environmental contamination. The details of the potential biological damage depend on where the contamination occurs.

The aquatic, particularly the marine, communities are the most vulnerable (Cairns and Buikema, 1984), and contamination comes from many sources (Table 9.4), although the Iraqi war of 1991 released, uniquely so far, as much as 240 million tonnes of oil into the Persian Gulf (Paine et al., 1996). Accidental oil spills from tankers are a relatively minor source, forming about 6% of the total oil released from routine or operational discharges. However, tanker accidents can release large amounts of oil rapidly in a small area, with the consequent risk of ecological impact. Organisms can then be affected in two ways (Perry, 1980), by direct toxicity (Anderson, 1979) and by physical smothering.

Oil released at sea forms a surface slick whose components can follow many pathways (Fig. 9.10). In favourable circumstances up to half of the surface oil may be lost by evaporation within the first day or so (Whittle et al., 1982). Of the rest, some compounds pass into the mass of seawater, where the available evidence suggests that they may persist for some weeks (Wolfe, 1987), although Howarth (1989) suggests that dissolved oil can be much more persistent. The slick becomes more viscous and discontinuous as the more volatile and soluble compounds are lost, forms water-in-oil emulsions (mousse) and tar balls, and residues that come ashore can be far more persistent than at sea.

To assess the environmental impact of an oil spill we need first a complete mass balance (Jordan and Payne, 1980). The most thorough followed the release of 36 000 tonnes of oil from the grounding of the oil tanker *Exxon*

Table 9.4 Sources and estimated amounts of crude oil and its products released into the marine environment

Source	Amount (kilotonnes)
Transport of oil	
Tanker operation (washing-out tanks)	158
Cleaning of tankers in dry docks	4
Releases in ports during routine handling	30
Bilge water and fuel oil	252
Accidents with tankers	121
Other accidents	20
Total	585
Offshore production platforms	50
Coastal refineries	100
Atmospheric (includes release from fuel combustion)	300
Municipal and industrial wastes, and surface run-off	1060
Ocean dumping of dredging spoil	20
Natural seeps/erosion	250
Sum	2365

For comparison, the annual world production of crude oil is about 3×10^6 kilotonnes (Whittle *et al.*, 1982), of which half is transported by sea.

Adapted from Clark (1992). Original data published in 1985 and 1991.

Valdez on a reef off the coast of Alaska (at 61°02′N—a cold environment) in March 1989 (Wolfe *et al.*, 1994). Figure 9.11 and Table 9.5 indicate the speed and amounts involved in the various processes: biodegradation and evaporation were the major pathways. To then determine the biological effects is less straightforward, because individual compounds differ in their toxicity, although compounds of medium molecular weight tend to be the most toxic.

On sheltered beaches, in estuaries, marshes and lagoons, where there is little wave energy and the oil can penetrate into the sediments, oil residues can persist for years, with loss principally by microbial degradation and dissolution (Prince, 1992). On rocky shores with a high input of wave energy stranded oil can be lost much more rapidly. Gundlach and Hayes (1978) used these two characteristics, of turbulence from waves and type of substrate, to devise a ten-point scale of vulnerability (Table 9.6): that is, the likelihood of oil being deposited on different types of shore, and the persistence of deposited oil. Their scale also used biomass and number of species as two additional characteristics, on the assumption that, as the biomass and number of species in a habitat increase, so does vulnerability. However, they were confusing two separate attributes of a community, its vulnerability and its

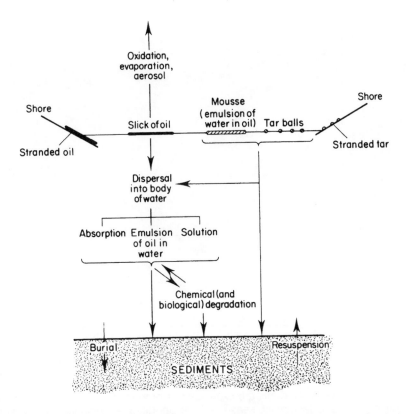

Fig. 9.10 Abiotic pathways of oil from the sea surface. (Modified from Gunkel and Gassmann, 1980, adapted from Whittle *et al.*, 1978.)

Table 9.5 Mass balance for oil spilt from the *Exxon Valdez* off the coast of Alaska

Fate (after 3½ years)	Estimated proportion of total tonnage split
Evaporated and oxidized	0.20
Biodegradation in sea and on beaches	0.50
Dispersed in sea	<0.01
Recovered	0.14
In marine sediments	0.13
On beaches	0.02
Total	~1.00

Adapted from Wolfe *et al.* (1994).

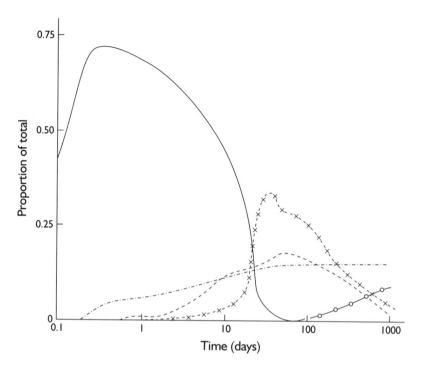

Fig. 9.11   Fate of oil spilt from the *Exxon Valdez*, which ran aground off the coast of Alaska in March 1989. Amounts are expressed as proportions of the total mass spilt: —, floating on the surface of the sea; – · – · –, evaporated; – –, dispersed in seawater; x– –x, deposited on the beach; ○—○, in the sediments. About 13% of the spilt oil was burnt or recovered. (Data from Wolfe *et al.*, 1994.)

susceptibility (see p. 250) and, as Gundlach and Hayes comment, "total prediction of biologic response to oil contamination is extremely difficult".

Possibly the most noticeable, and publicized, effect of oil spills is from the oil slick itself: seabirds become covered with oil and die. It is difficult to estimate numbers but it is reasonably certain that, in the seas around northwestern Europe spilt oil kills at least some tens of thousands of seabirds every year (Dunnet, 1982). There is little evidence to suggest that the populations of these seabirds are being reduced, even though most of these species have very low breeding rates and do not breed until they are several years old. In fact, populations of several species have increased during recent decades. If environmental conditions, apart from oil, become less favourable oil pollution might then have an adverse effect on these species. Where species have declined, factors such as climatic change appear to be responsible (Clark, 1984) and there is little evidence to suggest that any species, as distinct from local populations, is endangered at present by oil spills (Wolfe, 1985).

Table 9.6   Vulnerability of shores to contamination by oil. The higher the index, the more vulnerable the biota

Index of vulnerability	Type of shoreline	Speed with which oil disappears naturally
1	Exposed rocky headlands	Rapid—little contamination occurs, because waves wash oil off
2	Eroding wave-cut platforms	Usually within a few weeks
3	Fine-grained sandy beaches (particle diameter 0.06–0.25 mm)	Often several months
4	Coarse-grained sandy beaches (particle diameter 0.25–2.0 mm)	Several months, longer if lack of high-energy waves, because oil can penetrate deeply into the sand
5	Exposed compacted tidal flats[a]	Little penetration of, or adherence to, surface. Rapid degradation.
6	Mixed sand and gravel beaches	If lack of high-energy waves, may persist for years
7	Gravel beaches (particle diameter > 2 mm)	As above
8	Sheltered rocky coasts	May persist for years
9	Sheltered tidal flats	May persist for years
10	Salt marshes and mangroves	May persist for years

Modified from Gundlach and Hayes (1978).
[a] Given an index value of 5 by Gundlach and Hayes because the biota may be "severely damaged". In other words, this type of shoreline has low vulnerability and high susceptibility. By the more restricted criteria of deposition and persistence of oil, an index of 2–3 is more appropriate.

Apart from seabirds, most major observed impacts occur where oil comes ashore. The conventional view is that the degree of effect on shores is related to their vulnerability. Thus rocky headlands suffer least long-term damage, sheltered shores the most. This is perhaps over-simple. Studies on sites subject to repeated spills or chronic inputs of oil show that effects depend also on the composition of the oil, and on the susceptibility of the species present (Dicks and Hartley, 1982).

Mangroves do fit the conventional picture. They are both vulnerable and susceptible (Odum and Johannes, 1975) because they are adapted to live in anaerobic muds. They have either lenticels on aerial roots (e.g. *Rhizophora* spp.) or special pneumatophores from underground roots (e.g. *Avicennia* spp.). Either way, the aerial roots or pneumatophores create a baffle in the water that reduces current velocities and traps sediments, and will also trap oil. Mangroves are susceptible because the respiratory surfaces are readily clogged by oil, which presumably then kills the plant by blocking supplies to the roots (Baker, 1982). A mangrove community can take many years to recover from exposure to oil. If death of mangroves allows erosion to occur, then the time for recovery is indefinite.

Communities at the other end of the spectrum for vulnerability, on rocky headlands, may sometimes be more susceptible than is usually supposed. The immediate effect of oil, in temperate zones at least, is to kill most of the invertebrates and some of the algae (Southward, 1982). Algae recolonize first, and grow luxuriantly within 2–3 years because there are few herbivores. This luxuriant growth is often interpreted as recovery, but recolonization by invertebrates is slower.

One of the best-documented studies is of the consequences of the first large oil spill, when the oil tanker *Torrey Canyon* went aground because of navigational errors off the south-western tip of England in March 1967. Smith (1968) gives a detailed account of the immediate emergency response. The tanker broke up and released into the sea 119 000 tonnes of crude oil from Kuwait, a relatively non-toxic crude oil. Extensive lengths of shore in Cornwall were virtually denuded of life, in part because of the oil that came ashore, but principally because of the dispersants used to remove the oil (Southward and Southward, 1978). Given the extent of the areas affected, recolonization occurred mostly by settlement of planktonic spores or larvae from unaffected areas. Recolonization was difficult for species without a planktonic dispersal stage in their life cycle.

Before this oil spill there were two types of community on Cornish rocky shores. Those fully exposed to waves had been dominated by barnacles, mussels and limpets, whereas sheltered shores had been dominated by large fucoid seaweeds. These seaweeds are controlled on exposed shores by the grazing of limpets, and shores with an intermediate degree of exposure exhibit an intermediate condition with considerable spatial and temporal variation in the distribution of *Fucus* spp., due in part to cyclic interactions between species (Hawkins and Southward, 1992).

The pattern of recovery after the oil spill followed the same general pattern on all rocky shores. An initial flush of the green algae *Enteromorpha* and *Ulva* was succeeded, in the late summer of 1967, by the large brown algae *Fucus vesiculosis* and *F. serratus*, and these two species gave virtually complete cover of the rocks during the period 1968–1971 (Fig. 9.12), when they eliminated those barnacles (*Chthamalus* spp.) that had survived exposure to oil and dispersants. Limpets (*Patella vulgata*)—the few survivors and incoming juveniles—grew very rapidly on the abundant algae, and reached a peak of numbers in 1972, by which time *Fucus* was declining rapidly. As the food supply diminished, so the population of *P. vulgata* declined to a more stable equilibrium level. Barnacles began to reappear in 1972, five years after the spill. The amount of *Fucus* started to increase again in 1979 and then oscillated during the next 11 years. The community appears to have regained its normal ecological balance 10 to 15 years after the oil spill, after a series of damped oscillations in species numbers (Hawkins *et al.*, 1994).

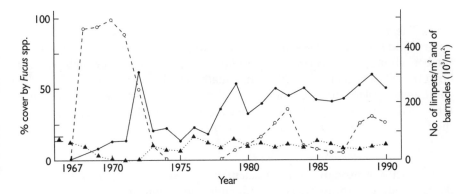

Fig. 9.12 Densities of algae, limpets and barnacles on rocks at mean tide level at Porthleven. Cornwall, after exposure to oil and dispersants in spring 1967: (O--O), *Fucus* spp.; (●—●), *Patella vulgata*; (▲ ⋯ ▲), *Chthamalus* spp., with the mean and twice the standard deviation (▲) for counts of *Chthamalus* spp. in the years 1955–1967, before the oil spill. (Adapted from Southward and Southward, 1978 and from Hawkins and Southward, 1992.)

Part of the reason for this slow return to the presumed original condition lies in the structure of the community: like the Scottish heather community (Fig. 3.12) it is a mosaic, with local patches at different stages of the succession after disturbance, be it heavy predation, abiotic factors or chance. An oil spill is a large-scale disturbance that synchronizes succession over a large area, thus reducing diversity and the amount of food available for some trophic levels. It then becomes difficult to define recovery after an oil spill. In this example, had recovery occurred when all species had returned and were, as individuals, functioning normally, or later when a mosaic of successional stages had developed with particular species densities and age structure?

A community has recovered when conditions are those that would have existed had no oil been spilt, but one has to determine the appropriate measures of a community. Part of the value judgement of the acceptability, or otherwise, of oil contamination on temperate rocky shores rests on the concept of naturalness. Not only is it desired, by some at least, that all species normally present should survive, but that their distribution be indistinguishable from that occurring before any significant pollution occurred. Statistical and biological difficulties, not least normal biological variability, can make it difficult to demonstrate damage, or recovery, unequivocally (Howarth, 1989; Paine *et al.*, 1996).

Habitat vulnerability and species susceptibility both influence the effects of an oil spill, but one of the basic difficulties in scientific studies of oil pollution is that one cannot readily quantify the exposure or the dose. Many individual compounds are not measured; there are, at least in theory, innumerable possibilities for interactions between individual compounds, and their relative abundances change with time.

Attempts are often made to reduce or prevent wide-scale damage from oil spills. The appropriate strategy depends on both the type of oil spilt and the habitat endangered. Briefly, apart from doing nothing, which is sometimes the best choice, one can physically remove, disperse with chemicals, sink or burn the oil. Any action, or inaction, usually has disadvantages, and ideally contingency plans for dealing with oil spills are discussed, agreed and rehearsed before the need arises (Doerffer, 1992). Accidents do appear to be the major environmental hazard from crude oil.

The capacity of human error, miscalculation or simple lack of awareness to damage our environment is considerable. Even within the limits of our knowledge, adequate assessment of problems demands a multidisciplinary approach.

10 Conclusions

Truhaut drew the parallel with toxicology when he defined ecotoxicology (see p. 1). Biochemistry and ecology respectively are the major underlying sciences, and it is relevant to note that those two sciences are at very different stages of development. Wöhler synthesized urea in 1828, and in so doing dispelled the vitalist theory, which can be traced back to Aristotle (384–322 BC), that organic compounds can only be synthesized by living organisms. Biochemistry has developed considerably since then, but vestiges of vitalism have only been dispelled far more recently from ecology. For example, many of the British chalk grasslands, renowned for their attractive flora, were thought to have existed for centuries, and possibly since neolithic times (Tansley, 1953), with the associated widely-held view that many of these species were present only as a result of many centuries of development without major disruption. We now know that we can create facsimiles of these communities artificially by sowing carefully selected seed mixtures which, within 10–15 years, may have a structure and floristic composition similar to that of long-established grasslands (Wells, 1983). In brief, the underlying science on which ecotoxicology rests is far less developed than is the scientific basis of toxicology.

Ecotoxicology is of course a multidisciplinary subject, and much time and effort has been wasted by framing questions in the wrong mode. The classic example is perhaps for the amounts of persistent pollutants in organisms, a question that has often been seen in ecological terms, of food chains, with little attention paid to the more physiological, and important, questions of rates of absorption and loss. This misapprehension led for example to considerable confusion about the high proportion of the total mercury in fish that occurs in the methyl form. It was commonly supposed that fish must methylate inorganic mercury within their tissues, because of the relatively low proportion of organic mercury in their diet. In fact, differential rates of absorption for inorganic and organic mercury from food appear to provide an adequate explanation (Miller, 1978). Conversely, the effects of pollutants on species of wildlife need to be considered in ecological terms, of population dynamics and genetics, whereas toxicological appraisals are sometimes taken as the final, adequate, measure of effect.

It should be obvious by now that one cannot logically argue that the ecosystem must be affected if a pollutant affects some individual organisms. Holdgate (1979) suggested a crude cascade model in which an initial disturbance at the primary biochemical level of organization spills over successively, as exposure increases, into the physiological, population and ecosystem levels. This model implies that the greater the disturbance at the first level the greater the probability or degree of disturbance at subsequent levels. Later texts have sometimes adopted this approach, which may explain the over-reliance on LD_{50} and LC_{50} values for the prediction of effects. We have seen that sublethal effects can be more important ecologically than is death of some members of a population. The LD_{50} or LC_{50} is therefore biased towards underestimating exposures that can have an ecological impact. Not only is its seeming precision nullified by the effects of environmental changes and of experimental conditions, but the whole approach exemplifies what logicians call the fallacy of composition, that what is true of simpler levels of organization will also be true of more complex levels. It risks overlooking ecological effects when effects at lower levels appear to be relatively unimportant. Sublethal effects could usefully replace death as the measure of effect, when selection of the most appropriate effect is facilitated if there is some indication of the site or mode of action (Moriarty, 1968).

There is, understandably, considerable effort to discover simple relationships between things that are relatively easy to determine, such as partition coefficients and acute toxicities, and those characteristics which are more difficult to measure but are of more immediate ecological relevance, such as the distribution of a pollutant in the environment and chronic toxicity. Mount (1968) suggested, for example, the use of an application factor, defined as the maximum chronic exposure that did not affect growth or reproduction, divided by the 96-h LC_{50} value. He suggested that the value of this factor, for a range of fish species and water qualities, might be constant for any particular pollutant. There is no obvious reason why such a simple relationship should be valid, and Winner et al. (1977) showed it to be invalid for the effects of copper on the crustacean *Daphnia magna* when given different diets. This same search for useful generalizations underlies much of the work on themes such as diversity and stress. Useful generalizations may be possible, but it seems less misleading for the present to suppose that the complexities of ecosystems, with their many interactions, necessitate attention to the fine detail of each particular situation. The important underlying principles for ecotoxicology are that effects, as in toxicology, depend on the exposure and dose, but, unlike toxicology, the measure of effect is on the population. It is arguable that general predictions about the effects of pollutants are impossible, because each ecosystem is unique. To some extent, the truth or otherwise of this proposition rests on the degree of effect with which one is concerned. The smaller the total impact, the more this proposition is likely to be true.

We have to accept that prediction of ecological effects by pollutants is unlikely, for the foreseeable future, ever to be very precise. Many predictions for lack of significant effect can be made with a fair degree of confidence, but not for all compounds. One general point of interpretation needs to be noted. Even though predictions may be perfectly correct for many species, one or a few species may be significantly more vulnerable and/or susceptible than most other species. To some extent, our concern is not with how most species will be affected by a pollutant, but with the first-affected species.

Given the difficulties of prediction, there is considerable need for the ability to make sound judgements from inadequate data. The Third Zuckerman Working Party was established in the UK in the early 1950s "to investigate the possible risks to the natural fauna and flora of the countryside from the use in agriculture of toxic substances", and quickly concluded that they could not assess the risks for lack of evidence either way (Sheail, 1985). Subsequent experience supports two propositions:

(1) The effects of pollutants on wildlife often pass unnoticed.
(2) When there is an obvious impact on wildlife, it can be difficult to establish that pollutants are involved.

Many contaminants are unappreciated, and Holden (1981) has commented on the considerable number of synthetic chemicals that occur as contaminants in the sea. These are usually thought to have negligible biological effects (except perhaps near their sources), because there are no effects that are obvious to a casual observer and because levels are very low. Chlorinated paraffins are one such group of compounds, which nevertheless reach concentrations of several parts per million in the edible mussel (*Mytilus edulis*) around British coasts (Campbell and McConnell, 1980). Additional factors that militate against detection of effects, if they occur, include the insensitivity of some current monitoring methods, which will only detect severe widespread effects, and ecosystems sometimes cross national or administrative boundaries. Difficulties of this sort explain why many of the ecological effects by pollutants of which we are aware were detected first by chance events.

In the real world decisions have to be made, to control or not to control some form of environmental contamination for which evidence, albeit incomplete, suggests there is an undesirable ecological effect. Detached scientific judgement is sometimes made more difficult by strong sectional interests that tend to interpret data by reference to a preferred conclusion, rather than by the merits of the case. Moreover, unusually stringent criteria of proof are sometimes demanded for contrary conclusions. The scientist's instinctive approach, of "innocent until proved guilty", can be rephrased as "safe until proved dangerous", which is not the way most of us deal with potential personal dangers (Southwood, 1985). What should be our attitude to risk or, in other words, what level of proof should be required: absolute, beyond reasonable doubt, the balance of probability or the possibility? The conflicting aims are to avoid unnecessary costs of control, and to avoid unacceptable damage to wildlife. The pressure of public opinion can sometimes be important in deciding the issue, although one sometimes has cause to wonder about the scientific validity of that pressure (Ashby, 1978).

Personal attitudes to risk are influenced by many factors, such as differences of personality, the degree of personal control or choice, familiarity with the hazard, the benefit and, if a decision is imposed, the degree of trust in the authority (Adams, 1995).

Deciding the balance of probable advantage and disadvantage of two or more different possible decisions is made much more difficult by the lack of consensus about the value of wildlife. Attitudes range from the extreme environmentalists who regard man as an unwelcome intruder on the natural scene to those whose only concern is with company profits, the "bottom line".

Underlying much of the work on prediction and monitoring is an implied question: which species are we concerned with? Practical considerations restrict laboratory studies to relatively few species. Similarly, in the field, some species are much easier to observe and study than are others. So those species have necessarily to represent all of the others. Rarely though is the question asked: are we concerned to predict and detect the effects of pollutants on all species, or only on some? If we are more concerned about some species than others, what are the criteria for selection? Many criteria are possible, but in brief individual species may be valued for their aesthetic, economic, educational, scientific or sporting worth. The essential point is that efforts to predict or to monitor the effects of pollutants on wildlife derive, in practice usually implicitly, from a value judgement that we wish to minimize the impact of pollutants on some species. Lack of clarity on this judgement can lead to data that are difficult to interpret, as we saw in the previous three chapters.

One can argue that to concentrate on species—the structure of ecosystems—ignores the equally important measures of ecosystem function, such as energy flow and nutrient cycles (pp.50–54). Function should be measured as well as structure because the links between the two are neither simple nor direct (Römbke and Moltmann, 1996). Obviously if, say, the rate of soil mineralization is of interest in its own right, or is thought to be involved in a structural change, it should be measured. Otherwise, I would ask what evidence is there to suggest that functional changes indicate effects by pollutants sooner or more easily than do effects on populations?

To attach primary importance to measures of function suggests that the community is a supra-organism. Much the same criticism can be made of the use of single numbers, for example the EQI (p. 272), to describe communities. A simple analogy may illustrate the point. The size of rectangles can be measured by their areas—a single number—but one may well be interested in both their length and breadth, when one number is inadequate. In ecosystems, one is concerned with populations of individual species.

The question when one or more species is or may be affected by pollution is: what are we trying to achieve? The peregrine falcon was straightforward (pp. 255–260), but it is not always so simple. Moss *et al.* (1996) discussed whole communities for lakes in the UK and considered various possible standards, starting with conditions when lakes were first formed after the last glaciation. They argue that because the impact of organic pollution has been a major concern, present conditions should be judged against those prevailing in the period 1920–1940, when agriculture was determined by natural constraints. They preferred that standard to conditions just before the Industrial Revolution, when communities in some lakes began to be affected by acid rain. These are value judgements which merit debate.

Pollution used to be accepted as an inevitable part of industrial life. Now it is seen as a problem, and one historic reason for the difficulties of control is that the impact, or cost, of pollution is borne not by the producer but by the environment (Ward and Dubos, 1972). Cost–benefit analysis attempts to develop a rational solution. The cost and benefit of different degrees of abatement are calculated in monetary terms, and the optimum degree of abatement is that with the lowest net cost. Many would dispute that all benefits and costs can be adequately measured in this way (Adams, 1995). However, the principle that the cost of controlling unacceptable pollution should be borne by the polluter was adopted by the OECD in 1972. This "polluter-pays" principle has developed since then and is being incorporated into national, regional and international law (OECD, 1992). The limits of acceptable pollution are set by public authorities.

I have tried to relate the problems of ecotoxicology to their ecological context. Failure so to do has led to much muddled thinking and to unreliable conclusions. One should, finally, relate pollution to the wider aspects of man's impact on his environment. Pollutants are not the only way in which we affect wildlife. Indeed, it is commonly held that direct destruction of habitat, be it from building, agriculture, forestry, recreation or other activities, is the major human impact on wildlife. These impacts can, to a considerable extent, be controlled and mitigated, but this planet does have a finite carrying capacity for our own as well as for all other species. Below that limit, the more human beings, the greater will be our impact on wildlife, but this leads on to value judgements that are beyond the scope of this book.

References

References with more than four authors are given as, e.g. Brown, A., Smith, B. *et al.* No slight is intended to those whose names are omitted, but I wish to minimize the similarity to a telephone directory.

Abel, P. D. (1996). "Water Pollution Biology", 2nd edn. Taylor & Francis, London.

Abou-Donia, M. B. and Preissig, S. H. (1976). Delayed neurotoxicity from continuous low-dose oral administration of leptophos to hens. *Toxicology and Applied Pharmacology* **38**, 595–608.

Abrahamson, W. G. and Gadgil, M. (1973). Growth form and reproductive effort in goldenrods (*Solidago*, Compositae). *American Naturalist* **107**, 651–660.

Adams, J. (1995). "Risk". UCL Press, London.

Adams, S. M. and Giddings, J. M. (1982). Review and evaluation of microcosms for assessing effects of stress in marine ecosystems. *Environment International* **7**, 409–418.

Alabaster, J. S., Garland, J. H. N. *et al.* (1972). An approach to the problem of pollution and fisheries. *In* "Conservation and Productivity of Natural Waters" (R. W. Edwards and D. J. Garrod, Eds), Symposia of the Zoological Society of London, No. 29, 87–114. Academic Press, London.

Alabaster, J. S., Calamari, D. *et al.* (1988). Water quality criteria for European freshwater fish. *Chemistry and Ecology* **3**, 165–253.

Alabaster, J. S., Calamari, D. *et al.* (1994). Mixtures of toxicants. *In* "Water Quality for Freshwater Fish. Further Advisory Criteria" (G. Howells, Ed.), 145–205. Gordon & Breach, Switzerland.

Albert, A. (1985). "Selective Toxicity. The Physico-Chemical Basis of Therapy", 7th edn. Chapman & Hall, London.

Al-Hiyaly, S. A. K., McNeilly, T., Bradshaw, A. D. and Mortimer, A. M. (1993). The effect of zinc contamination from electricity pylons. Genetic constraints on selection for zinc tolerance. *Heredity* **70**, 22–32.

Allen, L. H. and Amthor, J. S. (1995). Plant physiological responses to elevated CO_2, temperature, air pollution, and UV-B radiation. *In* "Biotic Feedbacks in the Global Climatic System" (G. M. Woodwell and F. T. MacKenzie, Eds), 51–84. Oxford University Press, New York.

Allison, A. C. (1956). The sickle cell and haemoglobin C genes in some African populations. *Annals of Human Genetics* **21**, 67–89.

Allison, A. C. (1975). Abnormal haemoglobin and erythrocyte enzyme-deficiency traits. In "Human Variation and Natural Selection" (D. F. Roberts, Ed.), 13th Symposium of the Society for the Study of Human Biology, 101–122. Taylor & Francis, London.

Al-Shahristani, H. and Shihab, K. M. (1974). Variation of biological half-life of methylmercury in man. *Archives of Environmental Health* **28**, 342–344.

Anderson, D. W. and Hickey, J. J. (1976). Dynamics of storage of organochlorine pollutants in herring gulls. *Environmental Pollution* **10**, 183–200.

Anderson, J. M. and MacFadyen, A. (Eds) (1976). "The Role of Terrestrial and Aquatic Organisms in Decomposition Processes", 17th Symposium of the British Ecological Society. Blackwell Scientific, Oxford.

Anderson, J. W. (1979). An assessment of knowledge concerning the fate and effects of petroleum hydrocarbons in the marine environment. In "Marine Pollution: Functional Responses" (W. B. Vernberg, A. Calabrese, F. P. Thurberg and F. J. Vernberg, Eds), 3–21. Academic Press, New York.

Andreae, M. O. (1996). Raising dust in the greenhouse. *Nature (London)* **380**, 389–390.

Andrewartha, H. G. and Birch L. C. (1954). "The Distribution and Abundance of Animals". University of Chicago Press, Chicago.

Andrews, M. J. (1984). Thames estuary: pollution and recovery. In "Effects of Pollutants at the Ecosystem Level. SCOPE 22" (P. J. Sheehan, D. R. Miller, G. C. Butler and P. Bourdeau, Eds), 195–227. Wiley, Chichester.

Andrews, M. J. and Rickard, D. G. (1980). Rehabilitation of the inner Thames estuary. *Marine Pollution Bulletin* **11**, 327–332.

Andrews, M. J., Aston, K. F. A., Rickard, D. G. and Steel, J. E. (1982). The macrofauna of the Thames estuary. *London Naturalist*, **61**, 30–61.

Angel, M. V. and Rice, T. L. (1996). The ecology of the deep ocean and its relevance to global waste management. *Journal of Applied Ecology* **33**, 915–926.

Annau, Z. (1992). Neurobehavioral effects of organophosphorus compounds. In "Organophosphates. Chemistry, Fate, and Effects" (J. E. Chambers and P. E. Levi, Eds), 419–432. Academic Press, San Diego.

Anon. (1978). "Cleaning Our Environment. A Chemical Perspective", 2nd edn. American Chemical Society, Washington, D.C.

Antonovics, J., Bradshaw, A. D. and Turner, R. G. (1971). Heavy metal tolerance in plants. *Advances in Ecological Research* **7**, 1–85.

Ariëns, E. J., Simonis, A. M. and Offermeier, J. (1976). "Introduction to General Toxicology", revised printing. Academic Press, New York.

Armitage, P. D., Moss, D., Wright, J. F. and Furse, M. T. (1983). The performance of a new biological water quality score system based on macroinvertebrates over a wide range of unpolluted running-water sites. *Water Research* **17**, 333–347.

Armour, J. A. (1973). Quantitative perchlorination of polychlorinated biphenyls as a method for confirmatory residue measurement and identification. *Journal of the Association of Official Analytical Chemists* **56**, 987–993.

Ashby, E. (1978). "Reconciling Man with the Environment". Oxford University Press, London.

Ashenden, T. W. and Mansfield, T. A. (1977). Influence of wind speed on the sensitivity of ryegrass to SO_2. *Journal of Experimental Botany* **28**, 729–735.

Ashmore, M. R., Bell, J. N. B. and Reily, C. L. (1980). The distribution of phytotoxic ozone in the British Isles. *Environmental Pollution, B* **1**, 195–216.

Atkins, G. L. (1969). "Multicompartment Models for Biological Systems". Methuen, London.

Avery, O. T., MacLeod, C. M. and McCarty, M. (1944). Studies on the chemical nature of the substance inducing transformation of pneumococcal types. Induction of transformation by a desoxyribonucleic acid fraction isolated from pneumococcus type III. *Journal of Experimental Medicine* **79**, 137–158.

Avise, J. C. (1994). "Molecular Markers, Natural History and Evolution". Chapman & Hall, New York.

Baker, A. J. M. (1987). Metal tolerance. *New Phytologist* **106**, 93–111.

Baker, A. J. M. and Walker, P. L. (1989). Physiological responses of plants to heavy metals and the quantification of tolerance and toxicity. *Chemical Speciation and Bioavailability* **1**, 7–17.

Baker, A. J. M. and Walker, P. L. (1990). Ecophysiology of metal uptake by tolerant plants. *In* "Heavy Metal Tolerance in Plants: Evolutionary Aspects" (A. J. Shaw, Ed.), 155–177. CRC Press, Boca Raton.

Baker, J. M. (1982). Mangrove swamps and the oil industry. *Oil and Petrochemical Pollution* **1** 5–22.

Barker, R. J. (1958). Notes on some ecological effects of DDT sprayed on elms. *Journal of Wildlife Management* **22**, 269–274.

Barnes, J. M. (1975). Assessing hazards from prolonged and repeated exposure to low doses of toxic substances. *British Medical Bulletin* **31**, 196–200.

Barnola, J. M., Raynaud, D., Korotkevich, Y. S. and Lorius, C. (1987). Vostok ice core provides 160,000-year record of atmospheric CO_2. *Nature (London)* **329**, 408–414.

Baron, R. L. (1981). Delayed neurotoxicity and other consequences of organophosphate esters. *Annual Review of Entomology* **26**, 29–48.

Barrett, M. J., Mollowney, B. M. and Casapieri, P. (1978). The Thames model: an assessment. *Progress in Water Technology* **10**, 409–416.

Bates, T. S., Lamb, B. K. *et al.* (1992). Sulfur emissions to the atmosphere from natural sources. *Journal of Atmospheric Chemistry* **14**, 315–337.

Bateson, W. (1901). Experiments in plant hybridisation. *Journal of the Royal Horticultural Society* **26**, 1–32. (English translation of Mendel, 1865.)

Battaglia, B., Bisol, P. M. and Rodino, E. (1980). Experimental studies on some genetic effects of marine pollution. *Helgoländer Meeresuntersuchungen* **33**, 587–595.

Baughman, G. L. and Lassiter, R. R. (1978). Prediction of environmental pollutant concentration. *In* "Estimating the Hazard of Chemical Substances to Aquatic Life" (J. Cairns, K. L. Dickson and A. W. Maki, Eds), Special Technical Publications, No. 657, 35–54. American Society for Testing and Materials, Philadelphia.

Bayne, B. L. (1985). Cellular and physiological measures of pollution effect. *Marine Pollution Bulletin* **16**, 127–129.

Bayne, B. L., Moore, M. N. *et al.* (1979). An experimental approach to the determinants of biological water quality. *Philosophical Transactions of the Royal Society of London, B* **286**, 563–581.

Bayne, B. L., Brown, D. A. *et al.* (1985). "The Effects of Stress and Pollution on Marine Animals". Praeger, New York.

Bazzaz, F. A. (1990). The response of natural ecosystems to the rising global CO_2 levels. *Annual Review of Ecology and Systematics* **21**, 167–196.

Begon, M., Harper, J. L. and Townsend, C. R. (1996). "Ecology. Individuals, Populations and Communities", 3rd edn. Blackwell Science, Oxford.

Bell, J. N. B. and Clough, W. S. (1973). Depression of yield in ryegrass exposed to sulphur dioxide. *Nature (London)* **241**, 47–49.

Bell, J. N. B. and Mudd, C. H. (1976). Sulphur dioxide resistance in plants: a case study of *Lolium perenne*. In "Effects of Air Pollutants on Plants" (T. A. Mansfield, Ed.), 87–103. Cambridge University Press, Cambridge.

Bell, J. N. B., Rutter, A. J. and Relton, J. (1979). Studies on the effects of low levels of sulphur dioxide on the growth of *Lolium perenne* L. *New Phytologist* **83**, 627–643.

Bell, J. N. B., Ashmore, M. R. and Wilson, G. B. (1991). Ecological genetics and chemical modifications of the atmosphere. In "Ecological Genetics and Air Pollution" (L. F. Pitelka and M. T. Clegg, Eds), 33–59. Springer, New York.

Benet, L. Z. (1994). Foreword. In "Pharmacodynamics and Drug Development Perspectives in Clinical Pharmacology" (N. R. Cutler, J. J. Sramek and P. K. Narang, Eds), xiii–xv. Wiley, Chichester.

Bennett, B. G. (1981). The exposure commitment method in environmental pollutant assessment. *Experimental Monitoring and Assessment* **1**, 21–36.

Bennett, J. H. (Ed.) (1965). "Experiments in Plant Hybridisation". Oliver and Boyd, Edinburgh.

Benson, J. F. (1973). The biology of Lepidoptera infesting stored products, with special reference to population dynamics. *Biological Reviews* **48**, 1–26.

Beranek, A. P. and Oppenoorth, F. J. (1977). Evidence that the elevated carboxy-esterase (esterase 2) in organophosphorus-resistant *Myzus persicae* (Sulz.) is identical with the organophosphate-hydrolyzing enzyme. *Pesticide Biochemistry and Physiology* **7**, 16–20.

ten Berge, W. F. and Hillebrand, M. (1974). Organochlorine compounds in several marine organisms from the North Sea and the Dutch Wadden Sea. *Netherlands Journal of Sea Research* **8**, 361–368.

Bernard, R. F. (1963). Studies on the effects of DDT on birds. *Museum of the Michigan State University Biological Series* **2**, 155–191.

Berryman, A. A. (1986). "Forest Insects. Principles and Practice of Population Management". Plenum Press, New York.

Bess, H. A. (1961). Population ecology of the gypsy moth *Porthetria dispar* L. (Lepidoptera: Lymantridae). *Bulletin, Connecticut Agricultural Experiment Station* **646**, 4–43.

Betzer, S. B. and Pilson, M. E. Q. (1974). The seasonal cycle of copper concentration in *Busycon canaliculatum* L. *Biological Bulletin* **146**, 165–175.

Bick, H. (1963). A review of central European methods for the biological estimation of water pollution levels. *Bulletin of the World Health Organisation* **29**, 401–413.

Biddinger, G. R. and Gloss, S. P. (1984). The importance of trophic transfer in the bio-accumulation of chemical contaminants in aquatic ecosystems. *Residue Reviews* **91**, 103–145.

Bishop, J. A. (1972). An experimental study of the cline of industrial melanism in *Biston betularia* (L.) (Lepidoptera) between urban Liverpool and rural North Wales. *Journal of Animal Ecology* **41**, 209–243.

Bishop, J. A., Cook, L. M., Muggleton, J. and Seaward, M. R. D. (1975). Moths, lichens and air pollution along a transect from Manchester to North Wales. *Journal of Applied Ecology* **12**, 83–98.

Bitensky, L. (1963). The reversible activation of lysosomes in normal cells and the effects of pathological conditions. In "Lysosomes" (A. V. S. de Reuck and M. P. Cameron, Eds), Ciba Foundation Symposium, 362–383. Churchill, London.

Black, S. C. (1991). Data analysis and presentation. In "Instrumental Analysis of Pollutants" (C. N. Hewitt, Ed.), 335–355. Elsevier, London.

Blanck, H. and Dahl, B. (1996). Pollution-induced community tolerance (PICT) in marine periphyton in a gradient of tri-n-butyltin (TBT) contamination. *Aquatic Toxicology* **35**, 59–77.

Blanck, H. and Dahl, B. (in press). Recovery of marine periphyton communities around a Swedish marina after the ban of TBT use in antifouling paint. *Marine Pollution Bulletin.*

Blanck, H. and Wängberg, S-Å. (1988). Induced community tolerance in marine periphyton established under arsenate stress. *Canadian Journal of Fisheries and Aquatic Science* **45**, 1816–1819.

Blanck, H., Walling, G. and Wängberg, S-A. (1984). Species-dependent variation in algal sensitivity to chemical compounds. *Ecotoxicology and Environmental Safety* **8**, 339–351.

Bleasdale, J. K. A. (1973). Effects of coal-smoke pollution gases on the growth of ryegrass (*Lolium perenne* L.). *Environmental Pollution* **5**, 275–285.

Bloomquist, J. R. (1996). Ion channels as targets for insecticides. *Annual Review of Entomology* **41**, 163–190.

Blueweiss, L., Fox, H. *et al.* (1978). Relationships between body size and some life history parameters. *Oecologia* **37**, 257–272.

Bonga, S. E. W. and Balm, P. H. M. (1989). Endocrine responses to acid stress in fish. *In* "Acid Toxicity and Aquatic Animals" (R. Morris, E. W. Taylor, D. J. A. Brown and J. A. Brown, Eds), 243–263. Cambridge University Press, Cambridge.

Borg, K., Wanntorp, H., Erne, K. and Hanko, E. (1969). Alkyl mercury poisoning in terrestrial Swedish wildlife. *Viltrevy* **6**, 301–379.

Bormann, F. H. and Likens, G. E. (1981). "Pattern and Process in a Forested Ecosystem. Disturbance, Development and the Steady State Based on the Hubbard Brook Ecosystem Study". Springer, New York.

Boyce, H. S. (1984). Restitution of r- and K-selection as a model of density-dependent natural selection. *Annual Review of Ecology and Systematics* **15**, 427–447.

Boyd, C. M. (1981). Microcosms and experimental planktonic food chains. *In* "Analysis of Marine Ecosystems" (A. R. Longhurst, Ed.), 627–649. Academic Press, London.

Bradley, R. S. and Jones, P. D. (Eds) (1992). "Climate since A.D. 1500". Routledge, London.

Bradshaw, A. D. (1975). The evolution of metal tolerance and its significance for vegetation establishment on metal contaminated sites. *In* "International Conference on Heavy Metals in the Environment", 599–622. University of Toronto Press, Toronto.

Bradshaw, A. D. (1976). Pollution and evolution. *In* "Effects of Air Pollutants on Plants" (T. A. Mansfield, Ed.), 135–159. Cambridge University Press, Cambridge.

Bradshaw, A. D. (1982). Evolution of heavy metal resistance—an analogy for herbicide resistance? *In* "Herbicide Resistance in Plants" (H. M. LeBaron and J. Gressel, Eds), 293–307. Wiley, New York.

Bradshaw, A. D. and McNeilly, T. (1981). "Evolution and Pollution". Arnold, London.

Brakefield, P. M. (1987). Industrial melanism: do we have the answers? *Trends in Ecology and Evolution* **2**, 117–122.

Branson, D. R. (1978). Predicting the fate of chemicals in the aquatic environment from laboratory data. *In* "Estimating the Hazard of Chemical Substances to Aquatic Life" (J. Cairns, K. L. Dickson and A. W. Maki, Eds), Special Technical Publications, No. 657, 55–70. American Society for Testing and Materials, Philadelphia.

Braun-Blanquet, J. (1932). "Plant Sociology. The Study of Plant Communities". McGraw-Hill, New York.

Braune, B. M. (1987). Mercury accumulation in relation to size and age of Atlantic herring (*Clupea harengus harengus*) from the Southwestern Bay of Fundy, Canada. *Archives of Environmental Contamination and Toxicology* **16**, 311–320.

Brimblecombe, P. (1987). "The Big Smoke. A History of Air Pollution in London since Medieval Times". Methuen, London.

Brimblecombe, P. (1996). "Air Composition and Chemistry", 2nd edn. Cambridge University Press, Cambridge.

Brimblecombe, P., Hammer, C. *et al.* (1989). Human influence on the sulphur cycle. *In* "Evolution of the Global Biogeochemical Sulphur Cycle. SCOPE 39" (P. Brimblecombe and A. Y. Lein, Eds), 77–121. Wiley, Chichester.

Brooke, D. N., Dobbs, A. J. and Williams, N. (1986). Octanol:water partition coefficients (P): measurement, estimation, and interpretation, particularly for chemicals with P > 10⁵. *Ecotoxicology and Environmental Safety* **11**, 251–260.

Brown, G. C. and Skipsey, E. (1986). "Energy Resources. Geology, Supply and Demand". Open University Press, Milton Keynes.

Brown, K. A. (1982). Sulphur in the environment: a review. *Environmental Pollution, B* **3**, 47–80.

Brown, V. M. (1968). Calculation of the acute toxicity of mixtures of poisons to rainbow trout. *Water Research* **2**, 723–733.

de Bruijn, J., Busser, F., Seinen, W. and Hermens, J. (1989). Determination of octanol/water partition coefficients for hydrophobic organic chemicals with the "slow-stirring" method. *Environmental Toxicology and Chemistry* **8**, 499–512.

Bryan, G. W. (1976). Some aspects of heavy metal tolerance in aquatic organisms. *In* "Effects of Pollutants on Aquatic Organisms" (A. P. M. Lockwood, Ed.), 7–34. Cambridge University Press, Cambridge.

Bryan, G. W. (1979). Bioaccumulation of marine pollutants. *Philosophical Transactions of the Royal Society of London, B* **286**, 483–505.

Bryan, G. W. (1980). Recent trends in research on heavy-metal contamination in the sea. *Helgoländer Meeresuntersuchungen* **33**, 6–25.

Bryan, G. W. and Gibbs, P. E. (1991). Impact of low concentrations of tributyltin (TBT) on marine organisms: a review. *In* "Metal Ecotoxicology: Concepts and Applications" (M. C. Newman and A. W. McIntosh, Eds), 323–361. Lewis, Ann Arbor.

Bryan, G. W. and Hummerstone, L. G. (1973). Brown seaweed as an indicator of heavy metals in estuaries in south-west England. *Journal of the Marine Biological Association of the United Kingdom* **53**, 705–720.

Bryan, G. W., Langston, W. J. and Hummerstone, L. G. (1980). "The Use of Biological Indicators of Heavy Metal Contamination in Estuaries. With Special Reference to an Assessment of the Biological Availability of Metals in Estuarine Sediments from South-west Britain", Occasional Publication No. 1. Marine Biological Association of the United Kingdom, Plymouth.

Bryan, G. W., Langston, W. J., Hummerstone, L. G. and Burt, G. R. (1985). "A Guide to the Assessment of Heavy-Metal Contamination in Estuaries using Biological Indicators", Occasional Publication No. 4. Marine Biological Association of the United Kingdom, Plymouth.

Bryan, G. W., Gibbs, P. E., Hummerstone, L. G. and Burt, G. R. (1986). The decline of the gastropod *Nucella lapillus* around south-west England: evidence for the effect of tributyltin from antifouling paints. *Journal of the Marine Biological Association of the United Kingdom* **66**, 611–640.

Buffington, J. D. and Little, L. W. (1980). Research needs and priorities. *In* Biological Monitoring for Environmental Effects" (D. L. Worf, Ed.). 197–199. Heath, Lexington, Massachusetts.

Bulich, A. A. (1982). A practical and reliable method for monitoring the toxicity of aquatic samples. *Process Biochemistry* **17** (part 2), 45–47.

Bulich, A. A. (1984). Microtox—a bacterial toxicity test with several environmental applications. *In* "Toxicity Screening Procedures Using Bacterial Systems" (D. Liu and B. J. Dutka, Eds), 55–64. Dekker, New York.

Bull, K. R. (1991). The critical loads/levels approach to gaseous pollutant emission control. *Environmental Pollution* **69**, 105–123.

Bunyan, P. J. (1973). An approach to the detection of pesticide poisoning in wildlife. *Proceedings of the Society for Analytical Chemistry* **10**, 34–36.

Bunyan, P. J., Jennings, D. M. and Taylor, A. (1968). Organophosphorus poisoning. Some properties of avian esterases. *Journal of Agricultural and Food Chemistry* **16**, 326–331.

Burrows, C. J. (1990). "Processes of Vegetation Change". Unwin Hyman, London.

Burt, P. E., Gregory, G. E. and Molloy, F. M. (1966). A histochemical and electrophysiological study of the action of diazoxon on cholinesterase activity and nerve conduction in ganglia of the cockroach *Periplaneta americana* L. *Annals of Applied Biology* **58**, 341–354.

Burton, J. D. (1979). Physico-chemical limitations in experimental investigations. *Philosophical Transactions of the Royal Society of London, B* **286**, 443–456.

Burton, M. A. S. (1986). "Biological Monitoring of Environmental Contaminants (Plants)", MARC Report Number 32. Monitoring and Assessment Research Centre, University of London, London.

Busvine, J. R. (1951). Mechanism of resistance to insecticide in houseflies. *Nature (London)* **168**, 193–195.

Busvine, J. R. (1971). "A Critical Review of the Techniques for Testing Insecticides", 2nd edn. Commonwealth Agricultural Bureaux, Farnham Royal.

Butler, G. C. (1978). Estimation of doses and integrated doses. *In* "Principles of Ecotoxicology. SCOPE 12" (G. C. Butler, Ed.), 91–112. Wiley, Chichester.

Butler, G. C. (1980). Methods of dose evaluation for incorporation of radio-nuclides: applicability to environmental chemicals. *Ecotoxicology and Environmental Safety* **4**, 384–392.

Butler, P. A., Andrén, L. *et al.* (1971). Monitoring organisms. *In* "FAO Fisheries Reports", No. 99 (Suppl. 1), 101–112.

Cairns, J. (1980). Scenarios on alternative futures for biological monitoring, 1978–1985. *In* "Biological Monitoring for Environmental Effects" (D. L. Worf, Ed.), 11–21. Heath, Lexington, Massachusetts.

Cairns, J. and Buikema, A. L. (1984). "Restoration of Habitats Impacted by Oil Spills" (Ann Arbor Science). Butterworth, Boston.

Cairns, J. and Mount, D. I. (1990). Aquatic toxicology. *Environmental Science and Technology* **24**, 154–161.

Cairns, J. and Pratt, J. R. (1986). On the relation between structural and functional analyses of ecosystems. *Environmental Toxicology and Chemistry* **5**, 785–786.

Cairns, J., Dickson, K. L. and Maki, A. W. (Eds) (1978). "Estimating the Hazard of Chemical Substances to Aquatic Life", Special Technical Publications, No. 657. American Society for Testing and Materials, Philadelphia.

Calabrese, E. J. (1994). Primer on BELLE. *In* "Biological Effects of Low Level Exposures: Dose–Response Relationships" (E. J. Calabrese, Ed.), 27–42. CRC Press, Boca Raton.

Calabrese, E. J., McCarthy, M. E. and Kenyon, E. (1987). The occurrence of chemically induced hormesis. *Health Physics* **52**, 531–541.

Calamari, D. and Vighi, M. (1992). Role of evaluative models to assess exposures to pesticides. In "Methods to Assess Adverse Effects of Pesticides on Non-Target Organisms. SCOPE 49" (R. G. Tardiff, Ed.), 119–132. Wiley, Chichester.

Calow, P. (Ed.) (1993a). "Handbook of Ecotoxicology", Vol. 1. Blackwell Scientific, Oxford.

Calow, P. (1993b). General principles and overview. In "Handbook of Ecotoxicology", Vol. 1 (P. Calow, Ed.), 1–5. Blackwell Scientific, Oxford.

Campbell, I. and McConnell, G. (1980). Chlorinated paraffins and the environment. 1. Environmental occurrence. *Environmental Science and Technology* **14**, 1209–1214.

Capen, D. E. and Leiker, T. J. (1979). DDE residues in blood and other tissues of white-faced ibis. *Environmental Pollution* **19**, 163–171.

Carlson, R. W. and Drummond, R. A. (1978). Fish cough response—a method for evaluating quality of treated complex effluents. *Water Research* **12**, 1–6.

Caroli, S. (Ed.) (1996). "Element Speciation in Bioinorganic Chemistry". *Chemical Analysis* **135**. Wiley, New York.

Carson, R. (1962). "Silent Spring". Houghton Mifflin, Boston.

Cates, R. G. and Orians, G. H. (1975). Successional status and the palatability of plants to generalised herbivores. *Ecology* **56**. 410–418.

Caughley, G. and Sinclair, A. R. E. (1994). "Wildlife Ecology and Management". Blackwell Scientific, Boston.

Cecil, H. C., Fries, G. F. *et al.* (1972). Dietary p,p'-DDT, o,p'-DDT or p,p'-DDE and changes in eggshell characteristics and pesticide accumulation in egg contents and body fat of caged white leghorns. *Poultry Science* **51**, 130–138.

Chadwick G. G. and Brocksen, R. W. (1969). Accumulation of dieldrin by fish and selected fish-food organisms. *Journal of Wildlife Management* **33**, 693–700.

Chadwick, J. W. and Canton, S. P. (1984). Inadequacy of diversity indices in discerning metal mine drainage effects on a stream invertebrate community. *Water, Air and Soil Pollution* **22**, 217–223.

Chambers, H. W. (1992). Organophosphorus compounds: an overview. In "Organophosphates Chemistry, Fate, and Effects" (J. E. Chambers and P. E. Levi, Eds), 3–17. Academic Press, San Diego.

Charlson, R. J., Lovelock, J. E., Andreae, M. O. and Warren, S. G. (1987). Oceanic phytoplankton, atmospheric sulphur, cloud albedo and climate. *Nature (London)* **326**, 655–661.

Chipman, J. K. and Walker, C. H. (1979). The metabolism of dieldrin and two of its analogues: the relationship between rates of microsomal metabolism and rates of excretion of metabolites in the male rat. *Biochemical Pharmacology* **28**, 1337–1345.

Christensen, E. R. and Chen, C-Y. (1989). Modeling of combined toxic effects of chemicals. In "Hazard Assessment of Chemicals", Vol. 6 (J. Saxena, Ed.), 125–186. Hemisphere, New York.

Clark, C. W. (1990). "Mathematical Bioeconomics. The Optimal Management of Renewable Resources", 2nd edn. Wiley, New York.

Clark, J. R. and Cripe, C. R. (1993). Marine and estuarine multi-species test systems. In "Handbook of Ecotoxicology", Vol. 1 (P. Calow, Ed.), 227–247. Blackwell Scientific, Oxford.

Clark, R. B. (1984). Impact of oil pollution on seabirds. *Environmental Pollution A* **33**, 1–22.

Clark, R. B. (1992). "Marine Pollution", 3rd edn. Clarendon, Oxford.

Clarke, A. G. (1992). The atmosphere. *In* "Understanding Our Environment: An Introduction to Environmental Chemistry and Pollution" (R. M. Harrison, Ed.), 2nd edn, 5–51. Royal Society of Chemistry, Cambridge.

Clarke, C. A., Mani, G. S. and Wynne, G. (1985). Evolution in reverse: clean air and the peppered moth. *Biological Journal of the Linnaean Society* **26**, 189–199.

Clarke, C. A., Clarke, F. M. M. and Dawkins, H. C. (1990). *Biston betularia* (the peppered moth) in West Kirby, Wirral, 1959–1989: updating the decline of f. *carbonaria*. *Biological Journal of the Linnaean Society* **39**, 323–326.

Clarkson, T. W. (1972). Recent advances in the toxicology of mercury with emphasis on the alkylmercurials. *Critical Reviews in Toxicology* **1**, 203–234.

Cleary, J. J. and Stebbing, A. R. D. (1985). Organotin and total tin in coastal waters of southwest England. *Marine Pollution Bulletin* **16**, 350–355.

Clements, F. E. (1916). "Plant Succession. An Analysis of the Development of Vegetation", Publication 242. Carnegie Institute of Washington, Washington.

Cody, M. L. (1966). A general theory of clutch size. *Evolution* **20**, 174–184.

Colburn, W. A. and Eldon, M. A. (1994). Simultaneous pharmacokinetic/pharmacodynamic modeling. *In* "Pharmacodynamics and Drug Development Perspectives in Clinical Pharmacology" (N. R. Cutter, J. J. Sramek and P. K. Narang, Eds), 19–44. Wiley, Chichester.

Collander, R. (1954). The permeability of *Nitella* cells to non-electrolytes. *Physiologia Plantarum* **7**, 420–445.

Committee on Strategies for the Management of Pesticide Resistant Pest Populations (1986). "Pesticide Resistance: Strategies and Tactics for Management". National Research Council, Washington, D. C.

Connell, D. W. (1994). The octanol–water partition coefficient. *In* "Handbook of Ecotoxicology", Vol. 2 (P. Calow, Ed.), 311–320. Blackwell Scientific, Oxford.

Connell, D. W. and Miller, G. J. (1984). "Chemistry and Ecotoxicology of Pollution". Wiley, New York.

Connell, J. H. (1961). The influence of interspecific competition and other factors on the distribution of the barnacle *Chthamalus stellatus*. *Ecology* **42**, 710–723.

Connell, J. H. (1975). Some mechanisms producing structure in natural communities: a model and evidence from field experiments. *In* "Ecology and Evolution of Communities" (M. L. Cody and J. M. Diamond, Eds), 460–490. Belknap, Cambridge, Massachusetts.

Connell, J. H. and Slatyer, R. O. (1977). Mechanisms of succession in natural communities and their role in community stability and organization. *American Naturalist* **111**, 1119–1144.

Connell, J. H. and Sousa, W. P. (1983). On the evidence needed to judge ecological stability or persistence. *American Naturalist* **121**, 789–824.

Connor, M. S. (1984). Fish/sediment concentration ratios for organic compounds. *Environmental Science and Technology* **18**, 31–35.

Cooke, A. S. (1971). Selective predation by newts on frog tadpoles treated with DDT. *Nature (London)* **229**, 275–276.

Cooke, A. S. (1973). Shell thinning in avian eggs by environmental pollutants. *Environmental Pollution* **4**, 85–152.

Cooke, A. S. (1979a). Changes in egg shell characteristics of the sparrowhawk (*Accipiter nisus*) and peregrine (*Falco peregrinus*) associated with exposure to environmental pollutants during recent decades. *Journal of Zoology, London* **187**, 245–263.

Cooke, A. S. (1979b). Egg shell characteristics of gannets *Sula bassana*, shags *Phalacrocorax aristotelis* and great black-backed gulls *Larus marinus* exposed to DDE and other environmental pollutants. *Environmental Pollution* 19, 47–65.

Cooper, J. P. (Ed.) (1975). "Photosynthesis and Productivity in Different Environments". Cambridge University Press, Cambridge.

Cope, A. C. and Hancock, E. M. (1939a). Substituted vinyl barbituric acids. I. Isopropenyl derivatives. *Journal of the American Chemical Society* 61, 96–98.

Cope, A. C. and Hancock, E. M. (1939b). Substituted vinyl barbituric acids. II. (1-methylpropenyl) derivatives. *Journal of the American Chemical Society* 61, 353–354.

Corbett, J. R., Wright, K. and Baillie, A. C. (1984). "The Biochemical Mode of Action of Pesticides", 2nd edn. Academic Press, London.

Corner, E. D. S. (1978). Pollution studies with marine plankton Part I. Petroleum hydrocarbons and related compounds. *Advances in Marine Biology* 15, 289–380.

Coulson, J. C., Deans, I. R. *et al.* (1972). Changes in organochlorine contamination of the marine environment of eastern Britain monitored by shag eggs. *Nature (London)* 236, 454–456.

Cowling, D. W. and Koziol, M. J. (1978). Growth of ryegrass (*Lolium perenne* L.) exposed to SO_2. I. Effects on photosynthesis and respiration. *Journal of Experimental Botany* 29, 1029–1036.

Cowling, D. W. and Lockyer, D. R. (1976). Growth of perennial ryegrass (*Lolium perenne* L.) exposed to a low concentration of sulphur dioxide. *Journal of Experimental Botany* 27, 411–417.

Cowling, D. W., Jones, L. H. P. and Lockyer, D. R. (1973). Increased yield through correction of sulphur deficiency in ryegrass exposed to sulphur dioxide. *Nature (London)* 243, 479–480.

Cowling, D. W., Lockyer, D. R., Chapman, P. F. and Koziol, M. J. (1981). An assessment of the concentration of SO_2 to which plants are exposed in a system of chambers. *Environmental Pollution, A* 26, 1–13.

Crisp, D. J., Christie, A. O. and Ghobashy, A. F. O. (1967). Narcotic and toxic action of organic compounds on barnacle larvae. *Comparative Biochemistry and Physiology* 22, 629–649.

Crittenden, P. D. and Read, D. J. (1978). The effects of air pollution on plant growth with special reference to sulphur dioxide. II. Growth studies with *Lolium perenne* L. *New Phytologist* 80, 49–62.

Croft, B. A. (1990). "Arthropod Biological Control Agents and Pesticides". Wiley, New York.

Crosby, D. G. (1975). The toxicant-wildlife complex. *Pure and Applied Chemistry* 42, 233–253.

Crow, M. E. and Taub, F. B. (1979). Designing a microcosm bioassay to detect ecosystem level effects. *International Journal of Environmental Studies* 13, 141–147.

Cullen, J. J. and Lesser, M. P. (1991). Inhibition of photosynthesis by ultraviolet radiation as a function of dose and dosage rate: results for a marine diatom. *Marine Biology* 111, 183–190.

Cullis, C. F. and Hirschler, M. M. (1980). Atmospheric sulphur: natural and man-made sources. *Atmospheric Environment* 14, 1263–1278.

Cummings, J. G., Zee, K. T. *et al.* (1966). Residues in eggs from low level feeding of five chlorinated hydrocarbon insecticides to hens. *Journal of the Association of Official Analytical Chemists* 49, 354–364.

Dale, W. E., Gaines, T. B. and Hayes, W. J. (1962). Storage and excretion of DDT in starved rats. *Toxicology and Applied Pharmacology* 4, 89–106.

Dansgaard, W., Johnsen, S. J. *et al.* (1993). Evidence for general instability of past climate from a 250-kyr ice-core record. *Nature (London)* **364**, 218–220.

Darwin, C. (1859). "On the Origin of Species by Means of Natural Selection, or the Preservation of Favoured Races in the Struggle for Life". Murray, London.

Davidson, E. A. (1995). Linkages between carbon and nitrogen cycling and their implications for storage of carbon in terrestrial ecosystems. *In* "Biotic Feedbacks in the Global Climatic System" (G. M. Woodwell and F. T. MacKenzie, Eds), 219–230. Oxford University Press, New York.

Davies, A. G. (1978). Pollution studies with marine plankton Part II. Heavy metals. *Advances in Marine Biology* **15**, 381–508.

Davies, J. M. and Gamble, J. C. (1979). Experiments with large enclosed ecosystems. *Philosophical Transactions of the Royal Society of London, B* **286**, 523–544.

Davies, R. P. and Dobbs, A. J. (1984). The prediction of bioconcentration in fish. *Water Research* **18**, 1253–1262.

Davison, K. L. (1970). Dieldrin accumulation in tissues of the sheep. *Journal of Agricultural and Food Chemistry* **18**, 1156–1160.

Dearden, J. C. and Townend, M. S. (1977). A theoretical approach to structure–activity relationships—some implications for the concept of optimal lipophilicity. *In* "Herbicides and Fungicides. Factors Affecting their Activity" (N. R. McFarlane, Ed.), Special Publication No. 29, 135–141. Chemical Society, London.

Deichmann, W. B., Dressler, I., Keplinger, M. and MacDonald, W. E. (1968). Retention of dieldrin in blood, liver, and fat of rats fed dieldrin for six months. *Industrial Medicine and Surgery* **37**, 837–839.

Deichmann, W. B., MacDonald, W. E. and Cubit, D. A. (1971). DDT tissue retention: sudden rise induced by the addition of aldrin to a fixed DDT intake. *Science (New York)* **172**, 275–276.

Deichmann, W. B., Cubit, D. A., MacDonald, W. E. and Beasley, A. G. (1972). Organochlorine pesticides in the tissues of the great barracuda (*Sphyraena barracuda*) (Waldbaum). *Archiv für Toxikologie* **29**, 287–309.

Dempster, J. P. (1975a). "Animal Population Ecology". Academic Press, London.

Dempster, J. P. (1975b). Effects of organochlorine insecticides on animal populations. *In* "Organochlorine Insecticides: Persistent Organic Pollutants" (F. Moriarty. Ed.), 231–248. Academic Press, London.

Dempster, J. P. (1983). The natural control of populations of butterflies and moths. *Biological Reviews* **58**, 461–481.

Department of the Environment (1971). "Report of a River Pollution Survey of England and Wales 1970", Vol. I. HMSO, London.

Department of the Environment (1975). "River Pollution Survey of England and Wales. Updated 1973". HMSO, London.

Derouane, A., Verduyn, G., Goedertier, R. and Hallez, S. (1982). On the validity of the acidimetric method for sulfur dioxide survey networks. *Science of the Total Environment* **22**, 275–283.

Devonshire, A. L. and Field, L. M. (1991). Gene amplification and insecticide resistance. *Annual Review of Entomology* **36**, 1–23.

Dicks, B. and Hartley, J. P. (1982). The effects of repeated small oil spillages and chronic discharges. *Philosophical Transactions of the Royal Society of London, B* **297**, 285–307.

Dickson, K. L., Gruber, D., King, C. and Lubenski, D. (1980). Biological monitoring to provide an early warning of environmental contaminants. *In* "Biological Monitoring for Environmental Effects" (D. L. Worf, Ed.), 53–74. Heath, Lexington, Massachusetts.

Dickson, W. (1980). Properties of acidified waters. *In* "Ecological Impact of Acid Precipitation" (D. Drabløs and A. Tollan, Eds), 75–83. SNSF Project, Oslo-Ås.

Dickson, W. (1986). Acidification effects in the aquatic environment. *In* "Acidification and its Policy Implications" (T. Schneider, Ed.), Studies in Environmental Science **30**, 19–28. Elsevier, Amsterdam.

Digby, P. G. N. and Kempton, R. A. (1987). "Multivariate Analysis of Ecological Communities". Chapman & Hall, London.

Dixon, D. R., Simpson-White, R. and Dixon, L. R. J. (1992). Evidence for thermal stability of ribosomal DNA sequences in hydrothermal-vent organisms. *Journal of the Marine Biological Association of the United Kingdom* **72**, 519–527.

Doerffer, J. W. (1992). "Oil Spill Response in the Marine Environment". Pergamon, Oxford.

Donkin, P. (1994). Quantitative structure–activity relationships. *In* "Handbook of Ecotoxicology", Vol. 1 (P. Calow, Ed.), 321–347. Blackwell Scientific, Oxford.

Doust, J. L., Schmidt, M. and Doust, L. L. (1994). Biological assessment of aquatic pollution: a review, with emphasis on plants as biomonitors. *Biological Reviews*, **69**, 147–186.

Drury, W. H. and Nisbet, I. C. T. (1973). Succession. *Journal of the Arnold Arboretum, Harvard University* **54**, 331–368.

Duffey, E., Morris, M. G. *et al.* (1974). "Grassland Ecology and Wildlife Management". Chapman & Hall, London.

Dunnet, G. M. (1982). Oil pollution and seabird populations. *Philosophical Transactions of the Royal Society of London*, B **297**, 413–427.

Duplessy, J-C. (1978). Isotope studies. *In* "Climatic Change" (J. Gribbin, Ed.), 46–67. Cambridge University Press, Cambridge.

de Duve, C. (1963). General properties of lysosomes. *In* "Lysosomes" (A. V. S. de Reuck and M. P. Cameron, Eds), Ciba Foundation Symposium, 1–35. Churchill, London.

Earnest, R. D. and Benville, P. E. (1971). Residues in fish, wildlife, and estuaries. *Pesticides Monitoring Journal* **5**, 235–241.

Eaton, P. L. and Klaassen, C. D. (1996). Principles of toxicology. *In* "Casarett and Doull's Toxicology. The Basic Science of Poisons" (C. D. Klaassen, Ed.), 13–33. McGraw-Hill, New York.

Eberhardt, L. L. (1970). Correlation, regression, and density dependence. *Ecology* **51**, 306–310.

Edwards, C. A. and Heath, G. W. (1963). The role of soil animals in breakdown of leaf material. *In* "Soil Organisms" (J. Doeksen and J. van der Drift, Eds), 76–84. North-Holland, Amsterdam.

EEC Council (1976). Directive 76/464/EEC on pollution caused by certain dangerous substances discharged into the aquatic environment. *Official Journal* **L129/23**, 18 May 1976.

EEC Council (1979). Directive 79/831/EEC amending for the sixth time Directive 67/548/EEC on the approximation of the laws, regulations and administrative provisions relating to the classification, packaging and labelling of dangerous substances. *Official Journal* **L259/10**, 15 October 1979.

EEC Council (1992). Directive 92/32/EEC, amending for the seventh time Directive 67/548/EEC on the approximation of the laws, regulations and administrative provisions relating to the classification, packaging and labelling of dangerous substances. *Official Journal* **L154/1**, 5 June 1992.

EEC Council (1993). Regulation 793/93 of 23 March 1993 on the evaluation and control of the risks of existing substances.

Egerton, F. N. (1973). Changing concepts of the balance of nature. *Quarterly Review of Biology* **48**, 322–350.

EINECS (European Inventory of Existing Commercial Chemical Substances) (1987). Advanced version published in the English language September 1987. HMSO, London.

Elliott, J. M. (1994). "Quantitative Ecology and the Brown Trout". Oxford University Press, Oxford.

Elliott, J. R. and Haydon, D. A. (1986). Mapping of general anaesthetic target sites. *Nature (London)* **319**, 77–78.

Elsom, D. M. (1992). "Atmospheric Pollution. A Global Problem", 2nd edn. Blackwell, Oxford.

Elstner, E. F. (1987). Ozone and ethylene stress. *Nature (London)* **328**, 482.

Elton, C. S. (1927). "Animal Ecology". Sidgwick & Jackson, London.

Elton, C. S. (1958). "The Ecology of Invasions by Animals and Plants". Methuen, London.

Elzen, G. W. (1989). Sublethal effects of pesticides on beneficial parasitoids. *In* "Pesticides and Non-Target Invertebrates" (P. C. Jepson, Ed.), 95–104. Intercept, Andover.

Emlen, J. M. (1984). "Population Biology: the Co-Evolution of Population Dynamics and Behavior". Macmillan, New York.

Emon, H. (Ed.) (1997). Biological environmental specimen banking. *Chemosphere* **34**, 1867–2250.

Endler, J. A. (1977). "Geographic Variation, Speciation, and Clines". Princeton University Press, Princeton.

Endler, J. A. (1986). "Natural Selection in the Wild". Princeton University Press, New Jersey.

Enting, I. G. and Mansbridge, J. V. (1989). Seasonal sources and sinks of atmospheric CO_2. Direct inversion of filtered data. *Tellus* **41B**, 111–126.

Environment Agency (1996). "The Quality of Rivers in England and Wales (1990 to 1995)". Environment Agency, Bristol.

Ernst, W. (1985). Accumulation in aquatic organisms. *In* "Appraisal of Tests to Predict the Environmental Behaviour of Chemicals. SCOPE 25" (P. Sheehan, F. Korte, W. Klein and P. Bourdeau, Eds), 243–255.* Wiley, Chichester.

Ernst, W., Goerke, H., Eder, G. and Schaefer, R. G. (1976). Residues of chlorinated hydrocarbons in marine organisms in relation to size and ecological parameters 1. PCB, DDT, DDE and DDD in fishes and molluscs from the English Channel. *Bulletin of Environmental Contamination and Toxicology* **15**, 55–65.

Eskin, R. A. and Coull, B. C. (1984). A priori determination of valid control sites: an example using marine meiobenthic nematodes. *Marine Environmental Research* **12**, 161–172.

Esser, H. O. and Moser, P. (1982). An appraisal of problems related to the measurement and evaluation of bioaccumulation. *Ecotoxicology and Environmental Safety* **6**, 131–148.

Eto, M. (1974). "Organophosphorus Pesticides: Organic and Biological Chemistry". Chemical Rubber Company Press, Cleveland.

Evans, G. C. (1976). A sack of uncut diamonds: the study of ecosystems and the future resources of mankind. *Journal of Ecology* **64**, 1–39.

Evans, L. V. (1988). Marine biofouling. *In* "Algae and Human Affairs" (C. A. Lembi and J. R. Waaland, Eds), 433–453. Cambridge University Press, Cambridge.

---

* One table has been inadvertently transposed to page 260.

Farman, J. C., Gardiner, B. G. and Shanklin, J. D. (1985). Large losses of total ozone in Antarctica reveal seasonal ClO_x/NO_x interaction. *Nature (London)* **315**, 207–210.

Fenchel, T. (1974). Intrinsic rate of natural increase: the relationship with body size. *Oecologia* **14**, 317–326.

Ferguson, J. (1939). The use of chemical potentials as indices of toxicity. *Proceedings of the Royal Society of London, B* **127**, 387–404.

Fielder, R. J. and Martin, A. D. (1993). Regulation of industrial chemicals and pesticides in the EEC. In "General and Applied Toxicology", Vol. 2 (B. Ballantyne, T. Marrs and P. Turner, Eds), 1133–1149. Stockton Press, New York.

Fifield, F. W. and Haines, P. J. (Eds) (1995). "Environmental Analytical Chemistry". Blackie, London.

Findlay, G. M. and deFreitas, A. S. W. (1971). DDT movement from adipocyte to muscle cell during lipid utilization. *Nature (London)* **229**, 63–65.

Fisher, R. A. (1930). "The Genetical Theory of Natural Selection". Clarendon Press, Oxford.

Florence, T. M. (1982). The speciation of trace elements in waters. *Talanta* **29**, 345–364.

Ford, E. B. (1940). Polymorphism and taxonomy. In "The New Systematics" (J. Huxley, Ed.), 493–513. Clarendon Press, Oxford.

Ford, E. B. (1975). "Ecological Genetics", 4th edn. Chapman & Hall, London.

Förstner, U. and Wittmann, G. T. W. (1981). "Metal Pollution in the Aquatic Environment", 2nd edn. Springer, Berlin.

Foulkes, E. C. (1989). Factors determining target dose: extrapolation from a shifting basis. In "Hazard Assessment of Chemicals", Vol. 6 (J. Saxena, Ed.), 31–47. Hemisphere, New York.

Fowler, D. (1984). Transfer to terrestrial surfaces. *Philosophical Transactions of the Royal Society of London, B* **305**, 281–297.

Fowler, S. W. and Elder, D. L. (1978). PCB and DDT residues in a Mediterranean pelagic food chain. *Bulletin of Environmental Contamination and Toxicology* **19**, 244–249.

Frazer, A. C. and Sharratt, M. (1969). The value and limitations of animal studies in the prediction of effects in man. In "The Use of Animals in Toxicological Studies", 4–14, Universities Federation for Animal Welfare, Potters Bar.

Freedman, B. (1995). "Environmental Ecology. The Ecological Effects of Pollution, Disturbance, and other Stresses", 2nd edn. Academic Press, San Diego.

Friberg, L., Nordberg, G. F. and Vouk, V. B. (Eds). (1979). "Handbook on the Toxicology of Metals". Elsevier, Amsterdam.

Friend, M., Haegle, M. A. et al. (1979). Correlations between residues of dichloro-diphenylethane, polychlorinated biphenyl, and dieldrin in the serum and tissues of mallard ducks (*Anas platyrhynchos*). In "Animals as Monitors of Environmental Pollutants", 319–326. National Academy of Sciences, Washington, D.C.

Frydman, I. and Whittaker, R. H. (1968). Forest associations of southeast Lublin province, Poland. *Ecology* **49**, 896–908.

Furst, A. (1987). Hormetic effects in pharmacology: pharmacological inversions as prototypes for hormesis. *Health Physics* **52**, 527–530.

Futuyma, D. J. (1986). "Evolutionary Biology", 2nd edn. Sinauer, Sunderland.

Gabrielescu, E. (1970). The lability of lysosomes during the response of neurons to stress. *Histochemical Journal* **2**, 123–130.

Galassi, S. and Migliavacca, M. (1986). Organochlorine residues in River Po sediment: testing the equilibrium condition with fish. *Ecotoxicology and Environmental Safety* **12**, 120–126.

Gallo, M. A. and Lawryk, N. J. (1991). Organic phosphorus insecticides. In "Handbook of Pesticide Toxicology", Vol. 2 (W. J. Hayes and E. R. Laws, Eds), 917–1123. Academic Press, San Diego.

Gardiner, J. and Mance, G. (1984). "Proposed Environmental Quality Standards for List II Substances in Water. Introduction". Technical Report TR 206, Water Research Centre, Medmenham.

Garratt, J. R. (1992). "The Atmospheric Boundary Layer". Cambridge University Press, Cambridge.

Garsed, S. G. (1984). Uptake and distribution of pollutants in the plant and residence time of active species. In "Gaseous Air Pollutants and Plant Metabolism" (M. J. Koziol and F. R. Whatley, Eds), 83–103. Butterworths, London.

Gartside, D. W. and McNeilly, T. (1974). The potential for evolution of heavy metal tolerance in plants II. Copper tolerance in normal populations of different plant species. Heredity 32, 335–348.

Gause, G. F. (1934). "The Struggle for Existence". Republished, 1971. Dover Publications, New York.

Gause, G. F. and Witt, A. A. (1935). Behavior of mixed populations and the problem of natural selection. American Naturalist 69, 596–609.

Gaylor, D. W. (1994). Biostatistical approaches to low level exposures. In "Biological Effects of Low Level Exposures: Dose–Response Relationships" (E. J. Calabrese, Ed.), 87–98. CRC Press, Boca Raton.

Gearing, J. N. (1989). The role of aquatic microcosms in ecotoxicologic research as illustrated by large marine systems. In "Ecotoxicology: Problems and Approaches" (S. A. Levin, M. A. Harwell, J. R. Kelly and K. D. Kimball, Eds), 411–470. Springer, New York.

Georghiou, G. P. (1980). Insecticide resistance and prospects for its management. Residue Reviews 76, 131–145.

Georghiou, G. P. and Saito, T. (Eds) (1983). "Pest Resistance to Pesticides". Plenum Press, New York.

Getz, W. M. and Haight, R. G. (1989). "Population Harvesting. Demographic Models of Fish, Forest, and Animal Resources". Princeton University Press, Princeton.

Gibbs, P. E. (1993). A male genital defect in the dog-whelk, Nucella lapillus (Neogastropoda), favouring survival in a TBT-polluted area. Journal of the Marine Biological Association of the United Kingdom 73, 667–678.

Gibbs, P. E. and Bryan, G. W. (1994). Biomonitoring of tributyltin (TBT) pollution using the imposex response of neogastropod molluscs. In "Biomonitoring of Coastal Waters and Estuaries" (K. J. M. Kramer, Ed.), 205–226. CRC Press, Boca Raton.

Gibbs, P. E., Pascoe, P. L. and Burt, G. R. (1988). Sex change in the female dog-whelk, Nucella lapillus, induced by tributyltin from antifouling paints. Journal of the Marine Biological Association of the United Kingdom 68, 715–731.

Gibbs, P. E., Bryan, G. W. and Pascoe, P. L. (1991). TBT-induced imposex in the dog-whelk, Nucella lapillus: geographical uniformity of the response and effects. Marine Environmental Research 32, 79–87.

Giese, A. C. (1978). "Living with our Sun's Ultraviolet Rays". Plenum, New York.

Gil, L., Fine, B. C. et al. (1968). Biochemical studies on insecticide resistance in Musca domestica. Entomologia Experimentalis et Applicata 11, 15–29.

Giller, P. S. (1984). "Community Structure and the Niche". Chapman & Hall, London.

Gillespie, J. H. (1987). Molecular evolution and the neutral allele theory. *Oxford Surveys in Evolutionary Biology* 4, 10–37.

Gillespie, J. H. (1991). "The Causes of Molecular Evolution". Oxford Univesity Press, Oxford.

Gillett, J. W. (1989). The role of terrestrial microcosms and mesocosms in ecotoxicologic research. In "Ecotoxicology: Problems and Approaches" (S. A. Levin, M. A. Harwell, J. R. Kelly and K. D. Kimball, Eds), 367–410. Springer, New York.

Glover, H. G. (1975). Acidic and ferruginous mine drainages. In "The Ecology of Resource Degradation and Renewal" (M. J. Chadwick and G. T. Goodman, Eds), 15th Symposium of the British Ecological Society, 173–195. Blackwell Scientific, Oxford.

Glover, R. S. (1979). Natural fluctuations of populations. *Ecotoxicology and Environmental Safety* 3, 190–203.

Gobas, F. A. P. C. and MacKay, D. (1987). Dynamics of hydrophobic organic chemical bioconcentration in fish. *Environmental Toxicology and Chemistry* 6, 495–504.

Godfray, H. C. J., Partridge, L. and Harvey, P. H. (1991). Clutch size. *Anual Review of Ecology and Systematics* 22, 409–429.

Godfrey, K. (1983). "Compartmental Models and their Application". Academic Press, London.

Goldberg, E. D., Bowen, V. T. *et al.* (1978). The mussel watch. *Environmental Conservation* 5, 101–125.

Goldsmith, P., Smith, F. B. and Tuck, A. F. (1984). Atmospheric transport and transformation. *Philosophical Transactions of the Royal Society of London B* 305, 259–279.

Goldstein, A., Aronow, L. and Kalman, S. M. (1974). "Principles of Drug Action: The Basis of Pharmacology", 2nd edn. Wiley, New York.

Gorham, E. (1958). The influence and importance of daily weather conditions in the supply of chloride, sulphate and other ions to fresh waters from atmospheric precipitation. *Philosophical Transactions of the Royal Society of London, B* 241, 147–178.

Graney, R. L., Cherry, D. S. and Cairns, J. (1984). The influence of substrate, pH, diet and temperature upon cadmium accumulation in the asiatic clam (*Corbicula fluminea*) in laboratory artificial streams. *Water Research* 18, 833–842.

Grant, J. and Cranford, P. J. (1991). Carbon and nitrogen scope for growth as a function of diet in the sea scallop *Placopecten magellanicus*. *Journal of the Marine Biological Association of the United Kingdom* 71, 437–450.

Gray, J. S. (1981a). Detecting pollution induced changes in communities using the long-normal distribution of individuals among species. *Marine Pollution Bulletin* 12, 173–176.

Gray, J. S. (1981b). "The Ecology of Marine Sediments. An Introduction to the Structure and Function of Benthic Communities". Cambridge University Press, Cambridge.

Gray, J. S. and Mirza, F. B. (1979). A possible method for the detection of pollution-induced disturbance on marine benthic communities. *Marine Pollution Bulletin* 10, 142–146.

Gray, M. W. and Doolittle, W. F. (1982). Has the endosymbiont hypothesis been proven? *Microbiological Reviews* 46, 1–42.

Grime, J. P. (1979). "Plant Strategies and Vegetation Processes". Wiley, Chichester.

Grime J. P. (1988). The C-S-R model of primary plant strategies—origins, implications and tests. In "Plant Evolutionary Biology" (L. D. Gottlieb and S. K. Jain, Eds), 371–393. Chapman & Hall, London.

Grime, J. P. and Hunt, R. (1975). Relative growth-rate: its range and adaptive signifi-cance in a local flora. *Journal of Ecology* **63**, 393–422.

Grinnell, J. (1917). The niche-relationships of the California thrasher. *Auk* **34**, 427–433.

GRIP (Greenland Ice-Core Project) Members (1993). Climate instability during the last interglacial period recorded in the GRIP ice core. *Nature (London)* **364**, 203–207.

Gronwald, J. W. (1994). Resistance to photosystem II inhibiting herbicides. In "Herbicide Resistance in Plants" (S. B. Powles and J. A. M. Holtum, Eds), 27–60. Lewis, Boca Raton.

Grue, C. E., Powell, G. V. N. and McChesney, M. J. (1982). Care of nestlings by wild female starlings exposed to an organophosphate pesticide. *Journal of Applied Ecology* **19**, 327–335.

Grzenda, A. R., Paris, D. F. and Taylor, W. J. (1970). The uptake, metabolism, and elim-ination of chlorinated residues by goldfish (*Carassius auratus*) fed on ^{14}C-DDT contaminated diet. *Transactions of the American Fisheries Society* **99**, 385–395.

Grzenda, A. R., Taylor, W. J. and Paris, D. F. (1971). The uptake and distribution of chlo-rinated residues by goldfish (*Carassius auratus*) fed a ^{14}C-dieldrin contaminated diet. *Transactions of the American Fisheries Society* **100**, 215–221.

Guderian, R. (1977). "Air Pollution. Phytotoxicity of Acidic Gases and its Significance in Air Pollution Control". Springer, Berlin.

Guderian, R. (Ed.) (1985). "Air Pollution by Photochemical Oxidants". Springer, Berlin.

Gundlach, E. R. and Hayes, M. O. (1978). Vulnerability of coastal environments to oil spill impacts. *Marine Technology Society Journal* **12**, 18–27.

Gunkel, W. and Gassmann, G. (1980). Oil, oil dispersants and related substances in the marine environment. *Helgoländer Meeresuntersuchungen* **33**, 164–181.

Guthrie, F. E. and Perry, J. J. (Eds) (1980). "Introduction to Environmental Toxicology". Elsevier, New York.

Gydesen, H. (1984). Mathematical models of the transport of pollutants in eco-systems. *Ecological Bulletin* **36**, 17–25.

Haegele, M. A. and Hudson, R. H. (1973). DDE effects on reproduction of ring doves. *Environmental Pollution* **4**, 53–57.

Haegele, M. A. and Hudson, R. H. (1977). Reduction of courtship behaviour induced by DDE in male ringed turtle doves. *Wilson Bulletin* **89**, 593–601.

Haigh, N. (1992). "Manual of Environmental Policy: the EC and Britain". Longman, Harlow.

Haigh, N. (Ed.) (1995). "Legislation for the Control of Chemicals". Institute for European Environmental Policy, London.

Haldane, J. B. S. (1924). A mathematical theory of natural and artificial selection. Part I. *Transactions of the Cambridge Philosophical Society* **23**, 19–41.

Haldane, J. B. S. (1956a). The theory of selection for melanism in Lepidoptera. *Proceedings of the Royal Society of London, B* **145**, 303–306.

Haldane, J. B. S. (1956b). The relation between density regulation and natural selec-tion. *Proceedings of the Royal Society of London, B* **145**, 306–308.

Hama. H. (1983). Resistance to insecticides due to reduced sensitivity of acetyl-cholinesterase. In "Pest Resistance to Pesticides" (G. P. Georghiou and T. Saito, Eds), 299–331. Plenum Press, New York.

Hamelink, J. L., Waybrant, R. C. and Ball, R. C. (1971). A proposal: exchange equilib-ria control the degree chlorinated hydrocarbons are biologically magnified in lentic environments. *Transactions of the American Fisheries Society* **100**, 207–214.

Hansch, C. (1978). Recent advances in biochemical QSAR. *In* "Correlation Analysis in Chemistry. Recent Advances" (N. B. Chapman and J. Shorter, Eds), 397–438. Plenum Press, New York.

Hansch, C. and Fujita, T. (1964). ρ-σ-π Analysis. A method for the correlation of biological activity and chemical structure. *Journal of the American Chemical Society* **86**, 1616–1626.

Hansch, C. and Leo, A. J. (1979). "Substituent Constants for Correlation Analysis in Chemistry and Biology". Wiley, New York.

Hansch, C., Muir, R. M. *et al.* (1963). The correlation of biological activity of plant growth regulators and chloromycetin derivatives with Hammett constants and partition coefficients. *Journal of the American Chemical Society* **85**, 2817–2824.

Hansch, C., Steward. A. R., Anderson, S. M. and Bentley, D. (1968). The parabolic dependence of drug action upon lipophilic character as revealed by a study of hypnotics. *Journal of Medicinal (and Pharmaceutical) Chemistry* **11**, 1–11.

Harberd, D. J. (1967). Observation on natural clones in *Holcus mollis. New Phytologist* **66**, 401–408.

Harding, G. C., Vass, W. P. and Drinkwater, K. F. (1981). Importance of feeding, direct uptake from seawater, and transfer from generation to generation in the accumulation of an organochlorine (p,p'-DDT) by the marine planktonic copepod *Calanus finmarchicus. Canadian Journal of Fisheries and Aquatic Sciences* **38**, 101–119.

Hardy, G. H. (1908). Mendelian proportions in a mixed population. *Science (New York)* **28**, 49–50.

Harper, D. (1992). "Eutrophication of Freshwaters. Principles, Problems and Restoration". Chapman & Hall, London.

Harper, J. L. (1977). "Population Biology of Plants". Academic Press, London.

Harper, J. L. and Hawksworth, D. L. (1994). Biodiversity: measurement and estimation. Preface. *Philosophical Transactions of the Royal Society of London, B* **345**, 5–12.

Harrison, R. M. and de Mora, S. J. (1996). "Introductory Chemistry for the Environmental Sciences", 2nd edn. Cambridge University Press, Cambridge.

Harrison, R. M. and Perry, R. (Eds) (1986). "Handbook of Air Pollution Analysis", 2nd edn. Chapman & Hall, London.

Hart, A. D. M. (1993). Relationships between behavior and the inhibition of acetylcholinesterase in birds exposed to organophosphorus pesticides. *Environmental Toxicology and Chemistry* **12**, 321–336.

Harvey, G. R., Miklas, H. P., Bowen, V. T. and Steinhauer, W. G. (1974). Observations on the distribution of chlorinated hydrocarbons in Atlantic Ocean organisms. *Journal of Marine Research* **32**, 103–118.

Harvey, H. H. (1980). Widespread and diverse changes in the biota of North American lakes and rivers coincident with acidification. *In* "Ecological Impact of Acid Precipitation" (D. Drabløs and A. Tollan, Eds), 93–98. SNSF Project, Oslo-Ås.

Harvey, H. W. (1950). On the production of living matter in the sea off Plymouth. *Journal of the Marine Biological Association of the United Kingdom* **29**, 97–137.

Harwell, C. C. (1989). Regulatory framework for ecotoxicology. *In* "Ecotoxicology: Problems and Approaches" (S. A. Levin, M. A. Harwell, J. R. Kelly and K. D. Kimball, Eds), 497–516. Springer, New York.

Hassell, M. P. and Anderson, R. M. (1989). Predator–prey and host–pathogen interactions. *In* "Ecological Concepts" (J. M. Cherrett, Ed.), 147–196. Blackwell Scientific, Oxford.

Hassell, M. P., Southwood, T. R. E. and Reader, P. M. (1987). The dynamics of the viburnum whitefly (*Aleurotrachelus jelinekii*): a case study of population regulation. *Journal of Animal Ecology* **56**, 283–300.

Hawkins, S. J. (1981). The influence of season and barnacles on the algal colonization of *Patella vulgata* exclusion areas. *Journal of the Marine Biological Association of the United Kingdom* **61**, 1–15.

Hawkins, S. J. and Hartnoll, R. G. (1983). Changes in a rocky shore community: an evaluation of monitoring. *Marine Environmental Research* **9**, 131–181.

Hawkins, S. J., and Southward, A. J. (1992). The *Torrey Canyon* oil spill: recovery of rocky shore communities. *In* "Restoring the Nation's Marine Environment" (G. W. Thayer, Ed.), 583–631. Maryland Sea Grant, Maryland.

Hawkins, S. J., Proud, S. V., Spence, S. K. and Southward, A. J. (1994). From the individual to the community and beyond: water quality, stress indicators and key species in coastal ecosystems. *In* "Water Quality and Stress Indicators in Marine and Freshwater Ecosystems: Linking Levels of Organisation (Individuals, Populations, Communities)" (D. W. Sutcliffe, Ed.), 35–62. Freshwater Biological Association, Ambleside.

Hawksworth, D. L. and McManus, P. M. (1989). Lichen recolonization in London under conditions of rapidly falling sulphur dioxide levels, and the concept of zone skipping. *Botanical Journal of the Linnaean Society* **100**, 99–109.

Hawksworth, D. L. and Rose, F. (1970). Qualitative scale for estimating sulphur dioxide air pollution in England and Wales using epiphytic lichens. *Nature (London)* **227**, 145–148.

Hawksworth, D. L. and Rose, F. (1976). "Lichens as Pollution Monitors". Arnold, London.

Hayes, W. J. (1975). "Toxicology of Pesticides". Williams & Wilkins, Baltimore.

Haynes, K. F. (1988). Sublethal effects of neurotoxic insecticides on insect behavior. *Annual Review of Entomology* **33**, 149–168.

Heatwole, H. and Levins, R. (1972). Trophic structure stability and faunal change during recolonization. *Ecology* **53**, 531–534.

Hedgecott, S. (1994). Prioritization and standards for hazardous chemicals. *In* "Handbook of Ecotoxicology", Vol. 2 (P. Calow, Ed.), 368–393. Blackwell Scientific.

Hedrick, P. W., Ginevan, M. E. and Ewing, E. P. (1976). Genetic polymorphism in heterogeneous environments. *Annual Review of Ecology and Systematics* **7**, 1–32.

Heinz, G. H. (1980). Comparison of game-farm and wild-strain mallard ducks in accumulation of methylmercury. *Journal of Environmental Pathology and Toxicology* **3**, 379–386.

Heinz, G. H., Hill, E. F., Stickel, W. H. and Stickel, L. F. (1979). Environmental contaminant studies by the Patuxent Wildlife Research Center. *In* "Avian and Mammalian Wildlife Toxicology" (E. E. Kenaga, Ed.), Special Technical Publications, No. 693, 9–35. American Society for Testing and Materials, Philadelphia.

Hellawell, J. M. (1978). "Biological Surveillance of Rivers. A Biological Monitoring Handbook". Water Research Centre, Stevenage.

Hellawell, J. M. (1986). "Biological Indicators of Freshwater Pollution and Environmental Management". Applied Science, London.

Hermens, J. L. M. (1989). Quantitative structure–activity relationships of environmental pollutants. *In* "Handbook of Environmental Chemistry", Vol. 2E (O. Hutzinger, Ed.), 111–162. Springer, Berlin.

Hill, E. F. (1992). Avian toxicology of anticholinesterases. *In* "Clinical and Experimental Toxicology of Organophosphates and Carbamates" (B. Ballantyne and T. C. Marrs, Eds), 272–294. Butterworth-Heinemann, Oxford.

Hill, R. J. (1982). Taxonomy and biological considerations of herbicide-resistant and herbicide-tolerant biotypes. *In* "Herbicide Resistance in Plants" (H. M. LeBaron and J. Gressel, Eds), 81–98. Wiley, New York.

Hodgson, E., Silver, I. S. *et al.* (1991). Metabolism. *In* "Handbook of Pesticide Toxicology", Vol. 1 (W. J. Hayes and E. R. Laws, Eds), 107–167. Academic Press, San Diego.

Hoelzel, A. R. (Ed.) (1992). "Molecular Genetic Analysis of Populations. A Practical Approach". IRL Press, Oxford.

Hoelzel, A. R. and Dover, G. A. (1991). "Molecular Genetic Ecology". IRL Press, Oxford.

Holden, A. V. (1975). Monitoring persistent organic pollutants. *In* "Organochlorine Insecticides: Persistent Organic Pollutants" (F. Moriarty, Ed.), 1–27. Academic Press, London.

Holden, A. V. (1981). Organochlorines—an overview. *Marine Pollution Bulletin* **12**, 110–115.

Holden, C. (1979). Specimen bank set up. *Science (New York)* **206**, 1057.

Holdgate, M. W. (1979). "A Perspective of Environmental Pollution". Cambridge University Press, Cambridge.

Holland, H. D. (1984). "The Chemical Evolution of the Atmosphere and Oceans". Princeton University Press, Princeton.

Hopkin, S. P. (1989). "Ecophysiology of Metals in Terrestrial Invertebrates". Elsevier, London.

Hopkins, W. G. (1995). "Introduction to Plant Physiology". Wiley, New York.

Horsman, D. C., Roberts, T. M. and Bradshaw, A. D. (1978). Evolution of sulphur dioxide tolerance in perennial ryegrass. *Nature (London)* **276**, 493–494.

Horsman, D. C., Roberts, T. M. and Bradshaw, A. D. (1979a). Studies on the effect of sulphur dioxide on perennial ryegrass (*Lolium perenne* L.). II. Evolution of sulphur dioxide tolerance. *Journal of Experimental Botany* **30**, 495–501.

Horsman, D. C., Roberts, T. M., Lambert, M. and Bradshaw, A. D. (1979b). Studies on the effect of sulphur dioxide on perennial ryegrass (*Lolium perenne* L.). I. Characteristics of fumigation system and preliminary experiments. *Journal of Experimental Botany* **30**, 485–493.

Houghton, J. T. (1997). "Global Warming: the Complete Briefing", 2nd edn. Lion, Oxford.

Houghton, J. T., Filho, L. G. M. *et al.* (Eds) (1995). "Climate Change 1994. Radiative Forcing of Climate Change and an Evaluation of the IPCC IS92 Emission Scenarios". Cambridge University Press, Cambridge.

Houghton, J. T., Filho, L. G. M., *et al.* (Eds) (1996). "Climate Change 1995. The Science of Climate Change". Cambridge University Press, Cambridge.

Houston, J. B. and Wood, S. G. (1980). Gastrointestinal absorption of drugs and other xenobiotics. *Progress in Drug Metabolism* **4**, 57–129.

Howarth, R. W. (1989). Determining the ecological effects of oil pollution in marine ecosystems. *In* "Ecotoxicology: Problems and Approaches" (S. A. Levin, M. A. Harwell, J. R. Kelly and K. D. Kimball, Eds), 69–97. Springer, New York.

Howells, G. (1995). "Acid Rain and Acid Waters", 2nd edn. Ellis Horwood, New York.

Howlett, R. J. and Majerus, M. E. N. (1987). The understanding of industrial melanism in the peppered moth (*Biston betularia*) (Lepidoptera: Geometridae). *Biological Journal of the Linnaean Society* **30**, 31–44.

Hubby, J. L. and Lewontin, R. C. (1966). A molecular approach to the study of genic heterozygosity in natural populations. I. The number of alleles at different loci in *Drosophila pseudoobscura*. *Genetics* **54**, 577–594.

Huet, M., Paulet, Y. M. and Le Pennec, M. (1996). Survival of *Nucella lapillus* in a tributyltin-polluted area in west Brittany: a further example of a male genital defect (Dumpton syndrome) favouring survival. *Marine Biology* **125**, 543–549.

Huffaker, C. B. and Messenger, P. S. (1964). The concept and significance of natural control. *In* "Biological Control of Insect Pests and Weeds" (P. DeBach, Ed.), 74–117. Chapman & Hall, London.

Hunt, E. G. and Bischoff, A. I. (1960). Inimical effects on wildlife of periodic DDD applications to Clear Lake. *California Fish and Game* **46**, 91–106.

Hunter, B. A., Johnson, M. S. and Thompson, D. J. (1987). Ecotoxicology of copper and cadmium in a contaminated grassland ecosystem. *Journal of Applied Ecology* **24**, 587–599.

Hurlbert, S. H. (1971). The nonconcept of species diversity: a critique and alternative parameters. *Ecology* **52**, 577–586.

Hushon, J. M., Klein, A. W., Strachan, W. J. M. and Schmidt-Bleek, F. (1983). Use of OECD premarket data in environmental exposure analysis for new chemicals. *Chemosphere* **12**, 887–910.

Hutchinson, G. E. (1948). Circular causal systems in ecology. *Annals of the New York Academy of Sciences* **50**, 221–246.

Hutchinson, G. E. (1957). Concluding remarks. *Cold Spring Harbor Symposium on Quantitative Biology* **22**, 415–427.

Hutchinson, G. E. (1961). The paradox of the plankton. *American Naturalist* **95**, 137–145.

Hutchinson, G. E. (1978). "An Introduction to Population Ecology". Yale University Press, New Haven.

Hynes, H. B. N. (1960). "The Biology of Polluted Waters". Liverpool University Press, Liverpool.

Hynes, H. B. N. (1972). "The Ecology of Running Waters". Liverpool University Press, Liverpool.

Idso, K. E. and Idso, S. B. (1994). Plant responses to atmospheric CO_2 enrichment in the face of environmental constraints: a review of the last 10 years' research. *Agricultural and Forest Meteorology* **69**, 153–203.

Indorato, A. M., Snyder, K. B. and Usinowicz, P. J. (1984). Toxicity screening using Microtox® analyzer. *In* "Toxicity Screening Procedures Using Bacterial Systems" (D. Liu and B. J. Dutka, Eds), 37–53. Dekker, New York.

IRPTC (International Register of Potentially Toxic Chemicals) (1983). About chemicals and chemophobia. *IRPTC Bulletin* **6**, 2.

Jackson, P. J., Unkefer, P. J., Delhaize, E. and Robinson, N. J. (1990). Mechanisms of trace metal tolerance in plants. *In* "Environmental Injury to Plants" (F. Katterman, Ed.), 231–255. Academic Press, San Diego.

Jameson, C. W. and Walters, D. B. (Eds) (1984). "Chemistry for Toxicity Testing". Butterworth, Stoneham, Maryland.

Jarvinen, A. W., Hoffman, M. J. and Thorslund, T. W. (1977). Long-term toxic effects of DDT food and water exposure on fathead minnows (*Pimephales promelas*). *Journal of the Fisheries Research Board of Canada* **34**, 2089–2103.

Jasanoff, S. (1986). Comparative risk assessment—the lessons of cultural variation. *In* "Toxic Hazard Assessment of Chemicals" (M. Richardson, Ed.), 259–281. Royal Society of Chemistry, London.

Jefferies, D. J. (1969). Induction of apparent hyperthyroidism in birds fed DDT. *Nature (London)* **222**, 578–579.

Jefferies, D. J. (1972). Organochlorine insecticide residues in British bats and their significance. *Journal of Zoology* **166**, 245–263.

Jefferies, D. J. (1973). The effects of organochlorine insecticides and their metabolites on breeding birds. *Journal of Reproduction and Fertility* **19** (Suppl.), 337–352.

Jefferies, D. J. (1975). The role of the thyroid in the production of sublethal effects by organochlorine insecticides and polychlorinated biphenyls. *In* "Organochlorine Insecticides: Persistent Organic Pollutants" (F. Moriarty, Ed.), 131–230. Academic Press, London.

Jefferies, D. J. and Davis, B. N. K. (1968). Dynamics of dieldrin in soil, earthworms, and song thrushes. *Journal of Wildlife Management* **32**, 441–456.

Jefferies, D. J. and French, M. C. (1971). Hyper- and hypothyroidism in pigeons fed DDT: an explanation for the "thin eggshell phenomenon". *Environmental Pollution* **1**, 235–242.

Jefferies, D. J. and French, M. C. (1972). Changes induced in the pigeon thyroid by *p,p'*-DDE and dieldrin. *Journal of Wildlife Management* **36**, 24–30.

Jefferies, D. J. and Prestt, I. (1966). Post-mortems of peregrines and lanners with particular reference to organochlorine residues. *British Birds* **59**, 49–64.

Johannessen, M., Skartveit. A. and Wright, R. F. (1980). Streamwater chemistry before, during and after snowmelt. *In* "Ecological Impact of Acid Precipitation" (D. Drabløs and A. Tollan, Eds), 224–225. SNSF Project, Oslo-Ås.

Johnson, M. K. (1975a). Organophosphorus esters causing delayed neurotoxic effects. Mechanism of action and structure/activity studies. *Archives of Toxicology* **34**, 259–288.

Johnson, M. K. (1975b). The delayed neuropathy caused by some organophosphorus esters: mechanism and challenge. *Critical Reviews in Toxicology* **3**, 289–316.

Johnson, M. K. (1981). Initiation of organophosphate neurotoxicity. *Toxicology and Applied Pharmacology* **61**, 480–481.

Johnson, M. K. (1990). Organophosphates and delayed neuropathy—is NTE alive and well? *Toxicology and Applied Pharmacology* **102**, 385–399.

Jonas, P. R., Charlson, R. J. and Rodhe, H. (1995). Aerosols. *In* "Climate Change 1994" (J. T. Houghton, L. G. M. Filho *et al.*, Eds), 127–162. Cambridge University Press, Cambridge.

Jones, D. T. and Hopkin, S. P. (1991). Biological monitoring of metal pollution in terrestrial ecosystems. *In* "Terrestrial and Aquatic Ecosystems. Perturbation and Recovery" (O. Ravera, Ed.), 148–152. Ellis Horwood, New York.

Jones, J. R. E. (1949). A further ecological study of calcareous streams in the "Black Mountain" district of South Wales. *Journal of Animal Ecology* **18**. 142–159.

Jordan, R. E. and Payne, J. R. (1980). "Fate and Weathering of Petroleum Spills in the Marine Environment". Ann Arbor, Michigan.

Jouzel, J., Barkov, N. I. *et al.* (1993). Extending the Vostok ice-core record of paleoclimate to the penultimate glacial period. *Nature (London)* **364**, 407–412.

Karcher, W. (1998). Recent trends and developments in the EU in the environmental control and management of chemicals. *Ecotoxicology and Environmental Safety* **40**, 97–102.

Kareiva, P. (1990). Population dynamics in spatially complex environments: theory and data. *Philosophical Transactions of the Royal Society of London, B* **330**, 175–190.

Kays, S. and Harper, J. L. (1974). The regulation of plant and tiller density in a grass sward. *Journal of Ecology* **62**, 97–105.

Keddy, P. A. (1989). "Competition". Chapman & Hall, London.

Keeling, C. D., Whorf, T. P., Wahlen, M. and van der Plicht, J. (1995). Interannual extremes in the rates of rise of atmospheric carbon dioxide since 1980. *Nature (London)* **375**, 666–670.

Keiding, J. (1975). Problems of housefly (*Musca domestica*) control due to multiresistance to insecticides. *Journal of Hygiene, Epidemiology, Microbiology and Immunology* **19**, 340–355.

Kellogg, W. W., Cadle, R. D. *et al.* (1972). The sulfur cycle. *Science (New York)* **175**, 587–596.

Kelly, J. R. and Harwell, M. A. (1989). Indicators of ecosystem response and recovery. *In* "Ecotoxicology: Problems and Approaches" (S. A. Levin, M. A. Harwell, J. R. Kelly and K. D. Kimball, Eds), 9–35. Springer, New York.

Kenaga, E. E. (1980). Predicted bioconcentration factors and soil sorption coefficients of pesticides and other chemicals. *Ecotoxicology and Environmental Safety* **4**, 26–38.

Kershaw, K. A. and Looney, J. H. H. (1985). "Quantitative and Dynamic Plant Ecology", 3rd edn. Arnold, London.

Kerswill, C. J. (1967). Studies on effects of forest sprayings with insecticides, 1952–63, on fish and aquatic invertebrates in New Brunswick streams: introduction and summary. *Journal of the Fisheries Research Board of Canada* **24**, 701–708.

Kettlewell, H. B. D. (1955). Selection experiments on industrial melanism in the Lepidoptera. *Heredity* **9**, 323–342.

Kettlewell, H. B. D. (1956). Further selection experiments on industrial melanism in the Lepidoptera. *Heredity* **10**, 287–301.

Kettlewell, H. B. D. (1973). "The Evolution of Melanism. The Study of a Recurring Necessity. With Special Reference to Industrial Melanism in the Lepidoptera". Clarendon Press, Oxford.

Kilroy, A. and Gray, N. F. (1995). Treatability, toxicity and biodegradability test methods. *Biological Reviews* **70**, 243–275.

Kimura, M. (1968). Evolutionary rate at the molecular level. *Nature (London)* **217**, 624–626.

Kimura, M. (1983). "The Neutral Theory of Molecular Evolution". Cambridge University Press, Cambridge.

Kimura, M. (1991). Recent development of the neutral theory viewed from the Wrightian tradition of theoretical population genetics. *Proceedings of the National Academy of Sciences* **88**, 5969–5973.

Kimura, M. and Ohta, T. (1971). Protein polymorphism as a phase of molecular evolution. *Nature (London)* **229**, 467–469.

Kingsolver, J. G. (1996). Physiological sensitivity and evolutionary responses to climate change. *In* "Carbon Dioxide, Populations, and Communities" (C. Körner and F. A. Bazzaz, Eds), 3–12. Academic Press, San Diego.

Klaassen, C. D. (Ed.) (1996). "Casarett and Doull's Toxicology. The Basic Science of Poisons", 5th edn. McGraw Hill, New York.

Koch, G. W. and Mooney, H. A. (Eds) (1996). "Carbon Dioxide and Terrestrial Ecosystems". Academic Press, San Diego.

Kokkinn, M. J. and Davis, A. R. (1986). Secondary production: shooting a halcyon for its feathers. *In* "Limnology in Australia" (P. de Dekker and W. D. Williams, Eds), 251–261. Junk, Dordrecht.

Kolkwitz, R. and Marsson, M. (1908). Ökologie der pflanzlichen Saprobien. *Berichte der Deutschen botanischen Gesellschaft* **26A**. 505–519.

Kolkwitz, R. and Marsson, M. (1909). Ökologie der tierischen Saprobien. *Internationale Revue der gesamten Hydrobiologie und Hydrographie* **2**, 126–152.

Könemann, H. (1986). Quantitative structure–activity relationships in aquatic toxicology. In "Organic Micropollutants in the Aquatic Environment" (A. Bjørseth and G. Angeletti, Eds), 465–474. Reidel, Dordrecht.

Kooijman, S. A. L. M. (1987). A safety factor for LC_{50} values allowing for differences in sensitivity among species. *Water Research* **21**, 269–276.

Korte, F. (1977). Occurrence and fate of synthetic chemicals in the environment. In "The Evaluation of Toxicological Data for the Protection of Public Health" (W. J. Hunter and J. G. P. M. Smeets, Eds), 235–246. Pergamon Press, Oxford.

Korte, F., Freitag, D. *et al.* (1978). Ecotoxicologic profile analysis. *Chemosphere* **7**, 79–102.

Korte, F., Klein, W. and Sheehan, P. (1985). The role and nature of environmental testing methods. In "Appraisal of Tests to Predict the Environmental Behaviour of Chemicals. SCOPE 25" (P. Sheehan, F. Korte, W. Klein and P. Bourdeau, Eds), 1–11. Wiley, Chichester.

Krebs, C. J. (1994). "Ecology. The Experimental Analysis of Distribution and Abundance", 4th edn. HarperCollins, New York.

Krull, I. S. (Ed.) (1991). "Trace Metal Analysis and Speciation". Elsevier, Amsterdam.

Krupa, S. and Kickert, R. N. (1987). An analysis of numerical models of air pollutant exposure and vegetation response. *Environmental Pollution* **44**, 127–158.

Kuenzler, E. J. (1961). Phosphorus budget of a mussel population. *Limnology and Oceanography* **6**, 400–415.

Lacis, A. A. and Mishchenko, M. I. (1995). Climate forcing, climate sensitivity, and climate response: a radiative modeling perspective on atmospheric aerosols. In "Aerosol Forcing of Climate" (R. J. Charlson and J. Heintzenberg, Eds), 11–42. Wiley, Chichester.

Lamb, H. H. (1965). The early medieval warm epoch and its sequel. *Palaeogeography, Palaeoclimatology and Palaeoecology* **1**, 13–37.

Lane, P. I. and Bell, J. N. B. (1984). The effects of simulated urban air pollution on grass yield: Part I—description and simulation of ambient pollution. *Environmental Pollution, B* **8**, 245–263.

Laughlin, R. B., Ng, J. and Guard, H. E. (1981). Hormesis: a response to low environmental concentrations of petroleum hydrocarbons. *Science (New York)* **211**, 705–707.

Law, R. and Watkinson, A. R. (1989). Competition. In "Ecological Concepts" (J. M. Cherrett, Ed.), 243–284. Blackwell Scientific, Oxford.

LeBaron, H. M. (1991). Distribution and seriousness of herbicide-resistant weed infestations worldwide. In "Herbicide Resistance in Weeds and Crops" (J. C. Caseley, G. W. Cussans and R. K. Atkins, Eds), 27–43. Butterworth-Heinemann, Oxford.

LeBaron, H. M. and Gressel, J. (Eds) (1982). "Herbicide Resistance in Plants". Wiley, New York.

Lees, D. R. (1981). Industrial melanism: genetic adaptation of animals to air pollution. In "Genetic Consequences of Man Made Change" (J. A. Bishop and L. M. Cook, Eds), 129–176. Academic Press, London.

Lefohn, A. S. and Benedict, H. M. (1982). Development of mathematical index that describes ozone concentration, frequency and duration. *Atmospheric Environment* **16**, 2529–2532.

Leo, A., Hansch, C. and Elkins, D. (1971). Partition coefficients and their uses. *Chemical Reviews* **71**, 525–554.

Lepp, N. W. (1979). Cycling of copper in woodland ecosystems. *In* "Copper in the Environment" (J. O. Nriagu, Ed.), Part I, "Ecological Cycling", 289–323. Wiley, New York.

Lertzman, K. P. (1992). Patterns of gap-phase replacement in a subalpine, old-growth forest. *Ecology* **73**, 657–669.

Levin, S. A. and Kimball, K. D. (1984). New perspectives in ecotoxicology. *Environmental Management* **8**, 375–442.

Lewin, B. (1997). "Genes VI". Oxford University Press, Oxford.

Lewontin, R. C. (1965). Selection for colonizing ability. *In* "The Genetics of Colonizing Species" (H. G. Baker and G. L. Stebbins, Eds), 77–94. Academic Press, New York.

Lewontin, R. C. and Hubby, J. L. (1966). A molecular approach to the study of genic heterozygosity in natural populations. II. Amount of variation and degree of heterozygosity in natural populations of *Drosophila pseudoobscura*. *Genetics* **54**, 595–609.

Li, X., Maring, H. *et al.* (1996). Dominance of mineral dust in aerosol light-scattering in the North Atlantic trade winds. *Nature (London)* **380**, 416–419.

Liebert, T. G. and Brakefield, P. M. (1987). Behavioural studies on the peppered moth *Biston betularia* and a discussion of the role of pollution and lichens in industrial melanism. *Biological Journal of the Linnaean Society* **31**, 129–150.

Likens, G. E. and Bormann, F. H. (1974). Acid rain: a serious regional environmental problem. *Science (New York)* **184**, 1176–1179.

Lincer, J. L. (1975). DDE-induced eggshell-thinning in the American kestrel: a comparison of the field situation and laboratory results. *Journal of Applied Ecology* **12**, 781–793.

Lindeman, R. L. (1942). The trophic-dynamic aspect of ecology. *Ecology* **23**, 399–418.

Lindzen, R. S. (1994). On the scientific basis for global warming scenarios. *Environmental Pollution* **83**, 125–134.

Lipnick, R. L. (1989). Narcosis, electrophile and proelectrophile toxicity mechanisms: application of SAR and QSAR. *Environmental Toxicology and Chemistry* **8**, 1–12.

Liu, D. and Dutka, B. J. (Eds) (1984). "Toxicity Screening Procedures Using Bacterial Systems". Dekker, New York.

Livett, E. A. (1988). Geochemical monitoring of atmospheric heavy metal pollution: theory and applications. *Advances in Ecological Research* **18**, 65–177.

Livingstone, D. R., Donkin, P. and Walker, C. H. (1992). Pollutants in marine ecosystems: an overview. *In* "Persistent Pollutants in Marine Ecosystems" (C. H. Walker and D. R. Livingstone, Eds), 235–263. Pergamon, Oxford.

Lloyd, R. (1960). The toxicity of zinc sulphate to rainbow trout. *Annals of Applied Biology* **48**, 84–94.

Lockie, J. D., Ratcliffe, D. A. and Balharry, R. (1969). Breeding success and organochlorine residues in golden eagles in west Scotland. *Journal of Applied Ecology* **6**, 381–389.

Lockyer, D. R., Cowling, D. W. and Jones, L. H. P. (1976). A system for exposing plants to atmospheres containing low concentrations of sulphur dioxide. *Journal of Experimental Botany* **27**, 397–409.

Loewe, S. (1928). Die quantitativen Probleme der Pharmakologie. *Ergebnisse der Physiologie* **27**, 47–187.

Longhurst, A. R. (1991). Role of the marine biosphere in the global carbon cycle. *Limnology and Oceanography* **36**, 1507–1526.

Longhurst, A. R. and Harrison, W. G. (1989). The biological pump: profiles of plankton production and consumption in the upper ocean. *Progress in Oceanography* **22**, 47–123.

van Loon, J. and Beamish, R. J. (1977). Heavy-metal contamination by atmospheric fallout of several Flin Flon area lakes and the relation to fish populations. *Journal of the Fisheries Research Board of Canada* **34**, 899–906.

Lorius, C, Jouzel, J. *et al.* (1990). The ice-core record: climate sensitivity and future greenhouse warming. *Nature (London)* **347**, 139–145.

Lovelock, J. (1995). "The Ages of Gaia", 2nd edn. Oxford University Press, Oxford.

Lovelock, J. and Margulis, L. (1974). Atmospheric homeostasis by and for the biosphere: the gaia hypothesis. *Tellus* **26**, 2–10.

Lu, P.-Y. and Metcalf, R. L. (1975). Environmental fate and biodegradability of benzene derivatives as studied in a model aquatic ecosystem. *Environmental and Health Perspectives* **10**, 269–284.

Lu, P.-Y., Metcalf, R. L., Hirwe, A. S. and Williams, J. W. (1975). Evaluation of environmental distribution and fate of hexachlorocyclopentadiene, chlordane, heptachlor, and heptachlor epoxide in a laboratory model ecosystem. *Journal of Agricultural and Food Chemistry* **23**, 967–973.

Luepke, N.-P. (Ed.) (1979). "Monitoring Environmental Materials and Specimen Banking". Nijhoff, The Hague.

Lundholm, M. E. (1987). Thinning of eggshells in birds by DDE: mode of action on the eggshell gland. *Comparative Biochemistry and Physiology* **88C**, 1–22.

Luoma, S. N. (1983). Bioavailability of trace metals to aquatic organisms—a review. *Science of the Total Environment* **28**, 1–22.

Luoma, S. N. and Bryan, G. W. (1978). Factors controlling the availability of sediment-bound lead to the estuarine bivalve *Scrobicularia plana*. *Journal of the Marine Biological Association of the United Kingdom* **58**, 793–802.

Luoma, S. N. and Ho, K. T. (1993). Appropriate uses of marine and estuarine sediment bioassays. *In* "Handbook of Ecotoxicology", Vol. 1 (P. Calow, Ed.), 193–226. Blackwell Scientific, Oxford.

McArthur, M. L. B., Fox, G. A., Peakall, D. B. and Philogène, B. J. R. (1983). Ecological significance of behavioral and hormonal abnormalities in breeding ring doves fed an organochlorine chemical mixture. *Archives of Environmental Contamination and Toxicology* **12**, 343–353.

MacArthur, R. H. (1955). Fluctuations of animal populations, and a measure of community stability. *Ecology* **36**, 533–536.

MacArthur, R. H. (1958). Population ecology of some warblers of northeastern coniferous forests. *Ecology* **39**, 599–619.

MacArthur, R. H. and Wilson, E. O. (1967). "The Theory of Island Biogeography". Princeton University Press, Princeton.

McClusky, D. S., Bryant, V. and Campbell, R. (1986). The effects of temperature and salinity on the toxicity of heavy metals to marine and estuarine invertebrates. *Oceanography and Marine Biology* **24**, 481–520.

McCully, K. A., Villeneuve, D. C. *et al.* (1966). Metabolism and storage of DDT in beef cattle. *Journal of the Association of Official Analytical Chemists*, **49**, 966–973.

MacDonald, D. R. and Webb, F. E. (1963). Insecticides and the spruce budworm. *Memoirs of the Entomological Society of Canada* **31**, 288–310.

MacDowall, F. D. H. (1965). Predisposition of tobacco to ozone damage. *Canadian Journal of Plant Science*, **45**, 1–12.

MacDowall, F. D. H., Mukammal, E. I. and Cole, A. F. W. (1964). Direct correlation of air-polluting ozone and tobacco weather fleck. *Canadian Journal of Plant Science* **44**, 410–417.

McDowell, W. H. (1986). Power plant operation on the Hudson River. *In* "The Hudson River Ecosystem" (K. E. Limburg, M. A. Moran and W. H. McDowell, Eds), 40–82. Springer, New York.

McElroy, M. B. (1994). Climate of the Earth: an overview. *Environmental Pollution* **83**, 3–21.

MacFadyen, A. (1963). "Animal Ecology, Aims and Methods", 2nd edn. Pitman, London.

McGuire, A. D., Melillo, J. M., *et al.* (1992). Interactions between carbon and nitrogen dynamics in estimating net primary productivity for potential vegetation in North America. *Global Biogeochemical Cycles* **6**, 101–124.

McIntosh, R. P. (1995). H. A. Gleason's "individualistic concept" and theory of animal communities: a continuing controversy. *Biological Reviews* **70**, 317–357.

McIntyre, A. D. (1977). Effects of pollution on inshore benthos. *In* "Ecology of Marine Benthos" (B. C. Coull, Ed.), Bell W. Baruch Library in Marine Science. Vol. 6, 301–318. Columbia University, Columbia.

McIntyre, A. D. and Pearce, J. B. (Eds) (1980). "Biological Effects of Marine Pollution and the Problems of Monitoring", Publication 179. Conseil International pour l'Exploration de la Mer, Copenhagen.

MacKay, D. (1979). Finding fugacity feasible. *Environmental Science and Technology* **13**, 1218–1223.

MacKay, D. (1994). Fate models. *In* "Handbook of Ecotoxicology", Vol. 2 (P. Calow, Ed.), 348–367. Blackwell Scientific, Oxford.

MacKay, D. and Paterson, S. (1981). Calculating fugacity. *Environmental Science and Technology* **15**, 1006–1014.

McKenzie, J. A. and Batterham, P. (1994). The genetic, molecular and phenotypic consequences of selection for insecticide resistance. *Trends in Ecology and Evolution* **9**, 166–169.

McLoughlin, J. and Bellinger, E. G. (1993). "Environmental Pollution Control. An Introduction to Principles and Practice of Administration". Graham and Trotman, London; Martinus Nijhoff, Dordrecht.

MacNair, M. R. (1989). The genetics of metal tolerance in natural populations. *In* "Heavy Metal Tolerance in Plants: Evolutionary Aspects" (A. J. Shaw, Ed.), 235–253. CRC, Boca Raton.

McNeilly, T. (1968). Evolution in closely adjacent plant populations. III. *Agrostis tenuis* on a small copper mine. *Heredity* **23**, 99–108.

McNeilly, T. and Bradshaw, A. D. (1968). Evolutionary processes in populations of copper tolerant *Agrostis tenuis* Sibth. *Evolution* **22**, 108–118.

Majerus, M. E. N. (1989). Melanic polymorphism in the peppered moth *Biston betularia*, and other Lepidoptera. *Journal of Biological Education* **23**, 267–284.

Manahan, S. E. (1993). "Fundamentals of Environmental Chemistry". Lewis, Boca Raton.

Manahan, S. E. (1994). "Environmental Chemistry", 6th edn. Lewis, Boca Raton.

Mance, G. (1987), "Pollution Threat of Heavy Metals in Aquatic Environments". Elsevier Applied Science, London.

Mani, G. S. (1990). Theoretical models of melanism in *Biston betularia*—a review. *Biological Journal of the Linnaean Society* **39**, 355–371.

Mani, G. S. and Majerus, M. E. N. (1993). Peppered moth revisited: analysis of recent decreases in melanic frequency and predictions for the future. *Biological Journal of the Linnaean Society* **48**, 157–165.

Mann, K. H. and Clark, R. B. (1978). Long-term effects of oil spills on marine intertidal communities. *Journal of the Fisheries Research Board of Canada* **35**, 791–795.

Manning, W. J. and Feder, W. A. (1980). "Biomonitoring Air Pollutants with Plants". Applied Science, London.

Mansfield, T. A. and Freer-Smith, P. H. (1981). Effects of urban air pollution on plant growth. *Biological Reviews* **56**, 343–368.

Mantoura, R. F. C., Dickson, A. and Riley, J. P. (1978). The complexation of metals with humic materials in natural waters. *Estuarine and Coastal Marine Science* **6**, 387–408.

Marcus, A. H. (1982). Multicompartment kinetic models for cadmium I. Effects of zinc on cadmium retention in male mice. *Environmental Research* **27**, 46–51.

Marcus, A. H. (1983). Compartmental models for trace metals in mammals. *Science of the Total Environment* **28**, 307–316.

Martin, M. H. and Coughtrey, P. J. (1982). "Biological Monitoring of Heavy Metal Pollution. Land and Air". Applied Science, London.

Mason, C. F. (1996). "Biology of Freshwater Pollution", 3rd edn. Longman, Harlow.

Matis, J. H. and Tolley, H. D. (1979). Compartmental models with multiple sources of stochastic variability: the one-compartment, time invariant hazard rate case. *Bulletin of Mathematical Biology* **41**, 491–515.

Maxwell, B. D. and Mortimer, A. M. (1994). Selection for herbicide resistance. In: "Herbicide Resistance in Plants" (S. A. Powles and J. A. M. Holtum, Eds), 1–25. Lewis, Boca Raton.

May, R. M. (1974). "Stability and Complexity in Model Ecosystems". 2nd edn. Princeton University Press, Princeton.

May, R. M. (1975). Patterns of species abundance and diversity. In "Ecology and Evolution of Communities" (M. L. Cody and J. M. Diamond, Eds), 81–120. Belknap Press, Cambridge, Massachusetts.

May, R. M. (1979). Production and respiration in animal communities. *Nature (London)* **282**, 443–444.

May, R. M. (1981a). Models for single populations. In "Theoretical Ecology. Principles and Applications" (R. M. May, Ed.), 2nd edn, 4–25. Blackwell Scientific, Oxford.

May, R. M. (Ed.) (1981b). "Theoretical Ecology. Principles and Applications", 2nd edn. Blackwell Scientific, Oxford.

Mayfield, C. I. (1993). Microbial systems. In "Handbook of Ecotoxicology", Vol. 1 (P. Calow, Ed.), 9–27. Blackwell Scientific, Oxford.

Mayr, E. (1963). "Animal Species and Evolution". Belknap Press, Cambridge, Massachusetts.

Mayr, E. (1970). "Populations, Species, and Evolution. An Abridgement of Animal Species and Evolution". Belknap Press, Cambridge, Massachusetts.

Medawar, P. B. (1945). Size, shape, and age. In "Essays on Growth, and Form. Presented to d'Arcy Wentworth Thompson" (W. E. le Gros Clark and P. B. Medawar, Eds), 157–187. Clarendon Press, Oxford.

Meeks, R. L. (1968). The accumulation of ^{36}Cl ring-labelled DDT in a freshwater marsh. *Journal of Wildlife Management* **32**, 376–398.

Mehlhorn, H. and Wellburn, A. R. (1987). Stress ethylene formation determines plant sensitivity to ozone. *Nature (London)* **327**, 417–418.

Melander, A. L. (1914). Can insects become resistant to sprays? *Journal of Economic Entomology* **7**, 167–172.

Melillo, J. M., Callaghan, T. V. et al. (1990). Effects on ecosystems. In "Climate Change. The IPCC Scientific Assessment" (J. T. Houghton, G. J. Jenkins and J. J. Ephraums, Eds), 282–310. Cambridge University Press, Cambridge.

Melillo, J. M., Prentice, I. C. et al. (1996). Terrestrial biotic responses to environmental change and feedbacks to climate. In "Climate Change 1995. The Science of Climate Change" (J. T. Houghton, L. G. M. Filho et al., Eds), 445–481. Cambridge University Press, Cambridge.

Mendel, G. (1865). Versuche über Pflanzen Hybriden. Abhandlungen des naturforschenden Vereines in Brünn, 4.

Menser, H. A., Heggestad, H. E. and Street, O. E. (1963). Response of plants to air pollutants. II. Effects of ozone concentration and leaf maturity on injury to Nicotiana tabacum. Phytopathology 53, 1304–1308.

Menzel, D. B. (1979). From animals to man, the grand extrapolation of environmental toxicology. In "Assessing Toxic Effects of Environmental Pollutants" (S. D. Lee and J. B. Mudd, Eds), 1–14. Ann Arbor Science, Ann Arbor.

Metcalf, R. L. (1977). Model ecosystem approach to insecticide degradation: a critique. Annual Review of Entomology 22, 241–261.

Metcalf, R. L., Sangha, G. K. and Kapoor, I. P. (1971). Model ecosystem for the evaluation of pesticide biodegradability and ecological magnification. Environmental Science and Technology 5, 709–713.

Mikkola, K. (1984). On the selective forces acting in the industrial melanism of Biston and Oligia moths (Lepidoptera: Geometridae and Noctuidae). Biological Journal of the Linnaean Society 21, 409–421.

Miller, C. A. (1963). The spruce budworm. Memoirs of the Entomological Society of Canada 31, 12–19.

Miller, D. R. (1978). Models for total transport. In "Principles of Ecotoxicology. SCOPE 12" (G. C. Butler, Ed.), 71–90. Wiley, Chichester.

Miller, H. G. (1984). Deposition–plant–soil interactions. Philosophical Transactions of the Royal Society of London, B 305, 339–352.

Miller, H. G. and Miller, J. D. (1980). Collection and retention of atmospheric pollutants by vegetation. In "Ecological Impact of Acid Precipitation" (D. Drabløs and A. Tollan, Eds), 33–40. SNSF Project, Oslo-Ås.

Mills, E. L. (1969). The community concept in marine zoology, with comments on continua and instability in some marine communities: a review. Journal of the Fisheries Research Board of Canada 26, 1415–1428.

Mitchell, M. F. and Roberts, J. R. (1984). A case study of the use of fenitrothion in New Brunswick: the evolution of an ordered approach to ecological monitoring. In "Effects of Pollutants at the Ecosystem Level. SCOPE 22" (P. J. Sheehan, D. R. Miller, G. C. Butler and P. Bourdeau, Eds), 377–422. Wiley, Chichester.

Molina, M. J. and Rowland, F. S. (1974). Stratospheric sink for chlorofluoromethanes: chlorine atom-catalysed destruction of ozone. Nature (London) 249, 810–812.

Monitoring and Assessment Research Centre (1985). "Historical Monitoring". MARC report number 31, University of London, London.

Moore, M. N. (1981). Elemental accumulation in organisms and food chains. In "Analysis of Marine Ecosystems" (A. R. Longhurst, Ed.), 535–569. Academic Press, London.

Moore, M. N. and Stebbing, A. R. D. (1976). The quantitative cytochemical effects of three metal ions on a lysosomal hydrolase of a hydroid. Journal of the Marine Biological Association of the United Kingdom 56, 995–1005.

Moore, N. W. (1965). Pesticides in birds—a review of the situation in Great Britain in 1965. *Bird Study* **12**, 222–252.

Moore, N. W. (Ed.) (1966a). Pesticides in the environment and their effects on wildlife. *Journal of Applied Ecology* **3** (Suppl.).

Moore, N. W. (1966b). A pesticide monitoring system with special reference to the selection of indicator species. *Journal of Applied Ecology* **3** (Suppl.), 261–269.

Moore, N. W. (1975). Pesticide monitoring from the national and international points of view. *Agriculture and Environment* **2**, 75–83.

Moore, N. W. and Walker, C. H. (1964). Organic chlorine insecticide residues in wild birds. *Nature (London)* **201**, 1072–1073.

Moriarty, F. (1968). The toxicity and sublethal effects of p,p'-DDT and dieldrin to *Aglais urticae* (L.) (Lepidoptera: Nymphalidae) and *Chorthippus brunneus* (Thunberg) (Saltatoria: Acrididae). *Annals of Applied Biology* **62**, 371–393.

Moriarty, F. (1969). The sublethal effects of synthetic insecticides on insects. *Biological Reviews* **44**, 321–357.

Moriarty, F. (1971). Prediction of adverse effects on wildlife by pollutants. *Mededelingen Fakulteit Landbouwwetenschappen Gent* **36**, 27–33.

Moriarty, F. (1972). The effects of pesticides on wildlife: exposure and residues. *Science of the Total Environment* **1**, 267–288.

Moriarty, F. (1974). Residues in animals during chronic exposure to dieldrin. *Environmental Quality and Safety* **3**, 104–112.

Moriarty, F. (1975a). "Pollutants and Animals. A Factual Perspective". Allen & Unwin, London.

Moriarty, F. (1975b). Exposures and residues. *In* "Organochlorine Insecticides: Persistent Organic Pollutants" (F. Moriarty, Ed.), 29–72. Academic Press, London.

Moriarty, F. (1975c). The dispersal and persistence of p,p'-DDT. *In* "The Ecology of Resource Degradation and Renewal" (M. J. Chadwick and G. T. Goodman, Eds), 15th Symposium of the British Ecological Society, 31–47. Blackwell Scientific, Oxford.

Moriarty, F. (1977). Prediction of ecological effects by pesticides. *In* "Ecological Effects of Pesticides" (F. H. Perring and K. Mellanby, Eds), Linnaean Society Symposium Series, No. 5, 165–174. Academic Press, London.

Moriarty, F. (1984). Persistent contaminants, compartmental models and concentration along food-chains. *Ecological Bulletin* **36**, 35–45.

Moriarty, F. (1985a). Bioaccumulation in terrestrial food chains. *In* "Appraisal of Tests to Predict the Environmental Behaviour of Chemicals" (P. Sheehan, F. Korte, W. Klein and P. Bourdeau, Eds), 257–284.* Wiley, Chichester.

Moriarty, F. (1985b). Toxic pollutants in aquatic and terrestrial ecosystems: similarities and differences. *Vakblad voor Biologen* **65**(13/14), 5–9.

Moriarty, F. and Walker, C. H. (1987). Bioaccumulation in food chains—a rational approach. *Ecotoxicology and Environmental Safety* **13**, 208–215.

Moriarty, F., Hanson, H. M. and Freestone, P. (1984). Limitations of body burden as an index of environmental contamination: heavy metals in fish *Cottus gobio* L. from the River Ecclesbourne, Derbyshire. *Environmental Pollution A* **34**, 297–320.

Moriarty, F., Bell, A. A. and Hanson, H. (1986). Does p,p'-DDE thin eggshells? *Environmental Pollution A* **40**, 257–286.

---

* One table has been inadvertently transposed to page 248. There are also minor errors, mostly self-evident, in the text: authors did not see proofs of this book.

Morris, M. G. (1971). The management of grassland for the conservation of invertebrate animals. *In* "The Scientific Management of Animal and Plant Communities for Conservation" (E. Duffey and A. S. Watt, Eds), 11th Symposium of the British Ecological Society, 527–552. Blackwell Scientific, Oxford.

Morris, R. F. (Ed.) (1963). The dynamics of epidemic spruce budworm populations. *Memoirs of the Entomological Society of Canada* **31**.

Moss, B., Johnes, P. and Phillips, G. (1996). The monitoring of ecological quality and the classification of standing waters in temperate regions: a review and proposal based on a worked scheme for British waters. *Biological Reviews* **71**, 301–339.

Mount, D. I. (1968). Chronic toxicity of copper to fathead minnows (*Pimephales promelas*, Rafinesque). *Water Research* **2**, 215–223.

Mudd, J. B. (1975). Sulfur dioxide. *In* "Responses of Plants to Air Pollution" (J. B. Mudd and T. T. Kozlowski, Eds), 9–22. Academic Press, New York.

Mudd, J. B. and Kozlowski, T. T. (Eds) (1975). "Responses of Plants to Air Pollution". Academic Press, New York.

Mullin, C. A. and Scott, J. G. (Eds) (1992). "Molecular Mechanisms of Insecticide Resistance". American Chemical Society, Washington.

Muniz, I. P. (1984). The effects of acidification on Scandinavian freshwater fish fauna. *Philosophical Transactions of the Royal Society of London, B* **305**, 517–528.

Munn, R. E. (1981). "The Design of Air Quality Monitoring Networks—Air Pollution Problems", Vol. 2. Macmillan, London.

Murphy, P. G. (1970). Effects of salinity on uptake of DDT, DDE and DDD by fish. *Bulletin of Environmental Contamination and Toxicology* **5**, 404–407.

Murton, R. K., Westwood, N. J. and Isaacson, A. J. (1974). A study of wood-pigeon shooting: the exploitation of a natural animal population. *Journal of Applied Ecology* **11**, 61–81.

Narahashi, T. (1971). Effects of insecticides on excitable tissues. *Advances in Insect Physiology* **8**, 1–93.

National Rivers Authority (NRA) (1991). "The Quality of Rivers, Canals and Estuaries in England and Wales. Report of the 1990 Survey". NRA, Bristol.

National Rivers Authority (NRA) (1994). "The Quality of Rivers and Canals in England and Wales (1990 to 1992)". HMSO, London.

National Water Council (1981). "River Quality: the 1980 Survey and Future Outlook". National Water Council, London.

Nebert, D. W. and Gonzales, F. J. (1987). P450 genes: structure, evolution, and regulation. *Annual Review of Biochemistry* **56**, 945–993.

Neely, W. B. and Blau, G. E. (1977). The use of laboratory data to predict the distribution of chlorpyrifos in a fish pond. *In* "Pesticides in Aquatic Environments" (M. A. Q. Khan, Ed.), 145–163. Plenum Press, New York.

Neely, W. B., Branson, D. R. and Blau, G. E. (1974). Partition coefficient to measure bioconcentration potential of organic chemicals in fish. *Environmental Science and Technology* **8**, 1113–1115.

Neill, W. E. (1974). The community matrix and interdependence of the competition coefficients. *American Naturalist* **108**, 399–408.

Nevo, E., Beiles, A. and Ben-Shlomo, R. (1984). The evolutionary significance of genetic diversity: ecological, demographic and life history correlates. *In* "Evolutionary Dynamics of Genetic Diversity" (G. S. Mani, Ed.), 13–213. Springer, Berlin.

Newell, R. C. (1979). "Biology of Intertidal Animals", 3rd edn. Marine Ecological Surveys, Faversham.

Newman, M. C. and McIntosh, A. W. (1982). The influence of lead in components of a freshwater ecosystem on molluscan tissue lead concentrations. *Aquatic Toxicology* **2**, 1–19.

Newton, I. (1979). "Population Ecology of Raptors". T. & A. D. Poyser, Berkhamsted.

Newton, I. and Bogan, J. (1978). The role of different organo-chlorine compounds in the breeding of British sparrowhawks. *Journal of Applied Ecology* **15**, 105–116.

Nicholson, A. J. (1933). The balance of animal populations. *Journal of Animal Ecology* **2**, 131–178.

Nicholson, A. J. and Bailey, V. A. (1935). The balance of animal populations.—Part I. *Proceedings of the Zoological Society of London* **3**, 551–598.

Nicholson, I. A., Fowler, D. *et al.* (1980). Continuous monitoring of airborne pollutants. *In* "Ecological Impact of Acid Precipitation" (D. Drabløs and A. Tollan, Eds), 144–145. SNSF Project, Oslo-Ås.

Nirmalakhandan, N. and Speece, R. E. (1988). Structure–activity relationships. Quantitative techniques for predicting the behavior of chemicals in the ecosystem. *Environmental Science and Technology* **22**, 606–615.

Noble, I. R. and Slatyer, R. O. (1980). The use of vital attributes to predict successional changes in plant communities subject to recurrent disturbances. *Vegetatio* **43**, 5–21.

Norby, R. J., Gunderson, C. A. *et al.* (1992). Productivity and compensatory responses of yellow-poplar trees in elevated CO_2. *Nature (London)* **357**, 322–324.

Norstrom, R. J., McKinnon, A. E. and deFreitas, A. S. W. (1976). A bioenergetics-based model for pollutant accumulation by fish. Simulation of PCB and methylmercury residue levels in Ottawa River yellow perch (*Perca flavescens*). *Journal of the Fisheries Research Board of Canada* **33**, 248–267.

Nott, J. A. and Nicolaidou, A. (1990). Transfers of metal detoxification along marine food chains. *Journal of the Marine Biological Association of the United Kingdom* **70**, 905–912.

O'Brien, R. D. (1967). "Insecticides. Action and Metabolism". Academic Press, New York.

Odum, E. P. (1962). Relationships between structure and function in the ecosystem. *Japanese Journal of Ecology* **12**, 108–118.

Odum, E. P. (1969). The strategy of ecosystem development. *Science (New York)* **164**, 262–270.

Odum, E. P. (1971). "Fundamentals of Ecology", 3rd edn. Saunders, Philadelphia.

Odum, H. T. (1957). Trophic structure and productivity of Silver Springs, Florida. *Ecological Monographs* **27**, 55–112.

Odum, W. E. and Johannes, R. E. (1975). The response of mangroves to man-induced environmental stress. *In* "Tropical Marine Pollution" (E. J. F. Wood and R. E. Johannes, Eds), 52–62. Elsevier, Amsterdam.

OECD (1980). "Chemical Trends in Wildlife: an International Cooperative Study". Organisation for Economic Co-operation and Development, Paris.

OECD (1981). "OECD Guidelines for Testing of Chemicals". Organisation for Economic Co-operation and Development, Paris.

OECD (1986). "Existing Chemicals. Systematic Investigation. Priority Setting and Chemicals Reviews". Organisation for Economic Co-operation and Development, Paris.

OECD (1989a). "Report of the OECD Workshop on Ecological Effects Assessment", Environment Monograph no. 26. Organisation for Economic Co-operation and Development, Paris.

OECD (1989b). "Compendium of Environmental Exposure Assessment Methods for Chemicals", Environment Monograph no. 27. Organisation for Economic Co-operation and Development, Paris.

OECD (1992). "The Polluter-Pays Principle", OCDE/GD(92)81. Organisation for Economic Co-operation and Development, Paris.

OECD (1995a). "Guidance Document for Aquatic Effects Assessment", Environment Monograph no. 92. Organisation for Economic Co-operation and Development, Paris.

OECD (1995b). "Report of the OECD Workshop on Environmental Hazard/Risk Assessment", Environment Monograph no. 105. Organisation for Economic Co-operation and Development, Paris.

OECD (1998). "OECD Guidelines for the Testing of Chemicals, Plus the 9th Addendum". Organisation for Economic Co-operation and Development, Paris.

Oechel, W. C. and Strain, B. R. (1985). Native species responses to increased carbon dioxide concentration. In "Direct Effects of Increasing Carbon Dioxide on Vegetation" (B. R. Strain and J. D. Cure, Eds), 117–154. US Department of Energy, DOE/ER-0238, Washington.

Oechel, W. C. and Vourlitis, G. L. (1996). Direct effects of elevated CO_2 on arctic plant and ecosystem function. In "Carbon Dioxide and Terrestrial Ecosystems" (G. W. Koch and H. A. Mooney, Eds), 163–176. Academic Press, San Diego.

Oechel, W. C., Hastings, S. J. et al. (1993). Recent change of Arctic tundra ecosystems from a net carbon dioxide sink to a source. Nature (London) 361, 520–523.

Open University (1989). "Ocean Circulation". Pergamon, Oxford.

Oppenoorth, F. J. (1985). Biochemistry and genetics of insecticide resistance. In "Comprehensive Insect Physiology, Biochemistry and Pharmacology", Vol. 12 (G. A. Kerkut and L. I. Gilbert, Eds), 731–773. Pergamon, Oxford.

Opperhuizen, A. and Schrap, S. M. (1987). Relationships between aqueous oxygen concentration and uptake and elimination rates during bioconcentration of hydrophobic chemicals in fish. Environmental Toxicology and Chemistry 6, 335–342.

Opperhuizen, A., Sinnige, T. L., van der Steen, J. M. and Hutzinger, O. (1987). Differences between retentions of various classes of aromatic hydrocarbons in reversed-phase high-performance liquid chromatography. Journal of Chromatography 388, 51–64.

Orians, G. H. (1975). Diversity, stability and maturity in natural ecosystems. In "Unifying Concepts in Ecology" (W. H. van Dobben and R. H. Lowe-McConnell, Eds), 139–150. Junk, The Hague.

O'Riordan, T. (1989). Best practicable environmental option (BPEO): a case-study in partial bureaucratic adaptation. Environmental Conservation 16, 113–122, 162.

Osborn, D. (1979). Seasonal changes in the fat, protein and metal content of the liver of the starling Sturnus vulgaris. Environmental Pollution 19, 145–155.

O'Shea, T. J., Brownell, R. L. et al. (1980). Organochlorine pollutants in small cetaceans from the Pacific and South Atlantic Oceans, November 1968–June 1976. Pesticides Monitoring Journal 14, 35–46.

Paine, R. T. (1980). Food webs: linkage, interaction strength and community infra-structure. Journal of Animal Ecology 49, 667–685.

Paine, R. T., Ruesink, J. L. et al. (1996). Trouble on oiled waters: lessons from the Exxon Valdez oil spill. Annual Review of Ecology and Systematics 27, 197–235.

Park, T. (1954). Experimental studies of interspecies competition II. Temperature, humidity, and competition in two species of Tribolium. Physiological Zoology 27, 177–238.

Parsons, T. R. (1978). Controlled aquatic ecosystem experiments in ocean ecology research. *Marine Pollution Bulletin* 9, 203–205.

Paton, W. D. M. (1961). A theory of drug action based on the rate of drug–receptor combination. *Proceedings of the Royal Society of London, B* 154, 21–69.

Patrick, R. (1973). Use of algae, especially diatoms, in the assessment of water quality. *In* "Biological Methods for the Assessment of Water Quality" (J. Cairns and K. L. Dickson, Eds), Special Technical Publications, No. 528, 76–95. American Society for Testing and Materials, Philadelphia.

Peakall, D. B. and Gilman, A. P. (1979). Limitations of expressing organochlorine levels in eggs on a lipid-weight basis. *Bulletin of Environmental Contamination and Toxicology* 23, 287–290.

Pearson, T. H. (1975). The benthic ecology of Loch Linnhe and Loch Eil, a sea-loch system on the west coast of Scotland. IV. Changes in the benthic fauna attributable to organic enrichment. *Journal of Experimental Marine Biology and Ecology* 20, 1–41.

Penner, J. E., Charlson, R. J. *et al.* (1994). Quantifying and minimizing uncertainty of climate forcing by anthropogenic aerosols. *Bulletin of the American Meteorological Society* 75, 375–400.

Perkins, E. J. (1979). The need for sublethal studies. *Philosophical Transactions of the Royal Society of London, B* 286, 425–442.

Perry, J. J. (1980). Oil in the biosphere. *In* "Introduction to Environmental Toxicology" (F. E. Guthrie and J. J. Perry, Eds), 198–209. Elsevier, New York.

Peter, T. (1994). The stratospheric ozone layer—an overview. *Environmental Pollution* 83, 69–79.

Peters, R. A. (1969). The biochemical lesion and its historical development. *British Medical Bulletin* 25, 223–226.

Peterson, P. J. (1993). Plant adaptation to environmental stress: metal pollution tolerance. *In* "Plant Adaptation to Environmental Stress" (L. Fowden, T. Mansfield and J. Stoddart, Eds), 171–188. Chapman & Hall, London.

Phillips, D. J. H. (1978). Use of biological indicator organisms to quantitate organochlorine pollutants in aquatic environments—a review. *Environmental Pollution* 16, 167–229.

Phillips, D. J. H. (1980). "Quantitative Aquatic Biological Indicators. Their Use to Monitor Trace Metal and Organochlorine Pollution". Applied Science, London.

Phillips, D. J. H. and Rainbow, P. S. (1993). "Biomonitoring of Trace Aquatic Contaminants". Elsevier, London.

Phillipson, J. (1966). "Ecological Energetics". Arnold, London.

Pianka, E. R. (1981). Competition and niche theory. *In* "Theoretical Ecology. Principles and Applications" (R. M. May, Ed.), 2nd edn, 167–196. Blackwell Scientific, Oxford.

Pielou, E. C. (1975). "Ecological Diversity". Wiley, New York.

Pielou, E. C. (1984). "The Interpretation of Ecological Data". Wiley, New York.

Pillmoor, J. B. and Foster, S. G. (1994). Molecular approaches to the design of chemical crop protection agents. *In* "Molecular Biology in Crop Protection" (G. Marshall and D. Walters, Eds), 41–67. Chapman & Hall, London.

Pimm, S. L. (1984). The complexity and stability of ecosystems. *Nature (London)* 307, 321–326.

Pollard, E. (1981). Resource limited and equilibrium models of populations. *Oecologia* 49, 377–378.

Portmann, J. E. and Lloyd, R. (1986). Safe use of the assimilative capacity of the marine environment for waste disposal—is it feasible? *Water Science and Technology* **18**, 233–244.

Powers, D. A., Smith, M. *et al.* (1993). A multidisciplinary approach to the selectionist/ neutralist controversy using the model teleost, *Fundulus heteroclitus*. *Oxford Surveys in Evolutionary Biology* **9**, 43–107.

Prát, S. (1934). Die Erblichkeit der Resistenz gegen Kupfer. *Bericht der Deutschen botanischen Gesellschaft* **52**, 65–67.

Prather, M., Derwent, R. *et al.* (1995). Other trace gases and atmospheric chemistry. In "Climate Change 1994. Radiative Forcing of Climate Change and an Evaluation of the IPCC IS92 Emission Scenarios" (J. T. Houghton, L. G. M. Filho *et al.*, Eds), 73–126. Cambridge University Press, Cambridge.

Preston, A. (1979). Standards and environmental criteria: the practical application of the results of laboratory experiments and field trials to pollution control. *Philosophical Transactions of the Royal Society of London, B* **286**, 611–624.

Preston, A., Jefferies, D. F. and Pentreath, R. J. (1972). The possible contributions of radioecology to marine productivity studies. In "Conservation and Productivity of Natural Waters" (R. W. Edwards and D. J. Garrod, Eds), Symposium of the Zoological Society of London, No. 29, 271–284.

Preston, F. W. (1948). The commonness, and rarity, of species. *Ecology* **29**, 254–283.

Preston, F. W. (1962). The canonical distribution of commonness and rarity: part I. *Ecology* **43**, 185–215.

Price, D. R. H. (1978). Fish as indicators of water quality. *Water Pollution Control* **77**, 285–296.

Prince, R. G. (1992). Bioremediation of oil spills, with particular reference to the spill from the *Exxon Valdez*. In "Microbial Control of Pollution" (J. C. Fry, G. M. Gadd *et al.*, Eds), 48th Symposium of the Society for General Microbiology, 19–34. Cambridge University Press, Cambridge.

Prosi, F. (1979). Heavy metals in aquatic organisms. In "Metal Pollution in the Aquatic Environment" (U. Förstner and G. T. W. Wittmann, Eds), 271–323. Springer, Berlin.

Putman, R. J. (1994). "Community Ecology". Chapman & Hall, London.

Randall, D., Burggren, W. and French, K. (1997). "Eckert Animal Physiology. Mechanisms and Adaptations", 4th edn. Freeman, New York.

Ratcliffe, D. A. (1963). The status of the peregrine in Great Britain. *Bird Study* **10**, 56–90.

Ratcliffe, D. A. (1967). Decrease in eggshell weight in certain birds of prey. *Nature (London)* **215**, 208–210.

Ratcliffe, D. A. (1970). Changes attributable to pesticides in egg breakage frequency and eggshell thickness in some British birds. *Journal of Applied Ecology* **7**, 67–115.

Ratcliffe, D. (1993). "The Peregrine Falcon", 2nd edn. Poyser, London.

Ravera, O. (1969). Seasonal variation of the biomass and biocoenotic structure of plankton of the Bay of Ispra (Lago Maggiore). *Verhandlungen internationale Vereinigung für theoretische und angewandte Limnologie* **17**, 237–254.

Raynaud, D., Jouzel, J. *et al.* (1993). The ice record of greenhouse gases. *Science (New York)* **259**, 926–934.

Regier, H. A. and Cowell, E. B. (1972). Applications of ecosystem theory, succession, diversity, stability, stress and conservation. *Biological Conservation* **4**, 83–88.

Reinert, R. E. (1972). Accumulation of dieldrin in an alga (*Scenedesmus obliquus*), *Daphnia magna*, and the guppy (*Poecilia reticulata*). *Journal of the Fisheries Research Board of Canada* **29**, 1413–1418.

Reinert, R. E. and Bergman, H. L. (1974). Residues of DDT in Lake Trout (*Salvelinus namaycush*) and Coho Salmon (*Oncorhynchus kisutch*) from the Great Lakes. *Journal of the Fisheries Research Board of Canada* **31**, 191–199.

Remmert, H. (Ed.) (1991). "The Mosaic-Cycle Concept of Ecosystems". Springer, Berlin.

Renshaw, E. (1991). "Modelling Biological Populations in Space and Time". Cambridge University Press, Cambridge.

Resh, V. H. and Unzicker, J. D. (1975). Water quality monitoring and aquatic organisms: the importance of species identification. *Journal of the Water Pollution Control Federation* **47**, 9–19.

Reynolds, C. M. (1979). The heronries census: 1972–1977 population changes and a review. *Bird Study* **26**, 7–12.

Richardson, J. (1992). Interactions of organophosphorus compounds with neurotoxic esterase. *In* "Organophosphates. Chemistry, Fate, and Effects" (J. E. Chambers and P. E. Levi, Eds), 299–323. Academic Press, San Diego.

Richie, P. J. and Peterle, T. J. (1979). Effect of DDE on circulating luteinizing hormone levels in ring doves during courtship and nesting. *Bulletin of Environmental Contamination and Toxicology* **23**, 220–226.

Ridley, M. (1993). "Evolution". Blackwell Scientific, Boston.

Riebesell, U., Wolf-Gladrow, D. A. and Smetacek, V. (1993). Carbon dioxide limitation of marine phytoplankton growth rates. *Nature (London)* **361**, 249–251.

Riegelman, S., Loo, J. C. K. and Rowland, M. (1968). Shortcomings in pharmacokinetic analysis by conceiving the body to exhibit properties of a single compartment. *Journal of Pharmaceutical Sciences* **57**, 117–123.

Rigler, F. H. (1975). The concept of energy flow and nutrient flow between trophic levels. *In* "Unifying Concepts in Ecology" (W. H. van Dobben and R. H. Lowe-McConnell, Eds), 15–26. Junk, The Hague.

Ritz, K., Dighton, J. and Giller, K. E. (Eds) (1994). "Beyond the Biomass: Compositional and Functional Analysis of Soil Microbial Communities". Wiley, Chichester.

Roberts, J. R. and Marshall, W. K. (1980). Retentive capacity: an index of chemical persistence expressed in terms of chemical-specific and ecosystem-specific parameters. *Ecotoxicology and Environmental Safety* **4**, 158–171.

Robin, G. deQ. (1986). Changing the sea level. *In* "The Greenhouse Effect, Climatic Change, and Ecosystems. SCOPE 29" (B. Bolin, B. R. Döäs, J. Jäger and R. A. Warwick, Eds), 323–359. Wiley, Chichester.

Robinson, J. (1969). Organochlorine insecticides and bird populations in Britain. *In* "Chemical Fallout: Current Research on Persistent Pesticides" (M. W. Miller and G. C. Berg, Eds), 113–173. Thomas, Springfield, Illinois.

Robinson, J. and Roberts, M. (1968). Accumulation, distribution and elimination of organochlorine insecticides by vertebrates. *In* "Symposium on Physico-Chemical and Biophysical Factors Affecting the Activity of Pesticides", Monograph No. 29, 106–119. Society of Chemical Industry, London.

Robinson, J., Richardson, A. *et al.* (1967). Organochlorine residues in marine organisms. *Nature (London)* **214**, 1307–1311.

Robinson, J., Roberts, M., Baldwin, M. and Walker, A. I. T. (1969). The pharmacokinetics of HEOD (dieldrin) in the rat. *Food and Cosmetics Toxicology* **7**, 317–332.

Rodgers, J. H., Crossland, N. O. *et al.* (1996). Design and construction of model stream ecosystems. *Ecotoxicology and Environmental Safety* **33**, 30–37.

Römbke, J. and Moltmann, J. (1996). "Applied Ecotoxicology". Lewis, Boca Raton.

Ronis, M. J. J. and Walker, C. H. (1989). The microsomal monooxygenases of birds. *Reviews in Biochemical Toxicology* **10**, 301–384.

Rose, C. I. and Hawksworth, D. L. (1981). Lichen recolonization in London's cleaner air. *Nature (London)* **289**, 289–292.

Rosenberg, D. M. (1975a). Fate of dieldrin in sediment, water, vegetation, and invertebrates of a slough in central Alberta, Canada. *Quaestiones Entomologicae* **11**, 69–96.

Rosenberg, D. M. (1975b). Food chain concentration of chlorinated hydrocarbon pesticides in invertebrate communities: a re-evaluation, *Quaestiones Entomologicae*, **11**, 97–110.

Roush, R. T. and Daly, J. C. (1990). The role of population genetics in resistance research and management. *In* "Pesticide Resistance in Arthropods" (R. T. Roush and B. E. Tabashnik, Eds), 97–152. Chapman & Hall, New York.

Royal Commission on Environmental Pollution (1972). "Third Report. Pollution in some British Estuaries and Coastal Waters". HMSO, London.

Royal Commission on Environmental Pollution (1992). "Sixteenth Report. Freshwater Quality". HMSO, London.

Royal Society (1983). "Risk Assessment". Report of a Royal Society Study Group. The Royal Society, London.

Royal Society Study Group (1978). "Long-term Toxic Effects". Final report of a Royal Society Study Group. The Royal Society, London.

Rudd, R. L. (1964). "Pesticides and the Living Landscape". University of Wisconsin Press, Madison.

Runeckles, V. C. (1974). Dosage of air pollutants and damage to vegetation. *Environmental Conservation* **1**, 305–308.

Runeckles, V. C. (1984). Impact of air pollutant combinations on plants. *In* "Air Pollution and Plant Life" (M. Treshow, Ed.), 239–258. Wiley, Chichester.

Runeckles, V. C. and Krupa, S. V. (1994). The impact of UV-B radiation and ozone on terrestrial vegetation. *Environmental Pollution* **83**, 191–213.

Rygg, B. (1986). Heavy-metal pollution and log-normal distribution of individuals among species in benthic communities. *Marine Pollution Bulletin* **17**, 31–36.

Sagan, L. A. (1989). On radiation, paradigms, and hormesis. *Science (New York)* **245**, 574, 621.

Sagan, L. (1994). A brief history and critique of the low dose effects paradigm. *In* "Biological Effects of Low Level Exposures: Dose–Response Relationships" (E. J. Calabrese, Ed.), 15–26. CRC Press, Boca Raton.

Samiullah, Y. (1990a). "Prediction of the Environmental Fate of Chemicals". Elsevier Applied Science, London.

Samiullah, Y. (1990b). "Biological Monitoring of Environmental Contaminants: Animals", MARC Report Number 37. Monitoring and Assessment Research Centre, University of London, London.

Sanchez, J. C., Fossi, M. C. and Focardi, S. (1997). Serum "B" esterases as a non-destructive biomarker for monitoring the exposure of reptiles to organophosphorus insecticides. *Ecotoxicology and Environmental Safety* **37**, 45–52.

Sanders, C. J., Stark, R. W., Mullins, E. J. and Murphy, J. (Eds) (1985). "Recent Advances in Spruce Budworms Research. Proceedings of the CANUSA Spruce Budworms Research Symposium". Canadian Forestry Service, Ottawa.

Sanders, F. S. (1985). Use of large enclosures for perturbation experiments in lentic ecosystems: a review. *Environmental Monitoring and Assessment* **5**, 55–99.

Santone, K. S. and Powis, P. (1991). Mechanisms of and tests for injuries. *In* "Handbook of Pesticide Toxicology", Vol. I (W. J. Hayes and E. R. Laws, Eds), 169–214. Academic Press, San Diego.

Scheunert, I. and Klein, W. (1985). Predicting the movement of chemicals between environmental compartments (air–water–soil–biota). *In* "Appraisal of Tests to Predict the Environmental Behaviour of Chemicals. SCOPE 25" (P. Sheehan, F. Korte, W. Klein and P. Bourdeau, Eds), 285–332. Wiley, Chichester.

Schimel D., Enting, I. G. *et al.* (1995). CO_2 and the carbon cycle. *In* "Climate Change 1994. Radiative Forcing of Climate Change and an Evaluation of the IPCC IS92 Emission Scenarios" (J. T. Houghton, L. G. M. Filho *et al.*, Eds), 35–71. Cambridge University Press, Cambridge.

Schmidt-Nielsen, K. (1990). "Animal Physiology: Adaptation and Environment", 4th edn. Cambridge University Press, Cambridge.

Schneider, T. (Ed.) (1986). "Acidification and its Policy Implications". Studies in Environmental Science **30**. Elsevier, Amsterdam.

Schoener, T. W. (1974). Resource partitioning in ecological communities. *Science (New York)* **185**, 27–39.

Schoener, T. W. (1989). The ecological niche. *In* "Ecological Concepts" (J. M. Cherrett, Ed.), 79–113. Blackwell Scientific, Oxford.

Schultz, T. W., Holcombe, G. W. and Phipps, G. L. (1986). Relationships of quantitative structure–activity to comparative toxicity of selected phenols in the *Pimephales promelas* and *Tetrahymena pyriformis* test systems. *Ecotoxicology and Environmental Safety* **12**, 146–153.

Schultz, T. W., Jain, R., Cajina-Quezada, M. and Lin, D. T. (1988). Structure–toxicity relationships for selected benzyl alcohols and the polar narcosis mechanism of toxicity. *Ecotoxicology and Environmental Safety* **16**, 57–64.

Scura, E. D. and Theilacker, G. H. (1977). Transfer of the chlorinated hydrocarbon PCB in a laboratory marine food chain. *Marine Biology* **40**, 317–325.

Segar, D. A. and Stamman, E. (1986). A strategy for design of marine pollution monitoring studies. *Water Science and Technology* **18**, 15–26.

Seim, W. K., Curtis, L. R., Glenn, S. W. and Chapman, G. A. (1984). Growth and survival of developing steelhead trout (*Salmo gairdneri*) continuously or intermittently exposed to copper. *Canadian Journal of Fisheries and Aquatic Science* **41**, 433–438.

Shaffer, M. (1987). Minimum viable populations: coping with uncertainty. *In* "Viable Populations for Conservation" (M. E. Soulé, Ed.), 69–86. Cambridge University Press, Cambridge.

Sheail, J. (1985). "Pesticides and Nature Conservation. The British Experience 1950–1975". Clarendon Press, Oxford.

Sheehan, P. J. (1984). Functional changes in the ecosystem. *In* "Effects of Pollutants at the Ecosystem Level. SCOPE 22" (P. J. Sheehan, D. R. Miller, G. C. Butler and P. Bourdeau, Eds), 101–145. Wiley, Chichester.

Sheehan, P., Korte, F., Klein, W. and Bourdeau, P. (Eds) (1985). "Appraisal of Tests to Predict the Environmental Behaviour of Chemicals. SCOPE 25". Wiley, Chichester.

Sheiner, L. B., Stanski, D. R. *et al.* (1979). Simultaneous modeling of pharmacokinetics and pharmacodynamics: application to *d*-tubocurarine. *Clinical Pharmacology and Therapeutics* **25**, 358–371.

Shine, K. P., Fouquart, Y. *et al.* (1995). Radiative forcing. *In* "Climate Change 1994. Radiative Forcing of Climate Change and an Evaluation of the IPCC IS92 Emission Scenarios" (J. T. Houghton, L. G. M. Filho *et al.*, Eds), 163–203. Cambridge University Press, Cambridge.

Shorrocks, B. and Swingland, I. R. (Eds) (1990). "Living in a Patchy Environment". Oxford University Press, Oxford.

Shriner, C. and Gregory, T. (1984). Use of artificial streams for toxicological research. *CRC Critical Reviews in Toxicology* **13**, 253–281.

Siegenthaler, U. and Sarmiento, J. L. (1993). Atmospheric carbon dioxide and the ocean. *Nature (London)* **365**, 119–125.

Simon, J. L. and Kahn, H. (Eds) (1984). "The Resourceful Earth. A Response to Global 2000". Basil Blackwell, Oxford.

Sinclair, A. R. E. (1989). Population regulation in animals. *In* "Ecological Concepts" (J. M. Cherrett, Ed.), 197–241. Blackwell Scientific, Oxford.

Sládeček, V. (1973). System of water quality from the biological point of view. *Archiv für Hydrobiologie. Beihefte: Ergebnisse der Limnologie* **7**, 1–218.

Slobodkin, L. B. (1972). On the inconstancy of ecological efficiency and the form of ecological theories. *Transactions of the Connecticut Academy of Arts and Sciences* **44**, 291–305.

Slooff, W., van Oers, J. A. M. and de Zwart, D. (1986). Margins of uncertainty in eco-toxicological hazard assessment. *Environmental Toxicology and Chemistry*, **5**, 841–852.

Smith, B. S. (1981). Tributyl tin compounds induce male characteristics on female mud snails *Nassarius obsoletus* = *Ilyanassa obsoleta. Journal of Applied Toxicology* **1**, 141–144.

Smith, J. B. and Tirpak, D. A. (Eds) (1990). "The Potential Effects of Global Climate Change on the United States". Hemisphere, New York.

Smith, J. E. (Ed.) (1968). "'Torrey Canyon' Pollution and Marine Life". Cambridge University Press, Cambridge.

Smith, R. A. (1872). "Air and Rain. The Beginnings of a Chemical Climatology". Longmans, Green, London.

Smith, T. M. and Shugart, H. H. (1993). The transient response of terrestrial carbon storage to a perturbed climate. *Nature (London)* **361**, 523–526.

Smith, W. H. (1981). "Air Pollution and Forests. Interactions between Air Contaminants and Forest Ecosystems". Springer, New York.

Smyth, H. F. (1967). Sufficient challenge. *Food and Cosmetics Toxicology* **5**, 51–58.

Soderlund, D. M. and Bloomquist, J. R. (1990). Molecular mechanisms of insecticide resistance. *In* "Pesticide Resistance in Arthropods" (R. T. Roush and B. E. Tabashnik, Eds), 58–96. Chapman & Hall, New York.

Sokal, R. R. and Rohlf, F. J. (1995). "Biometry. The Principles and Practice of Statistics in Biological Research", 3rd edn. Freeman, New York.

Solomon, M. E. (1969). "Population Dynamics". Arnold, London.

Southam, C. M. and Ehrlich, J. (1943). Effects of extract of western red-cedar heartwood on certain wood-decaying fungi in culture. *Phytopathology* **33**, 517–524.

Southward, A. J. (1982). An ecologist's view of the implications of the observed physiological and biochemical effects of petroleum compounds on marine organisms and ecosystems. *Philosophical Transactions of the Royal Society of London, B* **297**, 241–255.

Southward, A. J. and Southward, E. C. (1978). Recolonization of rocky shores in Cornwall after use of toxic dispersants to clean up the Torrey Canyon spill. *Journal of the Fisheries Research Board of Canada* **35**, 682–706.

Southwood, T. R. E. (1976). Bionomic strategies and population parameters. *In* "Theoretical Ecology. Principles and Applications" (R. M. May, Ed.), 26–48. Blackwell Scientific, Oxford.

Southwood, T. R. E. (1978). "Ecological Methods, with Particular Reference to the Study of Insect Populations", 2nd edn. Chapman & Hall, London.

Southwood, T. R. E. (1985). The roles of proof and concern in the work of the Royal Commission on Environmental Pollution. *Marine Pollution Bulletin* **16**, 346–350.

Southwood, T. R. E. (1996). Natural communities: structure and dynamics. *Philosophical Transactions of the Royal Society of London, B* **351**, 1113–1129.

Southwood, T. R. E., Hassell, M. P., Reader, P. M. and Rogers, D. J. (1989). Population dynamics of the viburnum whitefly (*Aleurotrachelus jelenekii*). *Journal of Animal Ecology* **58**, 921–942.

Spehar, R. L., Holcombe, G. W. *et al.* (1979). Effects of pollution of freshwater fish. *Journal of the Water Pollution Control Federation* **51**, 1616–1694.

Spellerberg, I. F. (1991). "Monitoring Ecological Change". Cambridge University Press, Cambridge.

Spooner, N., Gibbs, P. E., Bryan, G. W. and Goad, L. J. (1991). The effect of tributyl tin upon steroid titres in the female dogwhelk, *Nucella lapillus*, and the development of imposex. *Marine Environmental Research* **32**, 37–49.

Sprague, J. B. (1970). Measurement of pollutant toxicity to fish. II. Utilizing and applying bioassay results. *Water Research* **4**, 3–32.

Sprague, J. B. (1971). Measurement of pollutant toxicity to fish—III. Sublethal effects and "safe" concentrations. *Water Research* **5**, 245–266.

Stanley, P. I. and Bunyan, P. J. (1979). Hazards to wintering geese and other wildlife from the use of dieldrin, chlorfenvinphos and carbophenothion as wheat seed treatments. *Proceedings of the Royal Society of London, B* **205**, 31–45.

Stanley, P. I. and Elliott, G. R. (1976). An assessment based on residues in owls of environmental contamination arising from the use of mercury compounds in British agriculture. *Agro-Ecosystems* **2**, 223–234.

Stearns, S. C. (1977). The evolution of life history traits: a critique of the theory and a review of the data. *Annual Review of Ecology and Systematics* **8**, 145–171.

Stebbing, A. R. D. (1982). Hormesis—the stimulation of growth by low levels of inhibitors. *Science of the Total Environment* **22**, 213–234.

Steele, J. H. (1979). The uses of experimental ecosystems. *Philosophical Transactions of the Royal Society of London, B* **286**, 583–595.

Stenseth, N. C. (1979). Where have all the species gone? On the nature of extinction and the Red Queen hypothesis. *Oikos* **33**, 196–227.

Stern, C. and Sherwood, E. R. (Eds) (1966). "The Origin of Genetics. A Mendel Source Book". Freeman, San Francisco.

Stickel, L. F. (1968). "Organochlorine Pesticides in the Environment", Special Scientific Report—Wildlife, No. 119. Bureau of Sport Fisheries and Wildlife, Washington.

Stickel, L. F. (1973). Pesticide residues in birds and mammals. *In* "Environmental Pollution by Pesticides" (C. A. Edwards, Ed.), 254–312. Plenum Press, London.

Stickel, W. H. (1975). Some effects of pollutants in terrestrial ecosystems. *In* "Ecological Toxicology Research. Effects of Heavy Metal and Organochlorine Compounds" (A. D. MacIntyre and C. F. Mills, Eds), 25–74. Plenum Press, New York.

Stiles, T. R. (1993). Quality assurance in toxicology studies. *In* "General and Applied Toxicology", Vol. I (B. Ballantyne, T. Marrs and P. Turner, Eds), 345–358. Stockton Press, New York.

Stoiber, R. E., Williams, S. N. and Huebert, B. (1987). Annual contribution of sulfur dioxide to the atmosphere by volcanoes. *Journal of Volcanology and Geothermal Research* **33**, 1–8.

van Straalen, N. M. and Denneman, C. A. J. (1989). Ecotoxicological evaluation of soil quality criteria. *Ecotoxicology and Environmental Safety* **18**, 241–251.

Strain, B. R. and Cure, J. D. (1985). "Direct Effects of Increasing Carbon Dioxide on Vegetation". United States Department of Energy, DOE/ER-0238, Washington.

Strickberger, M. W. (1985). "Genetics", 3rd edn. Macmillan, New York.

Stumm, W. and Morgan, J. J. (1995). "Aquatic Chemistry. An Introduction Emphasizing Chemical Equilibria in Natural Waters", 3rd edn. Wiley, New York.

Sugiura, K., Ito, N. et al. (1978). Accumulation of polychlorinated biphenyls and poly-brominated biphenyls in fish: limitation of "correlation between partition coefficients and accumulation factors". Chemosphere 7, 731–736.

Swift, M. J., Heal, O. W. and Anderson, J. M. (1979). "Decomposition in Terrestrial Ecosystems". Blackwell Scientific, Oxford.

Takahashi, F. (1956). On the effect of population density on the power of reproduction of the almond moth, Ephestia cautella. III. The maximum reproduction of population and the larval density. Researches on Population Ecology. Entomological Laboratory, Kyoto University. Kyoto 3, 27–35.

Tamarin, R. H. (Ed.) (1978). "Population Regulation. Benchmark Papers in Ecology", Vol. 7. Dowden, Hutchinson & Ross, Stroudsburg, Pennsylvania.

Tans, P. P., Fung, I. Y. and Takahashi, T. (1990). Observational constraints on the global atmospheric CO_2 budget. Science (New York) 247, 1431–1438.

Tansley, A. G. (1935). The use and abuse of vegetational concepts and terms. Ecology 16, 284–307.

Tansley, A. G. (1953). "The British Islands and their Vegetation". Cambridge University Press, Cambridge.

Task Group on Metal Accumulation (1973). Accumulation of toxic metals with special reference to their absorption, excretion and biological half-times. Environmental Physiology and Biochemistry 3, 65–107.

Taylor, G. E. (1978). Genetic analysis of ecotypic differentiation within an annual plant species, Geranium carolinianum L., in response to sulfur dioxide. Botanical Gazette 139, 362–368.

Taylor, G. E. and Murdy, W. H. (1975). Population differentiation of an annual plant species, Geranium carolinianum, in response to sulfur dioxide. Botanical Gazette 136, 212–215.

Taylor, H. J., Ashmore, M. R. and Bell, J. N. B. (1986). "Air Pollution Injury to Vegetation". Institution of Environmental Health Officers, London.

Taylor, L. R., Kempton, R. A. and Woiwod, I. P. (1976). Diversity statistics and the log-series model. Journal of Animal Ecology 45, 255–272.

Thibodeaux, L. J. (1996). "Environmental Chemodynamics. Movement of Chemicals in Air, Water and Soil", 2nd edn. Wiley, New York.

Thomas, P. T. (1989). Approaches used to assess chemically induced impairment of host resistance and immune function. In "Hazard Assessment of Chemicals", Vol. 6 (J. Saxena, Ed.), 49–83. Hemisphere, New York.

Thomas, S. C. and Jasieński, M. (1996). Genetic variability and the nature of micro-evolutionary responses to elevated CO_2. In "Carbon Dioxide, Populations, and Communities" (C. Körner and F. A. Bazzaz, Eds), 51–81. Academic Press, San Diego.

Thompson, C. M. (1992). Preparation, analysis, and toxicity of phosphorothiolates. In "Organophosphates. Chemistry, Fate and Effects" (J. E. Chambers and P. E. Levi, Eds), 19–46. Academic Press, San Diego.

Thornton, I. (1975). Geochemical parameters in the assessment of estuarine pollution. In "The Ecology of Resource Degradation and Renewal" (M. J. Chadwick and G. T. Goodman, Eds), 15th Symposium of the British Ecological Society, 157–169. Blackwell Scientific, Oxford.

Tingey, D. T. and Reinert, R. A. (1975). The effect of ozone and sulphur dioxide singly and in combination on plant growth. *Environmental Pollution* **9**, 117–125.

Tiwari, J. L. (1979). A modeling approach based on stochastic differential equations, the principle of maximum-entropy, and Bayesian inference for parameters. *In* "Compartmental Analysis of Ecosystem Models" (J. H. Matis, B. C. Patten and G. C. White, Eds), 167–194. International Co-operative Publishing House, Fairland, Maryland.

Treshow, M. (Ed.) (1984). "Air Pollution and Plant Life". Wiley, Chichester.

Treshow, M. and Anderson, F. K. (1989). "Plant Stress from Air Pollution". Wiley, New York.

Tripathi, R. K. and O'Brien, R. D. (1973). Effect of organophosphates *in vivo* upon acetylcholinesterase isozymes from housefly head and thorax. *Pesticide Biochemistry and Physiology* **2**, 418–424.

Truhaut, R. (1977). Ecotoxicology: objectives, principles and perspectives. *Ecotoxicology and Environmental Safety* **1**, 151–173.

Turner, L., Choplin, F. *et al.* (1987). Structure–activity relationships in toxicology and ecotoxicology: an assessment. *Toxicology in Vitro* **1**, 143–171.

Udvardy, M. F. D. (1959). Notes on the ecological concepts of habitat, biotope and niche. *Ecology* **40**, 725–728.

Underwood, A. J. (1994). Spatial and temporal problems with monitoring. *In* "The Rivers Handbook", Vol. 2 (P. Calow and G. E. Petts, Eds), 101–123. Blackwell Scientific, London.

Underwood, T. (1986). The analysis of competition by field experiments. *In* "Community Ecology: Pattern and Process" (J. Kikkawa and D. J. Anderson, Eds), 240–268. Blackwell Scientific, Oxford.

United States Council on Environmental Quality and Department of State (1980). "The Global 2000 Report to the President: Entering the Twenty-First Century", Vols I–III. United States Government Printing Office, Washington.

United States Environmental Protection Agency (1990). "Biological Criteria: National Program Guidance for Surface Waters". EPA-440/5–90–004.

Unsworth, M. H. and Mansfield, T. A. (1980). Critical aspects of chamber design for fumigation experiments on grasses. *Environmental Pollution, A* **23**, 115–120.

Ure, A. M. and Davidson, C. M. (Eds) (1995). "Chemical Speciation in the Environment". Blackie, Glasgow.

Usher, M. B., Davis, P. R., Harris, J. R. W. and Longstaff, B. C. (1979). A profusion of species? Approaches towards understanding the dynamics of the populations of the micro-arthropods in decomposer communities. *In* "Population Dynamics" (R. M. Anderson, B. D. Turner and L. R. Taylor, Eds), 20th Symposium of the British Ecological Society, 359–384. Blackwell Scientific, Oxford.

Vandermeer, J. H. (1972). Niche theory. *Annual Review of Ecology and Systematics* **3**, 107–132.

Varley, G. C. and Gradwell, G. R. (1960). Key factors in population studies. *Journal of Animal Ecology* **29**, 399–401.

Varley, G. C. and Gradwell, G. R. (1968). Population models for the winter moth. *In* "Insect Abundance" (T. R. E. Southwood, Ed.), 132–142. Blackwell Scientific, Oxford.

Verhaar, H. J. M., van Leeuwen, C. J. and Hermens, J. L. M. (1992). Classifying environmental pollutants. 1: Structure–activity relationships for prediction of aquatic toxicity. *Chemosphere* **25**, 471–491.

Verkleij, J. A. C. and Schat, H. (1989). Mechanisms of metal tolerance in higher plants. *In* "Heavy Metal Tolerance in Plants: Evolutionary Aspects" (A. J. Shaw, Ed.), 179–193. CRC, Boca Raton.

Vetter, R. D., Powell, M. A. and Somero, G. N. (1991). Metazoan adaptations to hydrogen sulphide. *In* "Metazoan Life Without Oxygen" (C. Bryant, Ed.), 109–128. Chapman & Hall, London.

Vosser, J. L. (1986). The European Community chemicals notification scheme and environmental hazard assessment. *In* "Toxic Hazard Assessment of Chemicals" (M. Richardson, Ed.), 117–132. The Royal Society of Chemistry, London.

Vouk, V. B., Butler, G. C. *et al.* (1987). "Methods for Assessing the Effects of Mixtures of Chemicals. SCOPE 30. SGOMSEC 3". Wiley, Chichester.

Vreeland, V. (1974). Uptake of chlorobiphenyls by oysters. *Environmental Pollution* **6**, 135–140.

Wagner, C. and Løkke, H. (1991). Estimation of ecotoxicological protection levels from NOEC toxicity data. *Water Research* **25**, 1237–1242.

Waldichuk, M. (1979). Review of the problems. *Philosophical Transactions of the Royal Society of London, B* **286**, 399–424.

Walker, C. H. (1975). Variations in the intake and elimination of pollutants. *In* "Organochlorine Insecticides: Persistent Organic Pollutants" (F. Moriarty, Ed.), 73–130. Academic Press, London.

Walker, C. H. (1978). Species differences in microsomal monooxygenase activity and their relationship to biological half-lives. *Drug Metabolism Reviews* **7**, 295–323.

Walker, C. H. (1981). The correlation between *in vivo* and *in vitro* metabolism of pesticides in vertebrates. *Progress in Pesticide Biochemistry* **1**, 247–285.

Walker, C. H. (1987). Kinetic models for predicting bioaccumulation of pollutants in ecosystems. *Environmental Pollution* **44**, 227–240.

Walker, C. H. (1992). The ecotoxicology of persistent pollutants in marine fish-eating birds. *In* "Persistent Pollutants in Marine Ecosystems" (C. H. Walker and D. R. Livingstone, Eds), 211–232. Pergamon, Oxford.

Walker, C. H., Hopkin, S. P., Sibly, R. M. and Peakall, D. B. (1996). "Principles of Ecotoxicology". Taylor and Francis, London.

Wallace, D. C. (1982). Structure and evolution of organelle genomes. *Microbiological Reviews* **46**, 208–240.

Wallace, K. B. (1992). Species-selective toxicity of organophosphorus insecticides: a pharmacodynamic phenomenon. *In* "Organophosphates. Chemistry, Fate and Effects" (J. E. Chambers and P. E. Levi, Eds), 79–105. Academic Press, San Diego.

Walley, K. A., Khan, M. S. I. and Bradshaw, A. D. (1974). The potential for evolution of heavy metal tolerance in plants 1. Copper and zinc tolerance in *Agrostis tenuis*. *Heredity* **32**, 309–319.

Ward, B. M. and Dubos, R. (1972). "Only One Earth. The Care and Maintenance of a Small Planet". Deutsch, London.

Ward, R. D., Skibinski, D. O. F. and Woodwark, M. (1992). Protein heterozygosity, protein structure, and taxonomic differentiation. *Evolutionary Biology* **26**, 73–159.

Warner, F. (1979). Sources and extent of pollution. *Proceedings of the Royal Society of London, B* **205**, 5–15.

Warren, C. E. and Davis, G. E. (1967). Laboratory studies on the feeding, bioenergetics, and growth of fish. *In* "The Biological Basis of Freshwater Fish Production" (S. D. Gerking, Ed.), 175–214. Blackwell Scientific, Oxford.

Warren, C. E. and Liss, W. J. (1977). "Design and Evaluation of Laboratory Ecological System Studies". US Environmental Protection Agency, Duluth, Minnesota.

Warrick, R. A. (1993). Climate and sea level change: a synthesis. *In* "Climate and Sea Level Change: Observations, Projections and Implications" (R. A. Warrick, E. M. Barrow and T. M. L. Wigley, Eds), 3–31. Cambridge University Press, Cambridge.

Warrick, R. and Oerlemans, J. (1990). Sea level rise. *In* "Climate Change. The IPCC Scientific Assessment" (J. T. Houghton, G. J. Jenkins and J. J. Ephraums, Eds), 257–281. Cambridge University Press, Cambridge.

Warrick, R. A., Shugart, H. H. *et al.* (1986). The effects of increased CO_2 and climatic change on terrestrial ecosystems. *In* "The Greenhouse Effect, Climatic Change, and Ecosystems. SCOPE 29" (B. Bolin, B. R. Döös, J. Jäger and R. A. Warrick, Eds), 363–392. Wiley, Chichester.

Washington, H. G. (1984). Diversity, biotic and similarity indices. A review with special relevance to aquatic ecosystems. *Water Research* **18**, 653–694.

Watson, A. J., Nightingale, P. D. and Cooper, D. J. (1995). Modelling atmosphere–ocean CO_2 transfer. *Philosophical Transactions of the Royal Society of London, B*, **348**, 125–132.

Watson, J. D. (1965). "Molecular Biology of the Gene". Benjamin, New York.

Watson, J. D. and Crick, F. H. C. (1953). A structure for deoxyribose nucleic acid. *Nature (London)* **171**, 737–738.

Watson, R. T., Rodhe, H., Oeschger, H. and Siegenthaler, U. (1990). Greenhouse gases and aerosols. *In* "Climate Change. The IPCC Scientific Assessment" (J. T. Houghton, G. J. Jenkins and J. J. Ephraums, Eds), 1–40. Cambridge University Press, Cambridge.

Watt, A. S. (1947). Pattern and process in the plant community. *Journal of Ecology* **35**, 1–22.

Way, M. J. (1968). Inter-specific mechanisms with special reference to aphid populations. *In* "Insect Abundance" (T. R. E. Southwood, Ed.), 18–36. Blackwell Scientific, Oxford.

Way, M. J. and Banks, C. J. (1967). Intra-specific mechanisms in relation to the natural regulation of numbers of *Aphis fabae* Scop. *Annals of Applied Biology* **59**, 189–205.

Way, M. J. and Bevan, D. (1977). Dilemmas in forest pest and disease management. *In* "Ecological Effects of Pesticides" (F. H. Perring and K. Mellanby, Eds), 95–110. Linnaean Society Symposium Series, number 5. Academic Press, London.

Wayne, R. P. (1991). "Chemistry of Atmospheres", 2nd edn. Clarendon Press, Oxford.

Weaver, R. F. and Hedrick, P. W. (1995). "Basic Genetics", 2nd edn. Brown, Dubuque.

Webb, F. E., Blais, J. R. and Nash, R. W. (1961). A cartographic history of spruce budworm outbreaks and aerial forest spraying in the Atlantic region of North America, 1949–1959. *Canadian Entomologist* **93**, 360–379.

Webb, T. (1992). Past changes in vegetation and climate: lessons for the future. *In* "Global Warming and Biological Diversity" (R. L. Peters and T. E. Lovejoy, Eds), 59–75. Yale University Press, New Haven.

Webb, W. L., Lavenroth, W. K., Szarek, S. R. and Kinerson, R. S. (1983). Primary production and abiotic controls in forests, grasslands, and desert ecosystems in the United States. *Ecology* **64**, 134–151.

Weinberg, A. M. (1978). Benefit—cost analysis and the linear hypothesis. *Nature (London)* **271**, 596.

Weinberg, W. (1908). Über den Nachweis der Verebung beim Menschen. *Jahresheft des Vereins für vaterländische Naturkunde in Württemberg* **64**, 369–382.

Weiner, J. (1996). Problems in predicting the ecological effects of elevated CO_2. *In* "Carbon Dioxide, Populations, and Communities" (C. Körner and F. A. Bazzaz, Eds), 431–441. Academic Press, San Diego.

Weiss, C. M. (1961). Physiological effect of organic phosphorus insecticides on several species of fish. *Transactions of the American Fisheries Society* **90**, 143–152.

Wells, T. C. E. (1983). The creation of species-rich grasslands. In "Conservation in Practice" (A. Warren and F. B. Goldsmith, Eds), 215–232. Wiley, Chichester.

Welz, B. (1985). "Atomic Absorption Spectrometry", 2nd English edn, from the 3rd German edn of 1983. VCH, Weinheim.

Westhoff, V. and van der Maarel, E. (1973). The Braun-Blanquet approach. In "Ordination and Classification of Communities" (R. H. Whittaker, Ed.), 617–726. Junk, The Hague.

Westlake, G. E., Bunyan, P. J. and Stanley, P. I. (1978). Variation in the response of plasma enzyme activities in avian species dosed with carbophenothion. *Ecotoxicology and Environmental Safety* **2**, 151–159.

Wheeler, A. (1969). Fish-life and pollution in the lower Thames: a review and preliminary report. *Biological Conservation* **2**, 25–30.

Whitfield M. (1975). The electroanalytical chemistry of sea water. In "Chemical Oceanography" (J. P. Riley and G. Skirrow, Eds), 2nd edn, 1–154. Academic Press, London.

Whitfield, M. and Turner, D. R. (1986). Chemical speciation in tropical waters—a cautionary tale. *Science of the Total Environment* **58**, 9–35.

Whittaker, R. H. (1962). Classification of natural communities. *Botanical Review* **28**, 1–239.

Whittaker, R. H. (1965). Dominance and diversity in land plant communities. *Science (New York)* **147**, 250–260.

Whittaker, R. H. (1967). Gradient analysis of vegetation. *Biological Reviews* **42**, 207–264.

Whittaker, R. H. (1975). "Communities and Ecosystems", 2nd edn. Macmillan, New York.

Whittaker, R. H., Levin, S. A. and Root, R. B. (1973). Niche, habitat, and ecotope. *American Naturalist* **107**, 321–338.

Whittle, K. J., Hardy, R. and McIntyre, A. D. (1978). "Scientific Studies at Future Oil Spill Incidents in the Light of Past Experience". Marine Environmental Quality Committee, International Council for the Exploration of the Sea.

Whittle, K. J., Hardy, R., Mackie, P. R. and McGill, A. S. (1982). A quantitative assessment of the sources and fate of petroleum compounds in the marine environment. *Philosophical Transactions of the Royal Society London, B* **297**, 193–218.

Whyte, A. V. and Burton, I. (Eds) (1980). "Environmental Risk Assessment. SCOPE 15". Wiley, Chichester.

Widdows, J. (1993). Marine and estuarine invertebrate toxicity tests. In "Handbook of Ecotoxicology", Vol.1 (P. Calow, Ed.), 145–166. Blackwell Scientific, Oxford.

Widdows, J. and Donkin, P. (1992). Mussels and environmental contaminants: bioaccumulation and physiological aspects. In "The Mussel *Mytilus*" (E. M. Gosling, Ed.), 383–424. Elsevier, Amsterdam.

Widdows, J. and Page, D. S. (1993). Effects of tributyltin and dibutyltin on the physiological energetics of the mussel, *Mytilus edulis*. *Marine Environmental Research* **35**, 233–249.

Wilkins, D. A. (1957). A technique for the measurement of lead tolerance in plants. *Nature (London)* **180**, 37–38.

Wilkins, D. A. (1960). "The Measurement and Genetical Analysis of Lead Tolerance in *Festuca ovina*". Report of the Scottish Plant Breeding Station, Edinburgh, 85–98.

Wilkins, D. A. (1978). The measurement of tolerance to edaphic factors by means of root growth. *New Phytologist* **80**, 623–633.

Williams, J. H., Kingham, H. G., Cooper, B. J. and Eagle D. J. (1977). Growth regulator injury to tomatoes in Essex, England. *Environmental Pollution* **12**, 149–157.

Wilson, B. W., Hooper, M. J., Hansen, M. E. and Nieberg, P. S. (1992). Reactivation of organophosphorus inhibited AChE with oximes. In "Organophosphates. Chemistry, Fate, and Effects" (J. E. Chambers and P. E. Levi, Eds), 107–137. Academic Press, San Diego.

Wilson, G. B. and Bell, J. N. B. (1985). Studies on the tolerance to SO_2 of grass populations in polluted areas III. Investigations on the rate of development of tolerance. *New Phytologist* **100**, 63–77.

Windebank, A. J. (1987). Peripheral neuropathy due to chemical and industrial exposure. In "Handbook of Clinical Neurology", Vol. 51 (P. J. Vinken, G. W. Bruyn, H. L. Klawans and W. B. Matthews, Eds), 263–292. Elsevier, Amsterdam.

Winner, R. W., Keeling, T., Yeager, R. and Farrell, M. P. (1977). Effect of food type on the acute and chronic toxicity of copper to *Daphnia magna*. *Freshwater Biology* **7**, 343–349.

Wolfe, D. A. (1985). Fossil fuels: transportation and marine pollution. In "Wastes in the Ocean", Vol. 4 (I. W. Duedall, D. R. Kester, P. K. Park and B. W. Ketchum, Eds), 45–93. Wiley, New York.

Wolfe, D. A. (1987). Interactions of spilled oil with suspended materials and sediments in aquatic systems. In "Fate and Effects of Sediment-Bound Chemicals in Aquatic Systems" (K. L. Dickson, A. W. Maki and W. A. Brungs, Eds), 299–316. Pergamon, New York.

Wolfe, D. A., Hameedi, M. J. *et al.* (1994). The fate of the oil spilled from the *Exxon Valdez*. *Environmental Science and Technology* **28**, 561A–568A.

Wood, L. B. (1982). "The Restoration of the Tidal Thames". Adam Hilger, Bristol.

Wood, R. J. (1981). Insecticide resistance: genes and mechanisms. In "Genetic Consequences of Man Made Change" (J. A. Bishop and L. M. Cook, Eds), 53–96. Academic Press, London.

Wood, R. J. and Bishop, J. A. (1981). Insecticide resistance: populations and evolution. In "Genetic Consequences of Man Made Change" (J. A. Bishop and L. M. Cook, Eds), 97–127. Academic Press, London.

Woodhead, D. S. (1980). Marine disposal of radioactive wastes. *Helgoländer Meeresuntersuchungen* **33**, 122–137.

Woods, J. D. (1984). The upper ocean and air–sea interactions in global climate. In "The Global Climate" (J. T. Houghton, Ed.), 141–187. Cambridge University Press, Cambridge.

Woods, J. and Barkmann, W. (1993). The plankton multiplier—positive feedback in the greenhouse. *Journal of Plankton Research* **15**, 1053–1074.

Woodward, F. I. (1992). Predicting plant responses to global environmental change. *New Phytologist* **122**, 239–251.

Woodward, F. I., Thompson, G. B. and McKee, I. F. (1991). The effects of elevated concentrations of carbon dioxide on individual plants, populations, communities and ecosystems. *Annals of Botany* **67** (supplement 1), 23–38.

Woodwell, G. M. (1970). Effects of pollution on the structure and physiology of ecosystems. *Science (New York)* **168**, 429–433.

World Commission on Environment and Development (1987). "Our Common Future". Oxford University Press, Oxford.

Worthington, E. B. (Ed.) (1975). "The Evolution of IBP". Cambridge University Press, Cambridge.

Wright, D. A., Mihursky, J. A. and Phelps, H. L. (1985). Trace metals in Chesapeake Bay oysters: intra-sample variability and its implications for biomonitoring. *Marine Environmental Research* **16**, 181–197.

Wright, J. F., Moss, D., Armitage, P. D. and Furse, M. T. (1984). A preliminary classification of running-water sites in Great Britain based on macro-invertebrate species and the prediction of community type using environmental data. *Freshwater Biology* **14**, 221–256.

Wu, L. (1989). Colonization and establishment of plants in contaminated sites. *In* "Heavy Metal Tolerance in Plants: Evolutionary Aspects" (A. J. Shaw, Ed.), 269–299. CRC, Boca Raton.

Wu, L., Bradshaw, A. D. and Thurman, D. A. (1975). The potential for evolution of heavy metal tolerance in plants III. The rapid evolution of copper tolerance in *Agrostis stolonifera*. *Heredity* **34**, 165–187.

Wu, R. S. S. (1982). Effects of taxonomic uncertainty on species diversity indices. *Marine Environmental Research* **6**, 215–225.

Yang, R. S. H. (1994a). Introduction to the toxicology of chemical mixtures. *In* "Toxicology of Chemical Mixtures" (R. S. H. Yang, Ed.), 1–10. Academic Press, San Diego.

Yang, R. S. H. (1994b). "Toxicology of Chemical Mixtures. Case Studies, Mechanisms, and Novel Approaches". Academic Press, San Diego.

Zaroogian, G. E., Heltshe, J. F. and Johnson, M. (1985). Estimation of bioconcentration in marine species using structure-activity models. *Environmental Toxicology and Chemistry* **4**, 3–12.

Index

Species are usually referred to by their Latin, binomial, names, with cross-references where appropriate from the common English names. However, I have conformed to the usual, inconsistent, practice of referring to some species, mostly mammals, by their common names only: most if not all of the original references omit the Latin name.